GENOME RESEARCH ADVANCES

GENOME RESEARCH ADVANCES

ALBERTO P. BELLINI
EDITOR

Nova Biomedical Books
New York

For permission to use material from this book please contact us:
Telephone 631-231-7269; Fax 631-231-8175
Web Site: http://www.novapublishers.com

LIBRARY OF CONGRESS CATALOGING-IN-PUBLICATION DATA

Genome research advances / Alberto P. Bellini (editor).
p. ; cm.
Includes bibliographical references and index.
ISBN-13: 978-1-60021-805-7 (hardcover)
ISBN-10: 1-60021-805-9 (hardcover)
1. Genomes. 2. Genomics.
[DNLM: 1. Genomics--methods. 2. Genome. QU 58.5 G3342 2007] I. Bellini, Alberto P.
QH447.G46514 2007
572.8'6--dc22

2007022577

Published by Nova Science Publishers, Inc. ✦ New York

Contents

PREFACE

The genomic approach of technology development and large-scale generation of community resource data sets has introduced an important new dimension in biological and biomedical research. Interwoven advances in genetics, comparative genomics, high throughput biochemistry and bioinformatics are combining to attack basic understanding of human life and disease and to develop strategies to combat disease. Genomic Research began with The Human Genome Project (HGP), the international research effort that determined the DNA sequence of the entire human genome, completed in April 2003. The HGP also included efforts to characterize and sequence the entire genomes of several other organisms, many of which are used extensively in biological research. Identification of the sequence or function of genes in a model organism is an important approach to finding and elucidating the function of human genes. Integral to the HGP are similar efforts to understand the genomes of various organisms commonly used in biomedical research, such as mice, fruit flies and roundworms. Such organisms are called "model organisms," because they can often serve as research models for how the human organism behaves. This book presents the latest advances in this fast-moving field.

Chapter I - Recent progress in high-throughput proteomics techniques resulted in large but sparse amounts of information on protein-protein interactions. Thus, one of the major goals of the modern bioinformatics research is to integrate these large-scale genomic and proteomic results to facilitate understanding of the inner workings of a cell governed by complex interaction networks.

The chapter "Bioinformatics Tools for Researching Protein Interaction Networks" reviews the current experimental and computational techniques and databases that are commonly used for researching protein-protein interactions. The overview of these state-of-the-art approaches outlines the ongoing need for the standardized computational platform that would enable accumulation and integration of the entire range of the current bioinformatics network tools.

The method of Semantic Networks (SN) has been suggested as a potential framework for the existing bioinformatics resources; several examples of effective simulation of protein interaction networks within the SN environment have been discussed.

Chapter II - Human Myoblast Genome Therapy (HMGT) is a platform technology of cell transplantation, nuclear transfer, and tissue engineering. Myoblasts are differentiated, immature cells destined to become muscles. Myoblasts cultured from muscle biopsy survive, develop and function to revitalize degenerative muscles upon transplantation. Transplant injury activates regeneration of host myofibers that fuse with the injected myoblasts, sharing their nuclei in a common gene pool of the syncytium. Thus, through nuclear transfer and complementation, human genome can be transferred into muscles of genetically-ill patients to achieve phenotype repair. Myoblasts are safe and efficient universal gene transfer vehicles endogenous to muscles that constitute 50% of the body. Myoblasts fuse among themselves to form new myofibers. Patients take only 2-month cyclosporine to immunosuppress allograft rejection because myofibers do not express MHC-1 antigens.

The first correction of human gene defect was published in the Lancet on July 14, 1990 when the therapeutic protein dystrophin was found in the myoblast-injected muscle of a Duchenne muscular dystrophy (DMD) patient. Results over 280 HMGT procedures on MD subjects in the past 15 years demonstrated absolute safety. Myoblast-injected DMD muscles showed improved histology. Strength increase at 18 months post-operatively averaged 123%. FDA-approved clinical trials progressed unto Phase III in USA with direct cost recovery.

Heart muscle degeneration is the leading cause of human debilitation and death. The first human myoblast transfer into porcine myocardium in May 2000 revealed that it was safe to administer one billion myoblasts at 10^8/ml using 20 injections inside the left ventricle. Surgical HMGT into 50 infarcted porcine hearts demonstrated three mechanisms as proof of concept. Some myoblasts trans-differentiated to become cardiomyocytes. Others transferred their nuclei into host cardiomyocytes through cell fusion. As yet others formed skeletal myofibers with satellite cells. *De novo* production of contractile filaments augmented heart contractility. Myoblasts transduced with VEGF165 and Angiopoeitin-1 genes produced six times more mature capillaries in infarcted porcine myocardium than placebo. Xenograft rejection was not observed for up to 30 weeks despite cyclosporine discontinuation at 6 weeks.

Clinical trials on approximately 150 severe heart patients in 12 countries demonstrated <10% mortality since June 2000. Most subjects received autologous cells implanted epicardially after CABG or endomyocardially. Two subjects each receiving 1-billion myoblast allograft and 2-month cyclosporine survived over one year. Significant increases in left ventricular ejection fraction, wall thickness and motion have been reported, with reduction in perfusion defective areas, angina and shortness of breath. Two European trials are in Phase II.

HMGT on two Type II diabetic subjects aimed to correct the GLUT4 genomic defects, producing insulin receptors on myofibers to facilitate glucose uptake for metabolism. Each subject received 24- billion allogenic myoblasts and 2-month cyclosporine. It has been 10 months and HMGT appears to be safe for both subjects.

Other HMGT targets include cosmetic enhancement, bone degeneration, cancer, anemia, hemophilia, human growth deficiency, muscle trauma, pain and depression. Transduced myoblasts and/or their derivatives act as stable bioreactor(s) to deliver long-term supplies of therapeutic protein(s) at optimal levels.

Chapter III - LET-dependent features of cell abnormality leading to death, lethal sectoring, genomic instability and cell cycle change were compared between α- and X-irradiated HeLa cells by means of a time-lapse photography. Cell abnormality induced by α irradiation tended to be manifested in the form of death without division (mitotic or interphase death), rather than incomplete division accompanied by anaphase bridge derived from dicentric chromosome (two-hit type damage) that was predominant for X irradiation. Frequency of incomplete division at irradiated generation was decreased remarkably (25.8% → 6.8%) by splitting X ray dose (4 Gy → 2 × 2 Gy), correlating well with the increase in cell survival (6.8% → 14.0%). However, in the case of α irradiation especially at low dose, two split doses (0.6 Gy → 2 × 0.3 Gy) significantly decreased cell survival (12.3% → 8.8%) with the accompanying increase in the frequency of incomplete division (7.2% → 13.5%) (inverse split dose effect). This suggests that an interaction between the damage by the 1st dose and the one by the 2nd dose gives rise to 2-hit type damage owing to irreparability of the damage and enhances cell killing at low dose where interphase cell death is not predominant.

Lethal sectoring is the process for survival by which lethal damage remaining in irradiated cells is eliminated as lethal sector (offspring without reproductive integrity). This process occurs through the postirradiation 1st to 4th divisions, resulting in the appearance of a clonogenic progenitor harboring no lethal damage. However, nonlethal damage remaining in the clonogenic progenitors led to an elevated incidence of delayed cell death in their progeny. From an analysis of the pedigrees of cells surviving α (0.45 Gy) or X irradiation (3 Gy) (20% survival dose) which were constructed by means of a long-term (~150 h) time-lapse observation, it became clear that the mean incidence was higher for α-particle (16.3%) than X-ray survivors (8.3%), indicating the greater potentiality for genomic instability by α particles. Evidence is available to suggest the intrinsic difference in the mechanisms for the induction of genomic instability by these two radiations; the association of misrepaired clustered DNA damage (CD) with α particles and the fixation of unrepaired potentially lethal damage (PLD) with X rays.

A novel phenomenon "delayed division delay (DDD)" was noticed, though occasionally (~10% per clonogenic progeny; ~30% per pedigree), during the postirradiation 1st – 3rd generations. DDD was much longer in α- (mean: ~11 h) than X-irradiated cells (~4 h). DDD might be triggered by delayed break and rejoining of DNA double-strand at the fragile misrepaired sites. This demonstrates the usefulness of DDD as a new endpoint of radiation-induced genomic instability. Contrary to DDD, a significant shortening (~5 h) of cell cycle time at the postirradiation 1st generation was found with the progeny of X-ray survivors harboring no lethal nor nonlethal damage.

Chapter IV - Some viruses are integrated into host genome. DNA integration of viruses is considered one of the most important steps in the development of several types of human cancers. However, few methods can efficiently detect viral integration sites. Recently developed techniques for polymerase chain reaction (PCR) have facilitated sequencing of the host DNA adjacent to an integrated retroviral provirus. *Alu*-PCR and cassette-ligation-mediated PCR can be used to selectively amplify integrated DNA, provided that a portion of a nearby gene has already been sequenced. Analysis of virus integration can reveal new functions of genes located near integration sites. Such genes may be candidates for new tumor genes or tumor suppressor genes. Hepatitis B virus (HBV) is a major cause of chronic hepatitis and hepatocellular carcinoma (HCC) in humans. HBV DNA is known to be integrated into the human genome of HCC. In this issue, the authors review recent reports and discuss our findings concerning HBV DNA integration and the pathogenesis of liver cancer.

Chapter V - Although gastric cancer (GC) incidence and mortality have recently decreased, GC still represents the fourth most common cancer and the second leading cause of cancer-related death worldwide. GC incidence shows a great geographical variation, with the highest frequencies observed in Eastern Asia and Southern America, and much lower rates in many Western Countries.

The aetiology of the disease and the mechanisms involved in its carcinogenesis are still poorly understood. The association between GC risk and positive family history of the disease suggests the involvement of genetic factors in GC pathogenesis, besides the influence of environmental factors.

GC develops through genomic instability and in particular, instability at simple repetitive sequences, called microsatellite instability (MSI), is observed in about 15% of GCs.

Hypermethylation of the promoter of *hMLH1*, one of the main mismatch repair (MMR) genes, has been found to be responsible for the development of the majority of GCs exhibiting MSI.

MSI is identified by analysing paired tumor and normal DNAs for microsatellite loci. Nowadays, guidelines are established to characterize MSI by using a standard panel of 5 microsatellite markers. Three levels of MSI can be identified: high-level MSI (MSI-H), generally defined as MSI in more than 30% of the standard markers; low-level MSI (MSI-L), when changes are exhibited in less than 30% of the markers and no-MSI (MSS), in the absence of microsatellite alterations. The highly sensitive BAT25 and BAT26 mononucleotide repeats can be used to specifically define the MSI-H phenotype and, recently, the role of MMR protein immunohistochemistry in identify MSI-H GCs has been pointed out.

Simple genetic and/or epigenetic inactivation of MMR genes is not by itself a transforming event and therefore additional genetic changes are believed to be necessary for progression to malignancy. MSI-H GCs progress through mutations in coding repetitive sequences in genes involved in cell growth regulation (*TGFβRII, IGFIIR, RIZ, TCF4, DP2*), in apoptosis (*BAX, BCL10, FAS, CASPASE5, APAF1*) and in DNA repair (*hMSH6, hMSH3, MED1, RAD50, BLM, ATR, MRE11*). In most cases, these mutational events inactivate only one of the two alleles of these genes thus leading to haploinsufficiency. As a consequence, these mutations may tip the balance of regulatory pathways and drive cells to further progress in malignancy.

The occurrence of mutations in specific sets of cancer-related genes confers unique clinico-pathological features to MSI-H GCs. In fact, intestinal histotype, antral location, expanding type, lower prevalence of lymph-node metastases, *H. pylori* sieropositivity and improved survival have been associated with MSI-H GCs.

In conclusion, MSI is a promising screening tool because it allows for the identification of a specific GC subset characterized by peculiar molecular alterations and clinico-pathological features. This information may indicate improved therapeutic approaches for patients care.

Chapter VI - Genome sequencing was initially viewed by the vast majority of researchers as a technical endeavor, mostly meant to provide us with the blueprint of the Human Genome

[1]. Bacterial genetics, however, implied a much deeper vision because there were indications that genomes displayed some sort of organisation related to functional properties of the cell [2]. This latter vision prompted the creation of a consortium for the sequencing of the model bacterium *Bacillus subtilis* genome with the explicit underlying hypothesis that a genome is not simply a collection of genes, but an organised set that would, when understood, tell much about the nature of life itself [3]. As a matter of fact, as unexpected discoveries of genomics were made, it became self-evident that sequencing was just the first step in the new domain of the genetics of genomes. Perhaps the most seminal discovery in genomics was made in 1990 when, with the sequencing of the first yeast chromosome (chromosome III) and the first 100 kb of sequence of *B. subtilis*, it appeared that, contrary to expectations, a considerable number of genes (at the time, half of them) had completely been ignored until then [4,5]. In this context, and despite the world-wide importance of tuberculosis it nevertheless asked for a formidable effort to convince the relevant instances that sequencing the *Mycobacterium tuberculosis* genome should be a major priorityn [6] because sequencing genomes was still perceived as a waste of resources and not a research program *per se* [3]. It was not immediately understood that sequencing genomes is highly cost-effective, as we shall see later on, not only for gene discovery but also in the identification of drug resistance (though cost was high at the time). The genomics approach is however well accepted now [7].

In a book devoted to the fight against tuberculosis and related diseases, it seems essential to stress first the scientific reasons the authors have to sequence genomes (at the time of this article, approximately 1,500 public genome programs are completed or under way). The point of view I will hold is that of a microbiologist not particularly familiar with *M. tuberculosis*, but promoting the genetics of genomes as a basic tool, a new kind of microscope, for better understanding life processes and in particular those involved in pathogenesis, thereby improving our chances to target diseases with faster and better efficiency.

Chapter VII - The etiology and pathogenesis of tumors of the central nervous system are still inadequately explained. This study analyzes tumor suppressor gene—adenomatous polyposis coli (APC) in 28 patients with glioblastoma, the most aggressive form of glial tumors. APC protein has a structural role in adherens junctions, but also plays a signaling role as a negative regulator of the wnt pathway. The authors interest in the APC gene stemmed principally from the findings that the wild-type APC protein is highly expressed in the central nervous system, and upon the finding that it is critically involved in particular syndromes, among which brain tumors play a significant role. Glioblastoma samples were tested for gene instability by PCR/loss of heterozygosity using RFLP method. Two polymorphic markers were used: an Rsa I polymorphic site in exon 11, and an Msp I polymorphic site in exon 15. The results of the analysis for both markers showed allelic loss of the APC gene in 40% of our sample out of 25 heterozygous patients (informativeness 89%). Another 20% of samples demonstrated allelic imbalance of the APC allele in tumor tissue. Altogether, there were 15 samples (60%) demonstrating instability of this tumor suppressor gene. Despite increasing knowledge on glioma biology and genetics, the prognostic tools for glioblastoma still need improvement. The authors findings on genomic instability of the APC gene may contribute to a better understanding of glioblastoma genetic profile and could be used as a prognostic marker of disease evolution and progression.

In: Genome Research Advances
Editor: Alberto P. Bellini, pp. 1-67

ISBN: 978-1-60021-805-7

Chapter I

Bioinformatics Tools for Researching Protein Interaction Networks

Michael Hsing[1] and Artem Cherkasov[2]

[1]CIHR/MSFHR Strategic Training Program in Bioinformatics, Genetics Graduate Program, Faculty of Graduate Studies, University of British Columbia, Vancouver, British Columbia, Canada;

[2]Division of Infectious Diseases, Department of Medicine, Faculty of Medicine, University of British Columbia, Vancouver, British Columbia, Canada.

Abstract

Recent progress in high-throughput proteomics techniques resulted in large but sparse amounts of information on protein-protein interactions. Thus, one of the major goals of the modern bioinformatics research is to integrate these large-scale genomic and proteomic results to facilitate understanding of the inner workings of a cell governed by complex interaction networks.

The chapter "Bioinformatics Tools for Researching Protein Interaction Networks" reviews the current experimental and computational techniques and databases that are commonly used for researching protein-protein interactions. The overview of these state-of-the-art approaches outlines the ongoing need for the standardized computational platform that would enable accumulation and integration of the entire range of the current bioinformatics network tools.

The method of Semantic Networks (SN) has been suggested as a potential framework for the existing bioinformatics resources; several examples of effective simulation of protein interaction networks within the SN environment have been discussed.

1. Introduction

One of the major challenges of the current bioinformatics research is to integrate the multidimensional biological data and to encompass the flow of 'biological' information from individual gene to protein(s), followed by pair-wise protein interactions resulting in complex protein interactomes. At the same time, to understand complex intracellular systems, it is critically important to place protein interaction networks (PIN-s) into the context of cell spatial organization and time-dependent stoichiometry.

Different components of these general research objectives can currently be achieved through a variety of bioinformatics approaches and tools. The chapter "Bioinformatics Tools for Researching Protein Interaction Networks" elaborates on those bioinformatics techniques and reviews the progress of computational biology in predicting, storing, visualizing, analyzing, integrating and simulating individual protein interactions and large-scale PIN-s.

The chapter begins with a review of high-throughput experimental techniques for detecting protein-protein interactions, followed by the description of the current computational methods for their prediction. Next, we discuss publicly available protein interaction- and pathway databases and review the current bioinformatics tools for visualization and analysis of protein interactomes. The following section reflects on the challenges of transferring conventional static protein interaction maps into adequate physicochemical models enabling simulation of dynamic protein interactions.

The chapter "Bioinformatics Tools for Researching Protein Interaction Networks" concludes with the discussion of the method of semantic networks that represents an effective integration environment for studying protein interactions. We elaborate on our recent results illustrating effective application of the semantic network analysis for modeling kinase signaling pathways in human macrophages.

2. Experimental Methods for Detecting Protein Interactions

For decades, physical interactions between proteins have been studied by a variety of biochemical methods such as protein affinity chromatography, affinity blotting, co-immunoprecipitation and chemical cross-linking (Phizicky and Fields 1995). Recent comprehensive sequencing efforts and advancements in genomic technology led to a number of high-throughput technologies for establishing protein-protein interactions at large scale. This following section will review some of such experimental approaches and will reflect on their particular strengths and disadvantages.

2.1. Yeast two-hybrid System

The yeast two-hybrid assay detects pairwise interactions between the 'bait' and the 'pray' proteins that represent two sequences under question respectively fused with the AD and

DBD domains (Fields and Song 1989). The AD is an activating domain of the yeast transcriptional activator GAL4, while the DBD represents its DNA-binding fragment. Both these domains are required for synthesizing β-galactosidase enzyme and, therefore, when interaction between the bait and prey proteins brings the two domains into spatial proximity, the resulting β-galactosidase expression can be biochemically detected.

The yeast two-hybrid systems have been applied for identification of pairwise protein-protein interactions in several organisms including *Escherichia coli* bacteriophage T7 (Bartel et al. 1996), *Helicobacter pylori* (Rain et al. 2001), *Saccharomyces cerevisiae* (Fromont-Racine, Rain, and Legrain 1997; Fromont-Racine et al. 2000; Uetz et al. 2000; Ito et al. 2001; Drees et al. 2001), *Caenorhabditis elegans* (Li et al. 2004), *Drosophila melanogaster* (Giot et al. 2003; Stanyon et al. 2004) and some viral species (Uetz et al. 2004). Thus, Figure 1 features organism-wide protein interaction network of the *Caenorhabditis elegans* identified by the yeast-two hybrid experiments (Li et al. 2004).

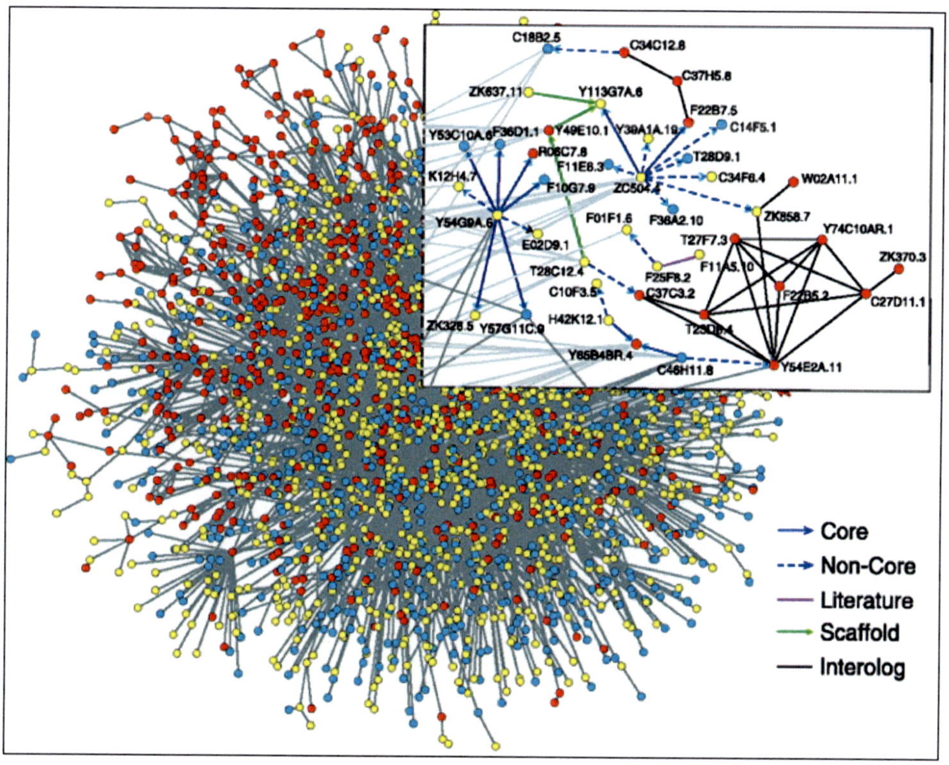

Figure 1. The network graph represents the Worm Interactome, determined from the yeast two-hybrid techniques, literature or in silico interolog method. The network contains 2898 proteins (nodes) in C. elegans and 5460 protein-protein interactions (edges). Reprinted with permission from Li et al. (2004). A map of the interactome network of the metazoan C. elegans. Science, 303, 540-543. Copyright 2005 AAAS.

An important advantage of the yeast two-hybrid assay is its simplicity and broad applicability (Fields and Song 1989). On the other hand, the assay has several technical limitations related to the expression of foreign proteins in yeast cells that may lack necessary chaperons and allosteric modulators and may cause non-adequate post-translational modifications (Fields and Sternglanz 1994; Uetz 2002) or result in untypical cellular localization (von Mering et al. 2002). There is also a potential danger that fusion between the target protein and yeast AD/DBD domains may block protein interaction surfaces (Fields and Sternglanz 1994).

2.2. Mass Spectrometry of Purified Protein Complexes

This approach consists of two independent components: *a)* purification of protein complexes occurring in a cell and *b)* identification of individual components of the complexes by mass spectrometry. The method utilizes sequence tags such as TAP (Rigaut et al. 1999; Gavin et al. 2002) and FLAG (Ho et al. 2002) that undergo fusion with the target genes at the 3' and/or 5' terminus and allow affinity-purification of the corresponding protein products. The purification step pulls down not only the tagged proteins but also their interacting partners and the isolated complexes further undergo mass spectrometry analysis (Neubauer et al. 1997; Mann, Hendrickson and Pandey 2001; Yates 2004).

To date, two mass spectrometry- constructed protein interaction maps have been reported for *Saccharomyces cerevisia* (Gavin et al. 2002; Ho et al. 2002); less detailed interactomes have also been established for certain human cell lines (Bouwmeester et al. 2004; Brajenovic et al. 2004). It is also expected that the use of high-throughout cloning techniques will soon enable systematic purification and identification of protein complexes for many other organisms and cell types (Marsischky and LaBaer 2004).

The important advantage of the affinity purification of protein complexes followed by mass spectrometry is that the method not only reveals composition of protein assemblies but also captures their actual stoichiometry (Rigaut et al. 1999; Gavin et al. 2002). The method also preserves and identifies post-translational modifications of the studied proteins (Pandey and Mann 2000) and allows their detection in very small quantities (Gavin et al. 2002). The limitations of protein tagging approach include possible interference of tags with protein complex assembly, localization and function (Gavin et al. 2002). In addition, some transient protein interactions that occur only under certain physiological conditions may go undetected by the method.

2.3. Phage Display

Phage display technology involves expressing a variety of foreign proteins on the surface of phage particles that get exposed to immobilized protein targets (Crameri and Kodzius 2001; Walter, Konthur and Lehrach 2001). After the interaction the bound phages are extracted, amplified and sequenced to identify proteins that possess binding affinity toward the targets .This approach is broadly used for large-scale screening of protein interactions as it can operate on very diverse protein sequences and detects protein interactions not only with other proteins, but also with short peptides, small molecules and drugs (Willats 2002; Sidhu, Fairbrother and Deshayes 2003). One the other hand, the phage display being an *in vitro* technology, also operates on proteins expressed in foreign environment that may lack native folding conditions and components.

2.4. Protein-Fragment Complementation Assays

The protein-fragment complementation assays (PCA) technique has the advantage over many other experimental approaches that it can detect protein-protein interactions in the living cells (Remy and Michnick 2001; Michnick 2003). This approach utilizes monomeric enzymes that can be expressed in the host cells and can be assayed *in vivo* for their catalytic activity (Pelletier, Remy and Michnick 1999; Galarneau et al. 2002). After the PCA enzyme (typically dihydrofolate reductase or β-lactamase) is rationally cleaved into two fragments, each of them undergoes fusion with a protein of interest and gets expressed in the cell. Thus, *in vivo* interaction between the 'bait' and the 'pray' proteins linked to the fragments of the PCA enzyme can be detected by assaying the corresponding enzymatic activity. Since both 'pray' and 'bait' are expressed under their native cellular conditions and at their actual cellular locations, the PCA can identify protein-protein interactions that may be missed by other *in vitro* techniques. In addition, the use of fluorescence-labeled enzymes in the PCA experiment allows real-time visualization of protein interactions. The drawback of the PCA method is common for all fusion-based techniques - the attached foreign enzymatic fragments can block certain protein-protein interactions.

2.5. ChIP-Chip

ChIP-chip technology combines chromatin immunoprecipitation method (ChIP) (Das et al. 2004) with DNA microarrays and detects *in vivo* protein-DNA interactions (Buck and Lieb 2004). The analysis begins with introduction of formaldehyde into the studied cells where the chemical cross-links proteins with their interacting DNA counterparts. After the cell is sonicated and protein-DNA complexes are extracted by immunoprecipitation or affinity chromatography (directed toward proteins), the DNA components can be released by heating. The purified DNA undergo further PCR amplification, fluorescent labeling and hybridization on the microarray base.

The ChiIP-chip technology has been successfully used in discovery of a broad range of interactions between genes and transcription factors in a number of whole-genome studies (Ren et al. 2000; Lieb et al. 2001; Iyer et al. 2001; Lee et al. 2002) and partial genome assays (Horak et al. 2002; Weinmann et al. 2002; Li et al. 2003; Wells et al. 2003). The ChIP-chip methodology has the advantage of detecting *in vivo* protein-DNA interactions and readily allows broad genomic coverage (Buck and Lieb 2004). It should be noted, however, that the ChIP-chip technology often covers only up to 1000 bp upstream gene regions and therefore, in order to study certain components of mammalian transcription machinery, it should be complemented with other methods such as the CpG islands sampling (Weinmann et al. 2002; Wells et al. 2003).

2.6. Protein Microarrays

The protein microarrays technology has been developed in recent years to enable direct measurement of protein expression levels and to detect protein-protein interactions in high-throughput fashion (Zhou et al. 2001; Kodadek 2002; Lopez and Pluskal 2003). Different protein capturing agents including specific antibodies, peptides, proteins or small molecules have been evaluated in various assays, typically employing fluorescent detection, followed by mass spectrometric analysis (MacBeath and Schreiber 2000; Zhou et al. 2001; Fung et al. 2001; Haab, Dunham, and Brown 2001; Borrebaeck et al. 2001; Cahill 2001; Paweletz et al. 2001; Williams and Cole 2001). Notably, such microarrays experiments allow large-scale evaluation of protein interactions in the whole cell lysates or for the phage display libraries (Pandey and Mann. 2000). Several protein microarrays systems have been specifically developed for detecting protein-protein interactions in yeast and human cell lines (MacBeath and Schreiber. 2000; Zhu et al. 2001; Paweletz et al. 2001).

2.7. Assessment of Interaction Data

Several comprehensive comparative studies have been carried out to evaluate the performance by the conventional high-throughput PIN identification approaches (von Mering et al. 2002; Bader and Hogue 2002; Edwards et al. 2002). It has been demonstrated that the current high-throughput techniques do not exhibit significant mutual overlap, some of them are biased toward higher abundance proteins while others suffer from under-representation of trans-membrane and translational proteins (von Mering et al. 2002). It has also been demonstrated that more than a half of the reported interactions detected by the yeast two-hybrid and complex purification experiments are, in fact, false-positive (von Mering et al. 2002). The rates of false-negative predictions have also been estimated as 45%-96% for yeast two hybrid system and as 15%-50% for mass spectrometry-based approaches (Edwards et al. 2002). It has been suggested that only comprehensive integration of high-throughput experimental results with other types of genomic and proteomic data, such as information on gene co-expression, proximity and protein co-localization can extrapolate true protein-protein interactions (von Mering et al. 2002; Kemmeren et al. 2002).

3. Prediction Methods for Protein Interactions

To complement costly and labor-intensive experimental high-throughput approaches, several computational models have been developed to enable forecasting hypothetical interaction between proteins using merely their sequences and/or structures. Some of the most popular conventional bioinformatics approach for predicting protein-protein coupling will be discussed in the following section.

3.1. Gene Proximity

Several previous large-scale bioinformatics analyses indicated that conservation of close-proximity of genes in multiple genomes may indicate functional association or physical interaction between the corresponding gene products (Dandekar et al. 1998; Overbeek et al. 1999). Thus, Dandekar et al. (1998) analyzed the conserved gene pairing in nine prokaryotic genomes and estimated that up to 75% of such gene pairs encode physically interacting proteins. Similarly, Overbeek et al. (1999) considered various prokaryotic genes separated by less than 300 bp in the studied genomes and demonstrated consistent functional association between the corresponding gene products. It should be mentioned, however, that genomic co-localization approximation is less informative for eukaryotic genes that can be organized in more complex manner (Enright et al. 1999).

3.2. Gene Fusion

Gene fusion approach utilizes the observation that the occurrence of two individual genes from one organism as parts of a single (fused) gene in another genome implies functional association and/or physical interaction between the encoded proteins (Marcotte et al. 1999; Enright et al. 1999). It has also been demonstrated that fusion may not only occur between two adjacent genes and that searching for fused gene components can merely rely on sequence homology (Enright and Ouzounis 2001a). Thus, the systematic identification of the fused genes in 24 complete prokaryotic and eukaryotic genomes revealed 39,730 possible pair-wise protein-protein associations (Enright and Ouzounis 2001a). Despite the fact that only small fraction of the predictions could be confirmed by the yeast two-hybrid results, good correlations with the gene co-expression data have been established (Enright and Ouzounis 2001a).

3.3. Phylogenetic Profiling

The phylogenetic profiling detects interacting proteins from the patterns of common genomic co-occurrence (Pellegrini et al. 1999). The method has been rigorously tested on the complete genome of *E. coli* where all 4,290 proteins have been associated with 16 binary (composed of ones and zeros) strings reflecting the presence of close homologues in other 16 complete genomes. The resulting Boolean vectors have then been used for clustering *E. coli* genes into groups implying possible physical interactions or similar function among the members. The resulting associations clearly demonstrated that similarly annotated sequences, proteins participating in the same metabolic pathway(s) or proteins localized in the same intracellular compartment(s) are more likely to cluster together according to their phylogenetic profiles (Ouzounis and Kyrpides 1996; Pellegrini et al. 1999; Marcotte et al. 2000).

3.4. Interologs

The notion of interologs is based on the observation that some pairwise protein interactions are conserved among organisms and, therefore, can be inferred from protein homology (Walhout et al. 2000). Thus, 1,195 interactions between *S. cerevisiae* proteins identified by the yeast two-hybrid experiments (Uetz et al. 2000; Ito el al. 2001) have been used to predict 257 interactions among 282 orthologous proteins in *C. elegans*, and 16% of these prediction could be confirmed experimentally (Matthews et al. 2001). In the work by Yu *et al.* (Yu et al 2004a), authors investigated interacting proteins from *S. cerevisiae, C. elegans, D. melanogaster, and H. pylori* and demonstrated that interlog inference is mostly feasible at >80% homology level. In other studies the interologs-based approach has been successfully used for predicting protein-protein interactions in *C. albicans, A. thaliana,* and *H. sapiens* (Matthews et al. 2001; Yu et al. 2004a; Lehner and Fraser 2004).

3.5. Gene Co-Expression

It is a broadly accepted notion, that co-expression of two genes over multiple experimental conditions implies possible interaction between the corresponding gene products. Numerous studies reported correlations between genes expression patterns and results of protein interaction experiments (Ge et al. 2001; Grigoriev 2001; Jansen, Greenbaum and Gerstein 2002). Thus, in the work (Ge et al. 2001) authors used clustering analysis to demonstrate that similarly expressed proteins either interact or have similar functions. The Bayesian Network approach has also been successfully applied to predict yeast protein-protein interactions from the gene expression data (Jansen et al. 2003).

3.6. '*In Silico*' two-hybrid Systems

'*In silico*' two-hybrid approach aims to predict protein-protein interactions by analyzing correlated mutations in protein sequences (Pazos and Valencia 2002). It has been shown that sequence mutations accumulated at the interaction interface of one protein are typically compensated by sequence changes in the binding partner. Such correlated mutations can often be identified using multiple sequence alignment of interacting proteins (Gobel et al. 1994; Pazos et al. 1997) and by computing the "interaction index" representing ratios of sequence correlations (Pazos and Valencia 2002). This approach has been successfully applied to *E. coli* proteome where it produced a number of high scores for known protein interactions (Pazos and Valencia 2002). The '*in silico*' two-hybrid system has the advantage of not only predicting protein-protein interactions but also suggesting potentially interacting residues. On the other hand, just like all other predictive approaches, the '*in silico*' two-hybrid system requires proper validation, while there are no 'golden standard' benchmark sets yet available for the protein interaction data (Pazos and Valencia 2002).

3.7. Interacting Domains and Motifs

Most protein-protein interactions are mediated by conserved binding domains or short sequence motifs, and therefore, experimentally verified domain-domain and domain-motif interactions can be used for developing general predictive approaches (Gomez and Rzhetsky 2002; Ng, Zhang, and Tan 2003; Obenauer and Yaffe 2004; Reiss and Schwikowski 2004). A large number of protein domains have been identified and catalogued into specialized databases that can be used for establishing hypothetical protein interactions (Sigrist et al. 2002; Bateman et al. 2004; Mulder et al. 2005). Thus, various protein interaction sources have been used to identify pairing among the entries of *Pfam* domain database (Bateman et al. 2004; Ng, Zhang, and Tan 2003; Finn, Marshall, and Bateman 2005). Another popular domain-based service *InterDom* can identify interacting proteins from their overall domain composition (Ng et al. 2003), while the *iPfam* predictions utilize spatial proximity between *Pfam* domains (Finn, Marshall, and Bateman 2005) (Figure 2 illustrates protein interactions identified for SH3 domain by the *iPfam* resource). The *Scansite* database catalogues position-specific scoring matrices for a number of known sequence motifs that can bind to common protein domains (Obenauer, Cantley, and Yaffe 2003), while Reiss and Schwikowski (2004) utilized probabilistic model for discovering domain binding motifs. Numerous other theoretical approaches for predicting protein-protein interactions from the primary sequence information are currently under way (Bock and Gough 2001; Martin, Roe, and Faulon 2005).

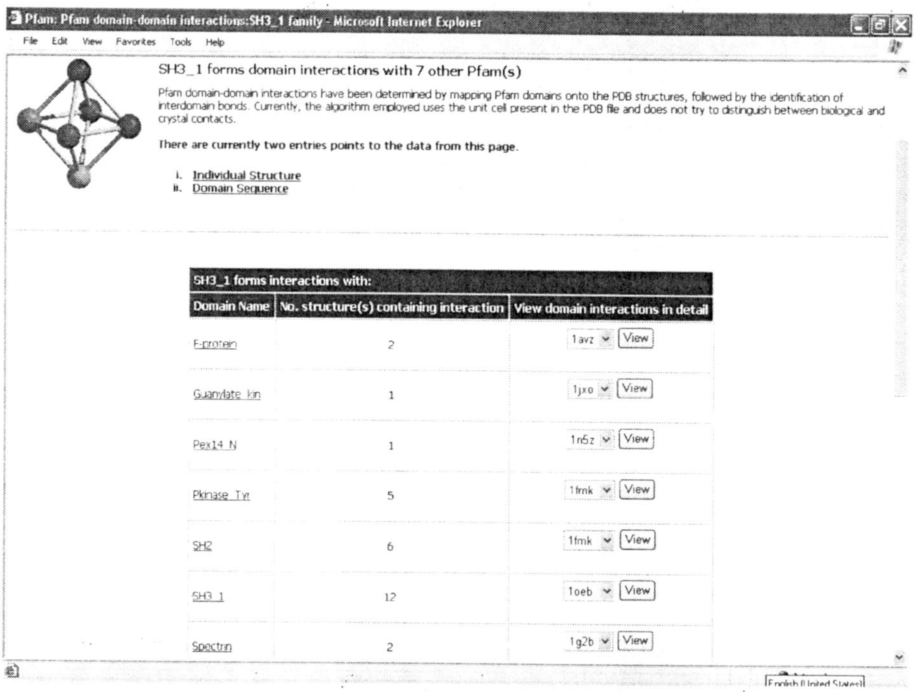

Figure 2. iPfam determines domain-domain interactions based on protein complex structures in PDB (Berman et al. 2000). The 7 protein domains that interact with SH3 domain are shown. The screenshot was reprinted with permission from the iPfam website.

3.8. Protein Docking and Threading

Protein docking and threading approaches differ from all computational techniques described above, as they predict interactions between proteins from their tertiary structures. The concept of protein docking relies on the search for the best geometrical and polar fit between two interacting protein surfaces (Janin 1995; Gabb et al. 1997). Thus, docking allows studying various physical-chemical aspects of protein interactions and can bring an important insight into configuration of protein binding complexes (Janin 2005). At the same time, the docking techniques are computationally intensive and not yet very reliable, and therefore they are less suitable for predicting protein interactions on a genome-wide scale (Lu, Lu and Skolnick 2002).

The multimeric threading approach represents a good alternative for macromolecular docking as it can predict protein interactions by fitting protein sequences into known protein complexes or standardized interacting folds (Lu, Lu and Skolnick 2002; Aloy and Russell 2002; Aloy and Russell 2003). The sequence-structure threading not only readily predicts protein-protein interactions, but it also provides adequate structural models for the resulting protein complexes and can identify contacting protein residues. When the multimeric threading approach has been applied to the complete yeast genome, it identified 7321 pair-wise interaction between 1256 proteins, and 374 of the predicted interactions could be confirmed by experiments (Lu et al. 2003).

3.9. Text Mining

To date, large amounts of information on protein-protein interactions have been accumulated in the literature. Therefore, the text mining methods can be very effective in extracting protein interaction data and, not surprisingly, such approaches are broadly used in the current bioinformatics practice (Marcotte, Xenarios, and Eisenberg 2001; Temkin and Gilder 2003; Albert et al. 2003; Donaldson et al. 2003; Daraselia et al. 2004; Hoffmann et al. 2005). The text mining tools such as PreBIND (Donaldson et al. 2003), MedScan (Daraselia et al. 2004), *etc* use natural language processing algorithms to find protein names and interaction keywords in article abstracts. The identified text associations typically undergo further expert curation to produce hypothetical protein interactions. Many protein interaction databases described in the following sections have been established by the large-scale automated text mining.

3.10. Additional Resources for Interaction Prediction

Even the brief overview of the current experimental and theoretical methods for identification of protein-protein interactions clearly demonstrates that none of the existing approaches alone can provide reliable and general solutions. Not surprisingly, a number of integrative schemes have been developed to incorporate multiple bioinformatics sources into the unified schemes for assessing quality of putative protein interactions. Thus, the STRING methodology integrates information on gene proximity, gene fusion and sequence phylogeny to predict functional associations between proteins in 89 fully sequenced genomes (von Mering et al. 2005). To illustrate the approach capabilities, Figure 3 features PIK3R1 (phosphoinositide 3-kinase p85 subunit) protein and its associates determined by various STRING modules. Another integrative resource *Predictome* also utilizes gene proximity, gene fusion and phylogenetic profiling to establish functional links between proteins from 44 genomes (Mellor et al. 2002). This approach does not operate by the unified consensus score, but rather integrates multiple predictions in non-discriminative manner.

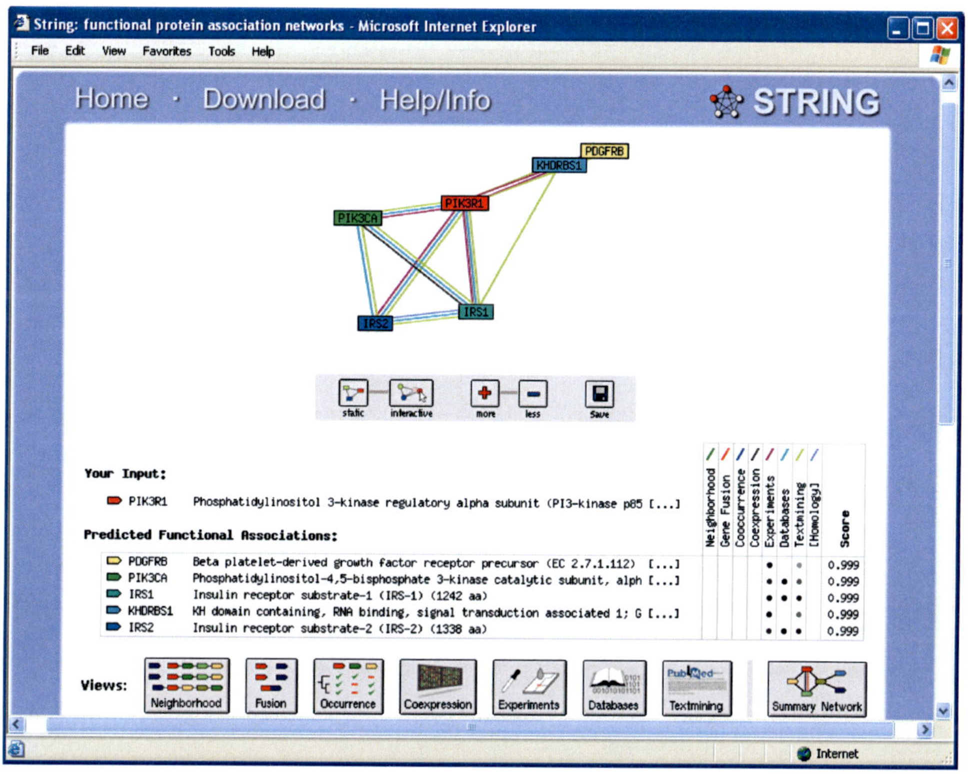

Figure 3. PIK3R1 protein (p85 subunit of a PI3K complex) in human is searched in STRING database for its functional associated partners, determined by methods including gene proximity, fusion, phylogenetic profiling, co-expression, experiments, other interaction databases and text-mining. The top 5 functional associated proteins with PIK3R1 and their interaction network are shown. The screenshot was reprinted with permission from the STRING website.

4. PROTEIN INTERACTION AND PATHWAY DATABASES

The current bioinformatics resources that store, integrate and disseminate large amounts of data on protein-protein interactions can be conventionally divided into two main classes of interaction and pathway databases. The formers archive individual protein interactions, while the pathway resources group them according to certain cellular functions such as metabolism, signal transduction or gene regulation. The next section will overview the most commonly used protein interaction databases and will briefly discuss their scope, structure, availability, coverage and data quality along with the accompanying bioinformatics tools.

4.1. Protein Interaction Databases

BIND

The Biomolecular Interaction Network Database (BIND, http://bind.ca/) contains information on more than 178,509 individual molecular interactions between proteins, DNA, RNA and small compounds organized into object-oriented data model (Bader, and Hogue 2000; Alfarano et al. 2005). Currently, BIND partially covers more that 1300 organisms and gets expert-curated daily updates encompassing the results of cross-linking, yeast two-hybrid, co-immunoprecipitation, chromatographic and crystallographic experiments.

The typical BIND interaction record encompasses information about the binding substances, their binding sites and chemical states. Figure 4 features BIND record for the query protein PIK3R1 where rows represent individual pairwise interactions retrieved from the database. The BIND is freely accessible through the web; it supports on-line searching and FTP downloads and is equipped with the protein interaction viewer and the text mining tool (PreBIND) among other bioinformatics services.

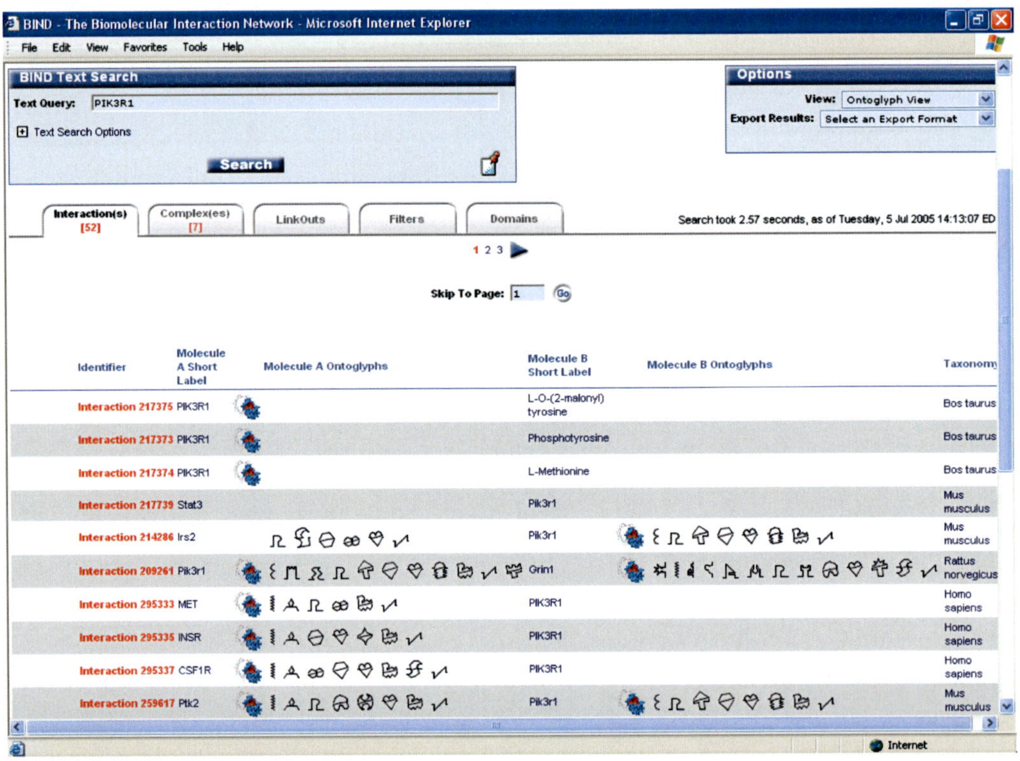

Figure 4. Protein interactions that involve a query protein, PIK3R1 are displayed in the separate rows in BIND. The binding molecules that have the gene ontology (GO) annotations are marked with the 'OntoGlyphs' icons. The screenshot was reprinted with permission from the BIND website.

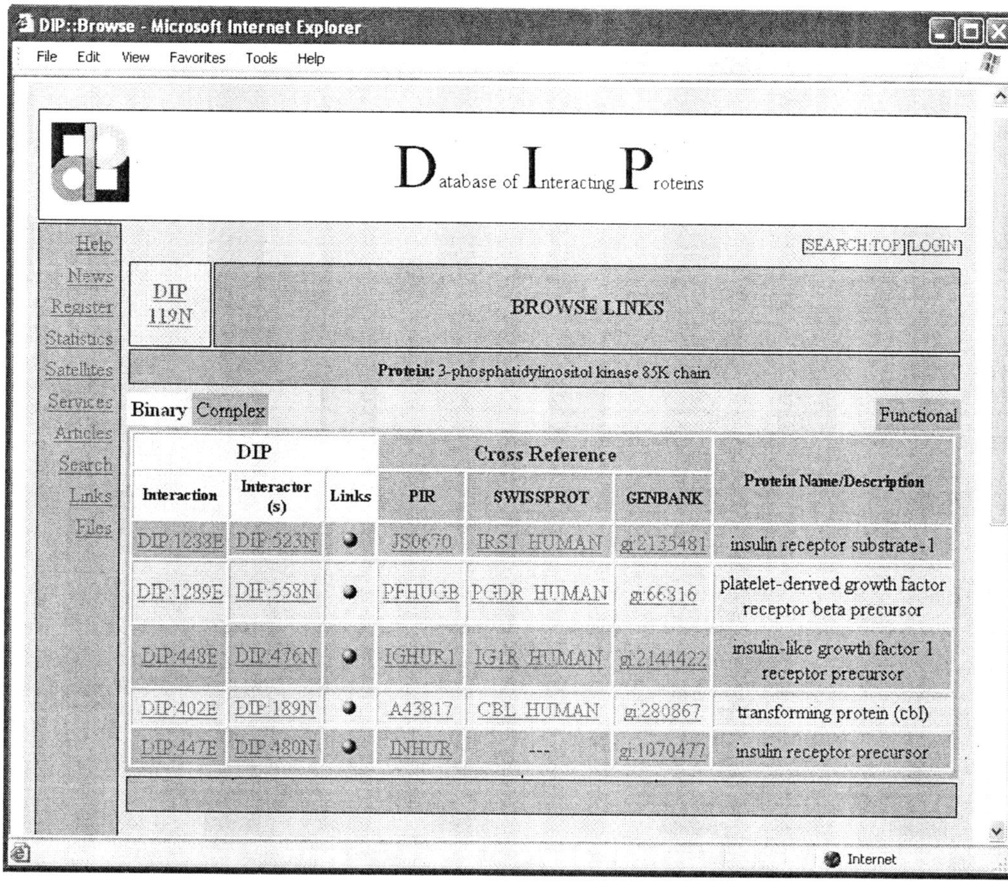

Figure 5. A screenshot of the DIP result page. Each interactor with 3-phosphatidylinositol kinase 85K chain (query protein) is shown on a separate row. The screenshot was reprinted with permission from the DIP website.

DIP

The Database of Interacting Proteins (DIP, http://dip.doe-mbi.ucla.edu/) catalogues 46,463 interactions among 17,556 proteins from 109 organisms; with the majority of data corresponding to *Drosophila melanogaster* and *Saccharomyces cerevisiae* species (Salwinski et al. 2004). The DIP is organized in a form of a relational database, composed of individual tables for proteins and interactions. The interaction data stored in the DIP is accessible via the web and can be downloaded in the XML, tab-delimited and PSI-MI formats. To illustrate the DIP capabilities, Figure 5 features the search record returned for 3-phosphatidylinositol kinase. This chart illustrates that the DIP resource offers several means for assessing the quality of protein interaction data such the EPR Index (standing for the Expression Profile Reliability) (Deane et al. 2002). The DIP also hosts the Paralogous Verification Method (PVM) and Domain Pair Verification (DPV) bioinformatics services as well as other complementary tools including *LiveDIP* and *JDIP* (Deane et al. 2002; Deng et al. 2002).

IntAct

An open source database *IntAct* (http://www.ebi.ac.uk/intact/index.jsp) contains 42,952 protein interactions abstracted from the literature and high-throughput experiments (Hermjakob et al. 2004b). The *IntAct* relational data model consists of three major components: *experiment*, *interactor* and *interaction* that are linked to the controlled vocabularies of the NCBI taxonomy and Gene Ontology (Wheeler et al. 2000, Harris et al. 2004). Figure 6 features a representative *IntAct* search record for *p85a_human* protein (a regulatory subunit of a PI3K complex) used as the query.

The *IntAct*-specific attributes ensure data integration and consistency; the content is monthly updated and can be accessed through the web and FTP downloads. The *IntAct* database provides a number of bioinformatics tools such as *ProViz* and MiNe that can visualize and analyze protein networks.

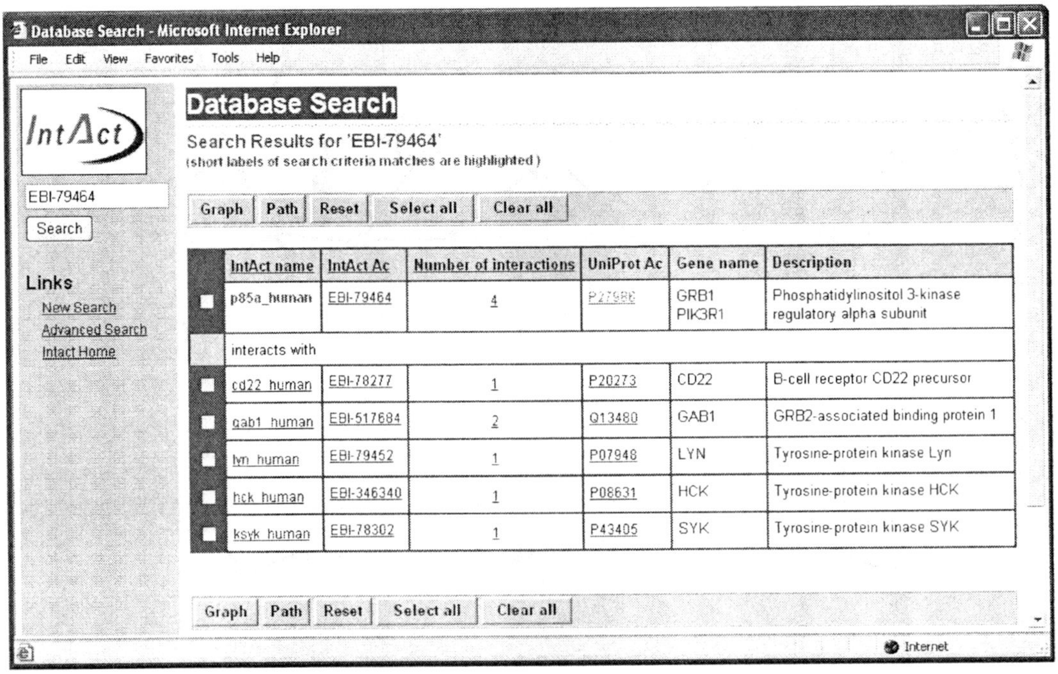

Figure 6. Snapshot of the IntAct result page for p85a_human query protein. The screenshot was reprinted with permission from the IntAct website.

GRID

The General Repository for Interaction Datasets (GRID: http://biodata.mshri.on.ca/grid/servlet/Index) stores 19,791 interactions from *Saccharomyces cerevisiae* (Yeast GRID), 4,453 interactions from *Caenorhabditis elegans* (Worm GRID) and 28,406 *Drosophila melanogaster* interactions (Fly GRID) (Breitkreutz, Stark, and Tyers 2003a). The GRID mostly contains protein-protein interactions that have been detected by the yeast two-hybrid assays and mass-spectrometry; the records are organized into periodically updated relational databases searchable through the web The GRID's graphic tool, Osprey allows visualization of individual protein interactions and protein networks (Breitkreutz,

Stark, and Tyers 2003b). Thus, Figure 7 demonstrates the GRID search results for the interacting partners of the Pi3K21B protein from *Drosophila melanogaster*.

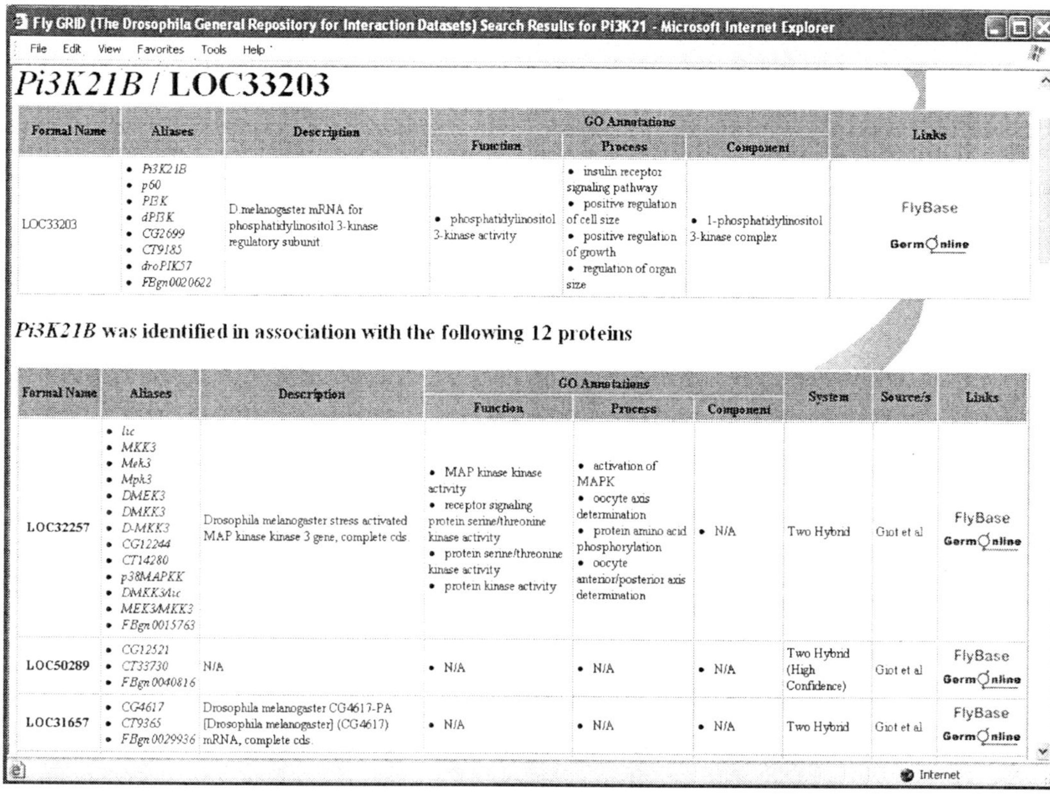

Figure 7. Pi3K21B protein of Drosophila is searched in Fly GRID, and each of its interacting proteins is displayed in an individual row of the table. The screenshot was reprinted with permission from the GRID website.

MINT

The MINT (http://mint.bio.uniroma2.it/mint/) - Molecular INTeraction database - stores data on protein-protein and protein-DNA interactions, enzymatic reactions and functional associations between proteins, genes and lipids (Zanzoni et al. 2002). Currently, the MINT contains 50,145 interactions between 20,675 proteins from 21 organisms. The MINT data has been derived from the yeast two-hybrid experiments, co-immunoprecipitation studies and from the literature. The interaction record of this relational database includes annotation, participating protein domains, kinetic constants and experimental source (Figure 8 features the MINT record for the human PI3K enzyme). The MINT is updated on a monthly basis and is offered freely in the XML-PSI and flat file formats. Two additional bioinformatics tools are available from the MINT resource: *MINT-Assistant* (for the text-mining) and *MINT-Viewer* (for visualizing protein interactions).

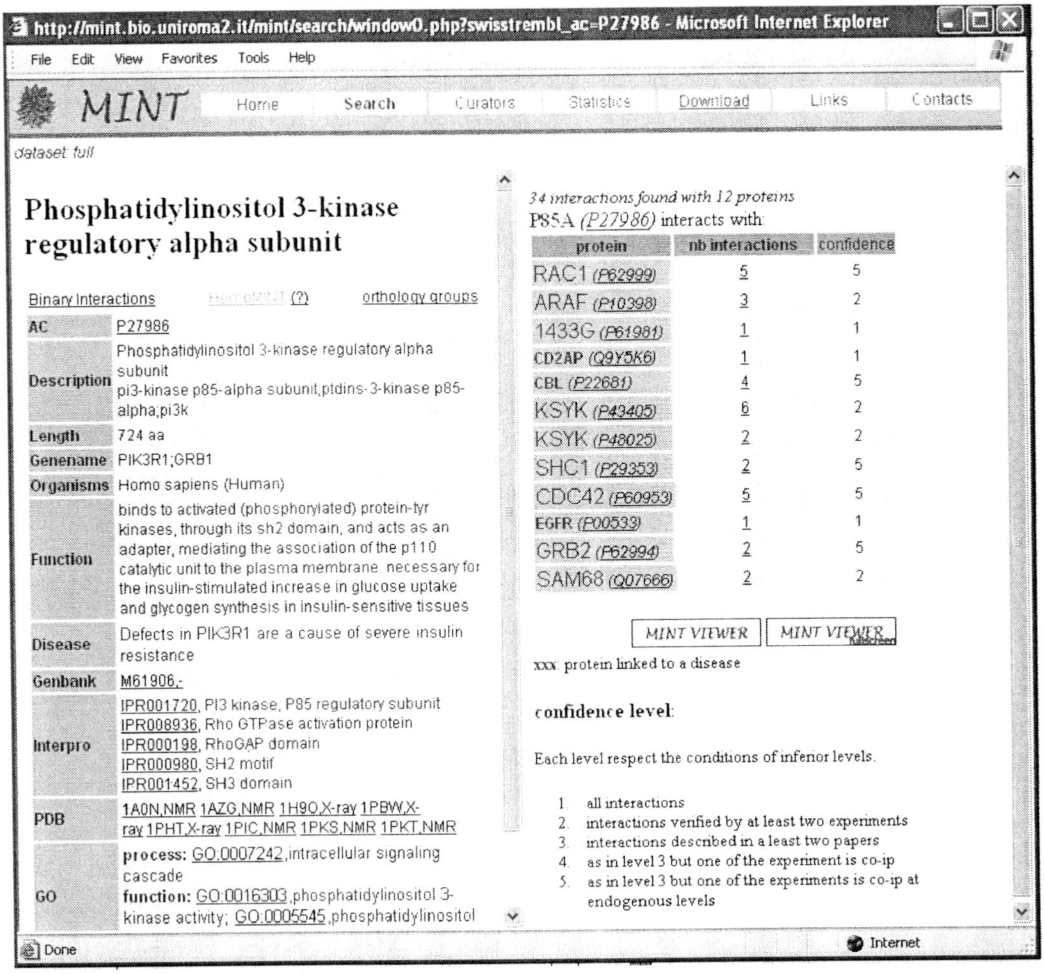

Figure 8. A protein record of PI3K alpha subunit in human displays the protein annotation on the left frame. The interacting proteins of PI3K and the number of experimental evidence are shown on the right frame of the window. The screenshot was reprinted with permission from the MINT website.

4.2 Pathway databases

KEGG

The Kyoto Encyclopedia of Genes and Genomes (KEGG, http://www.genome.jp/kegg/) is the knowledge-based resource that integrates genomic-, proteomic- and chemical information stored in several specialized and cross-referenced databases (Kanehisa et al. 2004). The genomic data are organized into GENES, SSDB and KO sub-databases, while the chemical information is catalogued into COMPOUND, GLYCAN, REACTION and ENZYME resources (Kanehisa et al. 2004). The protein interactions are represented by the PATHWAY database in a form of graphic diagrams (see Figure 9 for examples).

The KEGG pathway maps are organized into five major groups: metabolism, genetic information processing, environmental information processing, cellular processes and human diseases. Within each group, the maps are further classified by the corresponding cellular functions such as glycolysis, amino acid biosynthesis and lipid biosynthesis among others. The general reference map are manually drawn by the database curators, while the organism-specific diagrams are computationally generated and cross- linked to the records for orthologous genes.

Currently, the PATHWAYS database contains over 235 reference maps covering 74 organisms of eukaryotic origin, 210 bacteria and 21 archaea. The KEGG is daily updated; the stored pathway maps can be copied in GIF, HTML and KEGG Markup Language (KGML) formats.

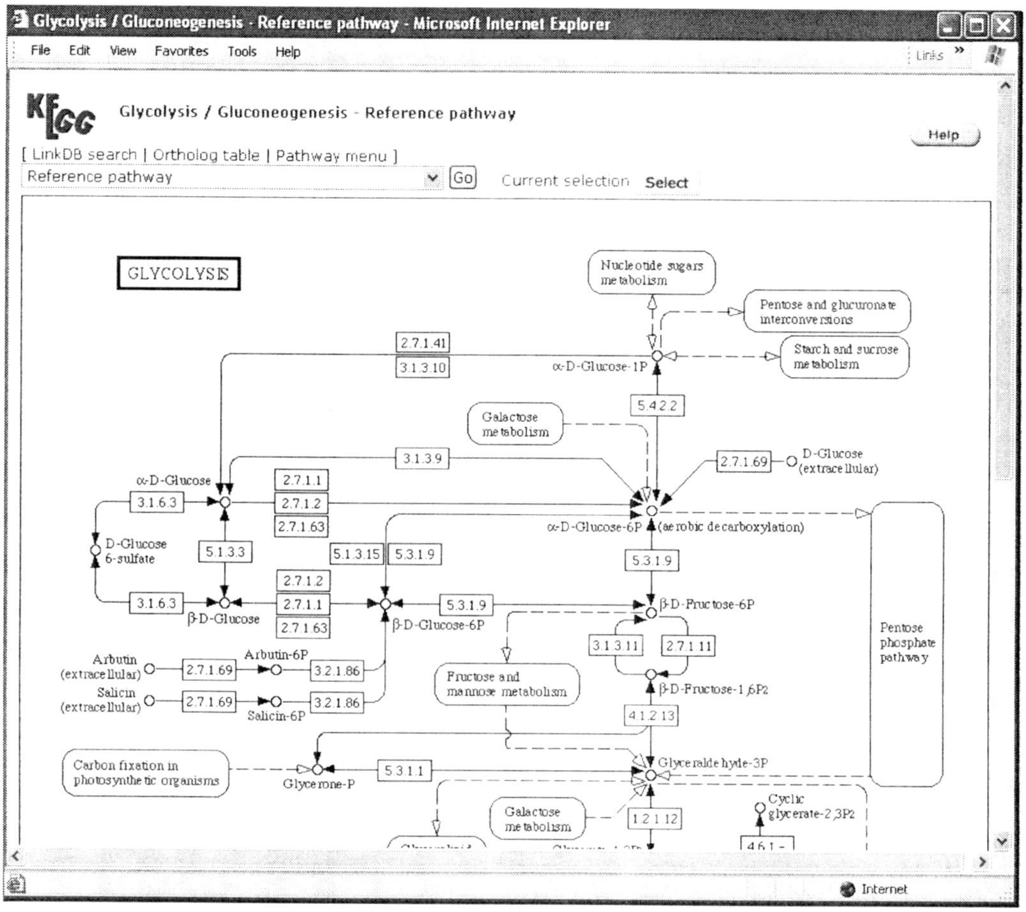

Figure 9. The reference pathway of glycolysis in KEGG is shown as a graph, where enzymes are displayed as boxes with EC numbers and chemical compounds are depicted as circles. Different arrows indicate interactions or relationships between biological entities. The screenshot was reprinted with permission from the KEGG website.

MetaCyc

The MetaCyc (http://metacyc.org/index.shtml) is a database of experimentally identified metabolic pathways for more than 240 organisms, mainly of microbial and plant origins (Krieger et al. 2004). Currently, there are over 500 metabolic pathways collected in the *MetaCyc* that encompass more than 4,900 enzymatic reactions. The stored pathways are classified according to their functions and represented as graphic diagrams composed of series of individual enzymatic reactions. The pathway maps are updated four times a year and contain great deal of useful additional information such as participating co-factors and ligands are hyper-linked to the corresponding literature references (an example of *MetaCyc* glycolysis pathway is demonstrated on Figure 10).

The *MetaCyc* is an extensively curated part of the *BioCyc* environment that incorporates computationally derived pathways from 159 organisms (BioCyc 2005). The *MetaCyc* is also equipped with the pathway prediction tool – the SRI's Pathway Tools software (Karp, Paley, and Romero 2002). The MetaCyc content is available for download in BioPAX, SBML, Pathway Tools and column-delimited formats; the BioCyc can also be installed locally.

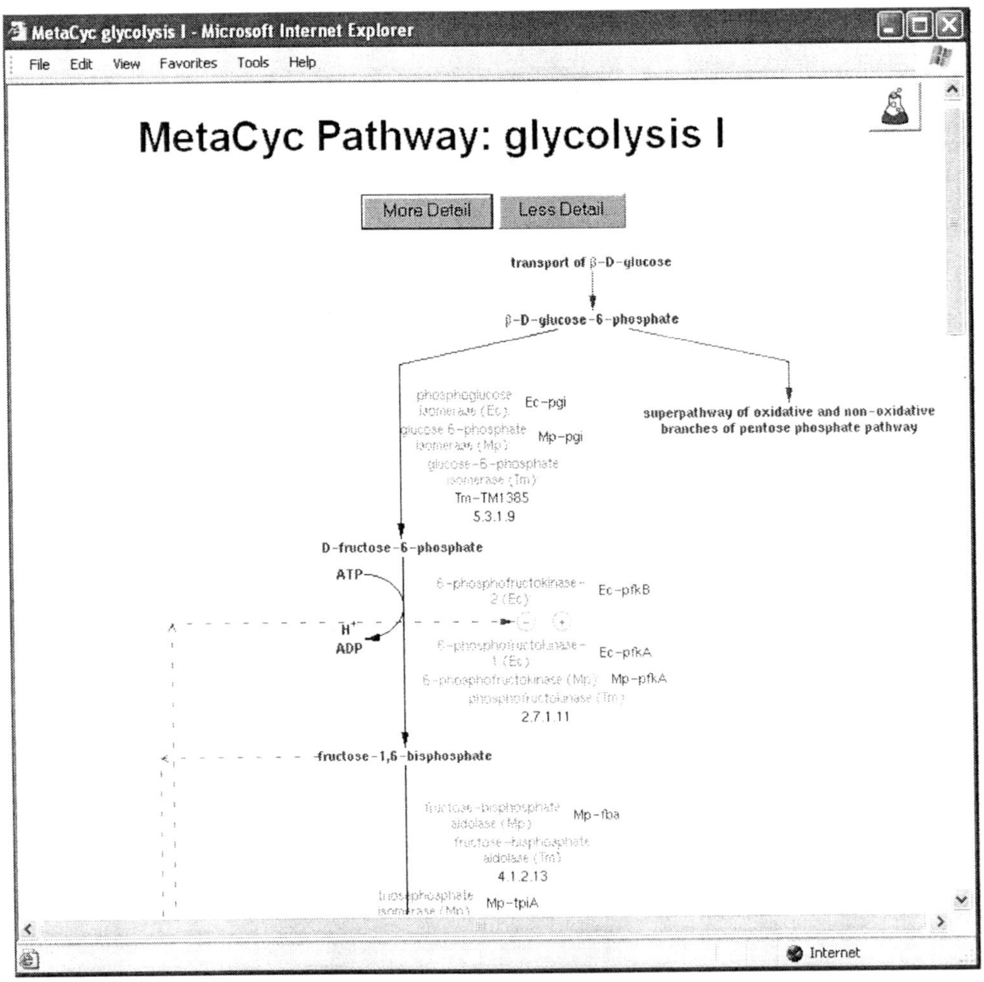

Figure 10. A glycolysis pathway is displayed as a graph in MetaCyc. Chemical compounds are shown in red texts and linked by blue arrows, representing chemical reactions. Enzymes are in orange texts and shown beside the blue arrows. Details about the pathways can be hided or displayed by clicking the corresponding buttons. The screenshot was reprinted with permission from the MetaCyc website.

STKE

The Signal Transduction Knowledge Environment (STKE, http://stke.sciencemag.org/) is a resource that classifies the signal transduction data into four major groups: literature, community, resources and connection maps (Gough 2002). The literature section of the STKE contains articles and reviews on the recent findings in the field of signal transduction (quarterly updated). The STKE connection maps are based on the expert-curated relational database that contains 39 canonical and 23 organism-specific signaling pathways. The map components are hyper-linked to multiple additional sources (for more details Figure 11 illustrating the PI3K pathway). Remarkably, the STKE contains numerous computational animations that illustrate signaling events across various intracellular compartments.

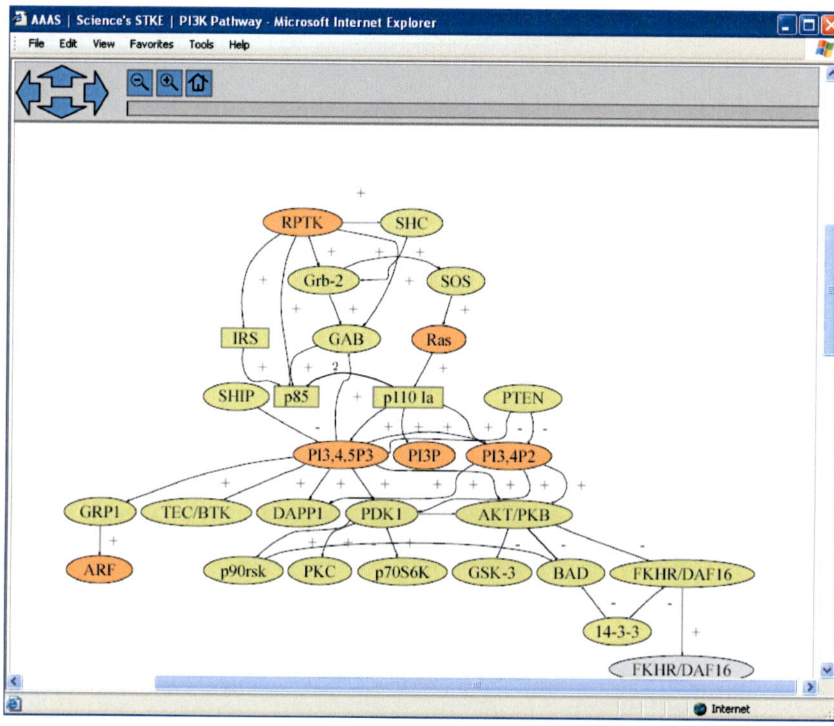

Figure 11. The PI3K pathway in STKE is displayed as a graph. The molecules are colored according to their intracellular location. A line with "+" sign represents stimulatory interaction, while a line with "-" is inhibitory. Reprinted with permission from Lewis C. Cantley, PI3K Pathway. Sci. STKE (Connection Map, as seen 14 Jun 2005, http://stke.sciencemag.org/cgi/cm/stkecm;CMP_6557).

TRANSPATH

The TRANSPATH (http://www.biobase.de/pages/products/transpath.html) is another large signal transduction database that is derived from more than 17,600 literature references (Choi et al. 2004b). This quarterly updated depository is an integral part of the TRANSFAC resource (Matys et al. 2003) that archives information on transcription factors and gene regulatory sites. The TRANSPATH catalogues 28,456 individual molecules, 10,962 genes and 52,007 reactions from a number of organisms (with an emphasis on the mammalians). The biochemical reactions are represented by the TRANSPATH at two levels: *mechanistic reactions* that capture detailed information on molecular interactions and *semantic reactions* representing the simplified view on conventional biochemical pathways (Choi et al. 2004a). The reactions stored in the TRANSPATH are assigned with the reliability values reflecting goodness of the corresponding experimental source(s). The TRANSPATH allows visualizing complex signaling networks with the PathwayBuilder application (Choi et al. 2004b). Another TRANSPATH utility - the ArrayAnalyzer can utilize the gene expression results to identify potential drug targets from the signaling pathways (Krull et al. 2003). In addition, the TRANSPATH offers 54 pre-generated signaling maps summarizing the current knowledge on the most studied cellular functions.

aMAZE

aMAZE (http://www.amaze.ulb.ac.be*)* is a relational database storing data on gene expression, gene regulation and metabolic reactions (Lemer et al. 2004). The data model of *aMAZE* is based on of three object types: *biochemical entities* (that include proteins, genes and small compounds), *interactions* (covering gene expression, protein assembly, translocation processes and chemical reactions) and *processes* objects representing biochemical pathways. Currently, the *aMAZE* contains over 118 pathways from *E. coli, S. cerevisiae* and *H. sapiens*. The *aMAZE* graphic web interface *LightBench* (Lemer et al. 2004) enables free access to the database content.

Currently, there is a significant number of other publicly available resources that store protein-protein interactions such as *Reactome* (Joshi-Tope et al. 2005) and *BioCarta* (BioCarta 2005) among others. The *Pathway Resource List* (Bader and Cary 2005) contains 181 interaction and pathway resources available on internet, sorted according to their web popularity. Many research groups and bioinformatics-oriented companies maintain in-house PIN databases and develop their own database formats.

4.3. Data Standards

It has already been outlined at several occasions that one of the most important tasks of the modern bioinformatics research is to integrate multiple sources on protein interaction data and to merge them with other types of genomic and proteomic information. This goal is particular challenging as there is no common standard utilized by the current interaction and pathway depositories including the above described resources. Thus, several collaborative bioinformatics projects aim to develop the universal data standard that would allow effective data sharing (Hermjakob et al. 2004a). Thus, the members of the Proteomics Standards Initiative (PSI) elaborated the PSI-MI standard representing the extensible markup language for biological interaction data. The PSI structured text file is composed of six major elements: *source*, *availabilityList*, *experimentList*, *interactorList*, *interactionList* and *attributeList*. These sections can be further divided into hierarchically controlled sub-elements describing more details on experimental techniques and conditions, classification of source organisms, *etc* (Hermjakob et al. 2004a) The PSI-MI data standard is already being adopted by several major protein interaction databases such as *BIND*, *DIP*, *MINT* and *IntAct*.

Numerous other database standards have also been proposed including *BioPAX* (BioPAX 2005), *SBML* (Hucka et al. 2003) and *CellML* (Lloyd, Halstead and Nielsen 2004) along with novel standards for molecular interactions involving RNA, DNA and small ligands (Cary, Bader, and Sander 2005).

5. Visualization and Analysis of Protein Interaction Networks

5.1. Visualization Algorithms and Tools

PIN visualization approaches represent an important mean for understanding complex biochemistry of living cells. Most of the current visualization algorithms map protein interactions in a form of two dimensional graphs consisting of nodes and connecting edges that can be associated with weights and directions (Basalaj and Eilbeck 1999; Uetz, Ideker, and Schwikowski 2002). Such data representation allows employing well-developed graph theory and to facilitate effective analysis of biological networks (Bollobas 1998).

In addition to the graph theory, various graphic symbols and notations have been proposed for detailed representation of biological molecules and their interactions (Kohn 1999; Cook, Farley, and Tapscott 2001; Kitano 2002). For example, the graphic notations developed by Kohn (1999), Cook, Farley, and Tapscott (2001) visualize proteins as a collection of binding domains and modification sites and represent non-covalent bindings and chemical reactions by directed arrows reflecting cause-effect relationships of biochemical interactions much like input and output signals of a computer chip (Cook, Farley, and Tapscott 2001; Kitano 2002). Some specialized graphic notations have been adopted by the KEGG database; some have been developed within other stand-alone visualization tools such as CADLIVE (Kurata, Matoba, and Shimizu 2003).

As it has been previously illustrated, the majority of the current interaction and pathway databases are equipped with some kind of PIN-s graphing tools. Thus, *BIND*, *DIP*, *IntAct* and *TRANSPATH* can automatically map interactions from a set of query substances, while other resources like *KEGG*, *MetaCyc*, *STKE* and *BioCarta* use manually drawn or precompiled diagrams of biochemical interactions and pathways. There are also a significant number of general-purpose visualization and analysis tools capable of taking arbitrary interaction dataset and converting it into two-dimensional graphs suitable for further analysis. These bioinformatics approaches include *Pajek* (Batagelj and Mrvar 1998), *BioLayout* (Enright and Ouzounis 2001b), *PATIKA* (Demir et al. 2002), *NetBuilder* (Brown et al. 2002), *Pathway Tools* (Karp, Paley, and Romero 2002), *PathwayAssist* (Nikitin et al. 2003) and *VitaPad* (Holford, Nadkarni, and Zhao 2005) among others. In the following section we will describe only the most commonly used or distinguished bioinformatics applications for visualizing large-scale protein interactomes.

Cytoscape

The *Cytoscape* (www.cytoscape.org) application is a broadly used visualization software that graphs molecular interactions into two-dimensional networks, where molecules are depicted as nodes and their interactions are represented by lines or arrows (Shannon et al. 2003). The *Cytoscape* utilizes interaction files that contain lists of protein-protein or protein-

DNA associations and transforms them into interaction maps, superimposable with other data types. For instance, various expression data can be used for color-coding of the network nodes, while the reaction kinetics parameters can be incorporated through the controlled thickness of the graph edges. Figure 12 illustrates a fragment of yeast PIN-s represented by the *Cytoscape*.

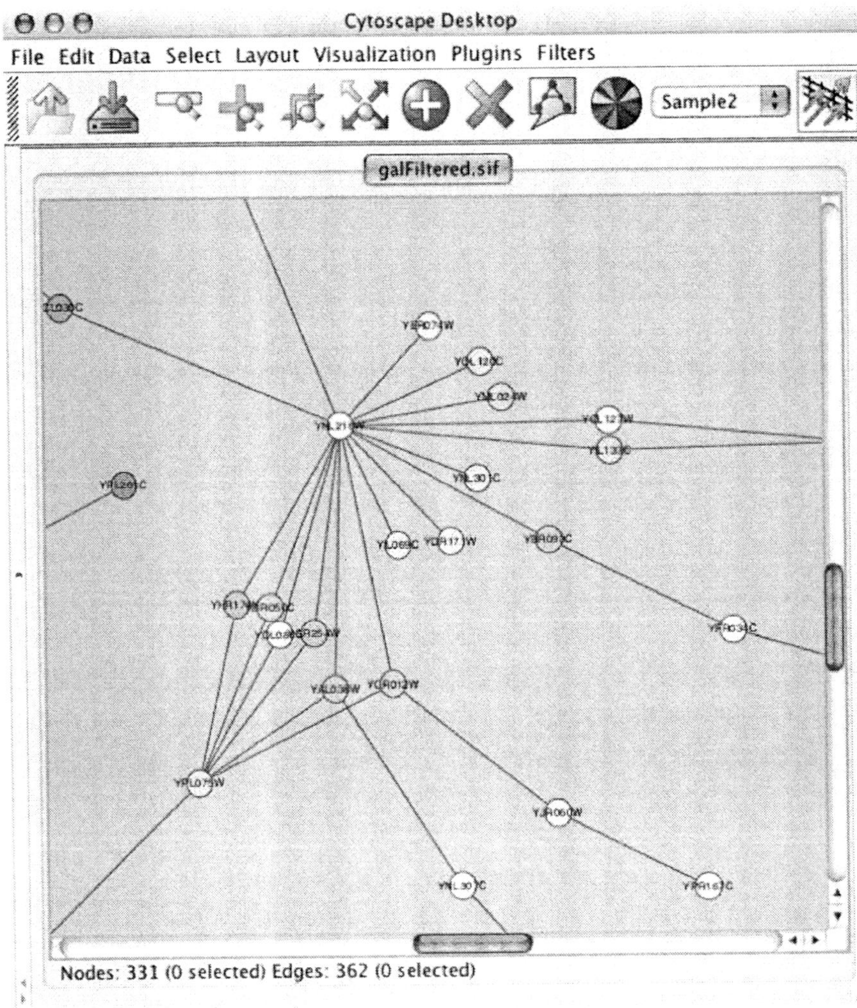

Figure 12. The yeast protein interaction network visualized by the Cytoscape. The nodes represent proteins and DNA molecule, connected by lines designating their interactions. The screenshot was reprinted with permission from the Cytoscape website.

The *Cytoscape* also allows selecting network interactions according to the number of neighbors, local distances, *etc*. Notably, the *Cytoscape* is an open-source software with scalable functionality; the third-party plug-ins such as *ActiveModules* (identifying distinct pathways within protein networks can be readily integrated into the application (Ideker et al. 2002).

Osprey

Unlike the *Cytoscape*, the *Osprey* (http://biodata.mshri.on.ca/osprey/) color-codes nodes of a gene network graph by the corresponding GO annotations, while the interaction arrows are colored according to experimental source (Breitkreutz, Stark and Tyers 2003b). An example of the protein interaction network visualized by Osprey is shown on Figure 13.

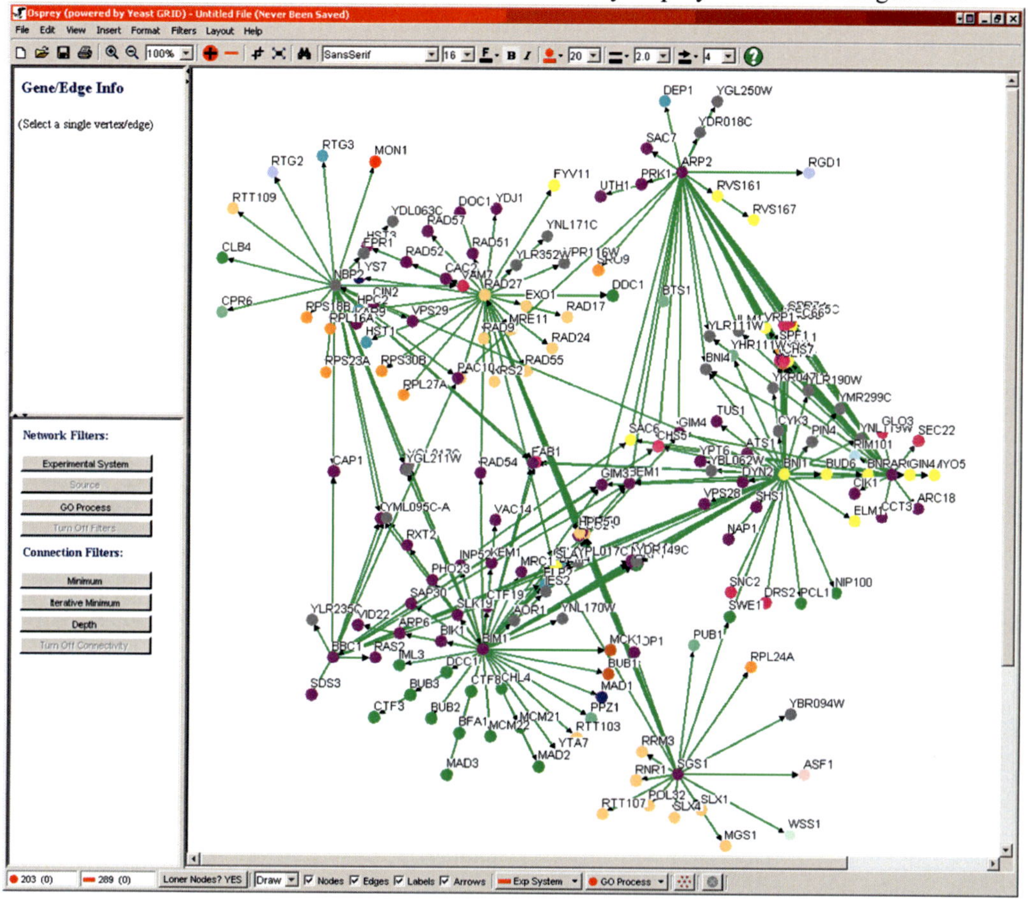

Figure 13. The Osprey visualizes protein interactions as a network graph composed of protein nodes and interaction lines. The noted are colored by the GO annotation, while the colors of lines depend on the type of experimental source. The screenshot was reprinted with permission from the Osprey website.

The interaction data can be imported into the *Osprey* in various formats including the GRID records that encompass gene names, synonyms, functional descriptions and GO annotations. Several graph layouts including circular, concentric circles, spoke and dual ring are implemented within the *Osprey's* graphic environment. Another important feature of the *Osprey* is that it can compare and analyze multiple datasets by superimposing the corresponding interaction maps and by finding common or specific network features.

VisANT

VisANT (http://visant.bu.edu) is a web-based visualization tool that utilizes the *Predictome* as a source of interaction data, including physical protein interactions, gene regulatory relationships and functional associations derived from phylogenetic profiling (Hu et al. 2004). Individual genes, represented as network nodes by the *VisANT* can be automatically cross-referenced with the Genbank (Benson et al. 2005), KEGG (Kanehisa et al. 2004) and Saccharomyces Genome Database (SGD) (Christie et al. 2004). The *VisANT* can color code the network edges according to the experimental source of interactions. An example of the yeast protein interactions visualized by the *VisANT* is shown on Figure 14.

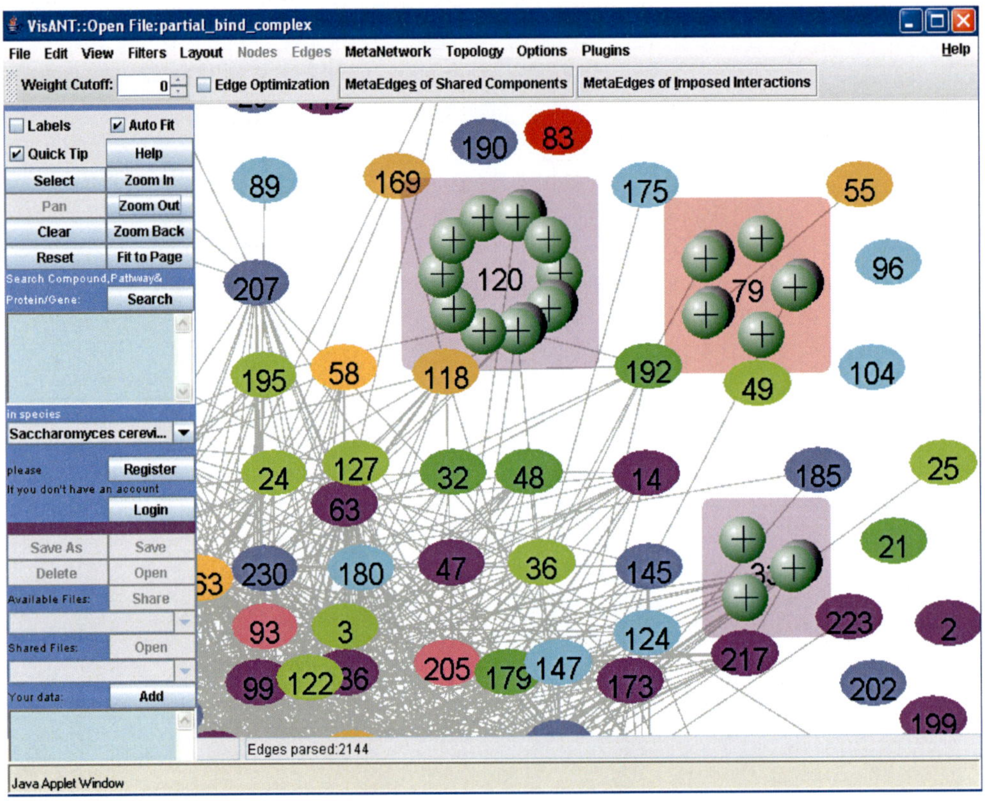

Figure 14. Yeast protein interactions determined from the mass spectrometry of purified protein complexes experiment (Gavin et al. 2002) is visualized by VisANT. In this example of network representation, a numbered node represents a protein complex that was purified in the experiment. A complex node can be expanded to show the individual protein components, visualized as nodes with "+" signs. A line connecting two complex nodes indicates the two protein complexes share at least one component. The screenshot was reprinted with permission from the ViaANT program, using the sample network data available on the VisANT website.

Like the previously described *Osprey* application, the *VisANT* allows comparing multiple interaction datasets through superimposition of the corresponding graphs and can compute various network properties including the degree distribution, shortest paths and feed-forward and feedback loops (Hu et al. 2004). More relevant discussion on these network-derived parameters will be presented in the following section *'Network Analysis and Integration'*.

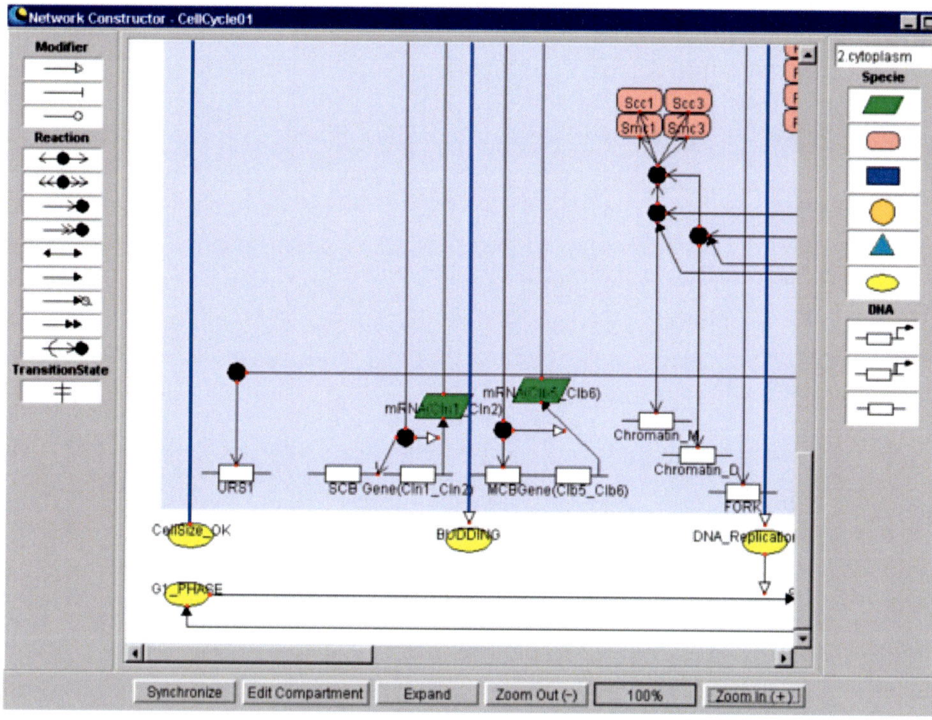

Figure 15. The graphic interface in CADLIVE allows constructing biochemical maps with various symbols and arrows, adopted from the Kohn's notations. The right panel of the screenshot features a list of available symbols for DNA, RNA, proteins and compounds, while the left panel contain a list of arrows to represent different interactions or relationships between the biological entities. The screenshot was reprinted with permission from the CADLIVE manual at its website.

CADLIVE

The Computer-Aided Design of Living Systems (CADLIVE, http://kurata21.bse.kyutech.ac.jp/cadlive/menu.htm) is another free bioinformatics tool capable of constructing and editing biochemical networks (Kurata, Matoba, and Shimizu 2003). The approach utilizes the Kohn's graphic notations for DNA, RNA, proteins, chemicals and molecular complexes (Kohn 1999). Within the CADLIVE environment, these biological symbols can be linked by different types of arrows representing physical-chemical processes (non-covalent binding, enzymatic reaction, gene transcription) or more abstract semantic concepts such as activation or inhibition (Kurata, Matoba, and Shimizu 2003). Figure 15 illustrates the CADLIVE graphic interface offering a choice of graphic notations for constructing and analyzing biochemical maps.

The interactions maps in the CADLIVE can also be converted into a set of the XML-formatted regulator-reaction equations that can further be used for dynamic simulations (Kurata et al. 2005). More details on simulating protein interaction networks will be discussed in the later sections.

CAVE

CAVE (http://cave.ucalgary.ca/) represents a unique three-dimensional virtual environment for visualizing complex genomics and proteomics data that are conventionally represented in the two dimensional space (Sensen 2005). The CAVE has been implemented at the University of Calgary as a room with the floor and three walls serving as projective surfaces. Using the Java 3D technology the CAVE server translates images onto these surfaces and allows users navigating through the three-dimensional visual objects using virtual reality goggles (featured on Figure 16). The CAVE has been successfully applied for visualizing and exploring human anatomy and various molecular structures. One prospective application for the CAVE is navigation through complex biological networks and real-time monitoring of simulated macromolecular interactions.

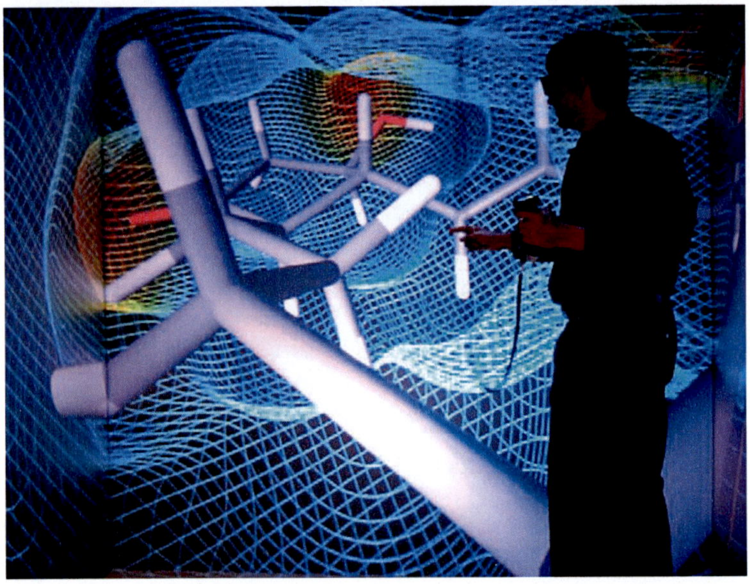

Figure 16. CAVE can visualize three-dimensional data by projecting images into three walls and a floor of a viewing room. A user with a virtual reality goggles can navigate the 3D objects with a remote control or by body movements. The picture was reprinted with permission from the CAVE website.

5.2. Network Analysis and Integration

Recent advances in networks theory have demonstrated a key role of uneven distributions occurring in many natural processes. It has been found that seemingly unrelated systems such as economic, professional, sexual and social networks, airline routing, power lines connections, language networks and internet hyperlinks all exhibit a power law decay of the cumulative degree distribution $P_X \approx X^{-\gamma}$, where x is the number of links connected to each network node as called a degree and γ is the value of the exponent typically varying in the range of 2-3 (Barabasi 2003).

The applicability of the scale-free methodology has been rigorously investigated in regard to the PIN-s including metabolic pathways (Jeong et al. 2000; Wagner and Fell 2001), experimentally observed protein-protein interactions (Jeong et al. 2001; Wagner 2001; Giot et al. 2003; Yook, Oltvai, and Barabasi 2004; Li et al. 2004) and genetic regulatory networks (Agrawal 2002; Featherstone and Broadie 2002). It has been suggested that the scale-free property of protein-protein interaction networks have emerged from the evolutionary cascade of gene duplication events (Pastor-Satorras, Smith, and Sole 2003).

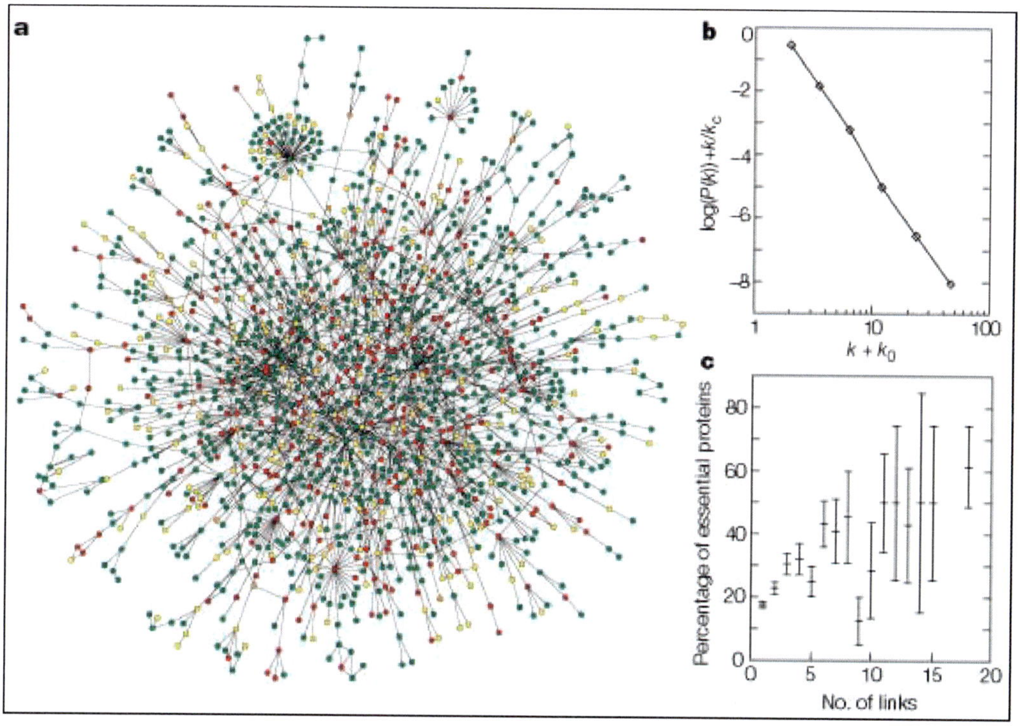

Figure 17. a) The yeast protein-protein interaction network is organized by a few highly connected proteins or hubs. The node is colored according to the phenotypic effect of removing the corresponding protein (red = lethal, green = non-lethal, orange = slow growth, yellow = unknown) (Jeong et al. 2001). b) The degree distribution, P(k), gives the probability of a node with k links in the network, and illustrates the scale-free topology of the yeast interaction network. c) The plot demonstrates that highly connected proteins are more likely to be essential than proteins with small number of connections. The figure is reproduced from Jeong, H., Mason, S. P., Barabasi, A. L., & Oltvai, Z. N. (2001). Lethality and centrality in protein networks. Nature, 411, 41-42. with permission from the Nature Publishing Group and the author.

Such heterogeneous architecture of protein interactomes imparts their robustness and error-tolerance from random perturbations (attacks) that would likely disable node(s) with small number of connections. On the other hand, a targeted attack directed toward protein network 'hubs' can efficiently impair system's topology and functionality (Albert, Jeong, and Barabasi 2000) (see Figure 17 and the legends for some details). This insight into topology and function of biological networks is particularly important for developing novel anti-infective therapies that can impact protein-protein interaction networks of target pathogens, without harming host's protein functions (as the correlation between the connectivity of proteins and their essentiality in a cell has already been confirmed (Jeong et al. 2001; Wuchty 2004)).

Other important features of biochemical networks include their modularity, hierarchy and clustering that can be analyzed at the sub-graph, motif and module levels (Milo et al. 2002; Barabasi and Oltvai 2004). A sub-graph is the basic building block of a network representing a subset of nodes connected in particular manner; while the motif can be characterized as a over-represented and often evolutionary conserved sub-graph in PIN; the network module is conventionally defined as a relatively large group of highly interconnected nodes that often correspond to distinct protein complex or cellular process (Wuchty, Oltvai, and Barabasi 2003, Spirin and Mirny 2003; Poyatos and Hurst 2004; Barabasi and Oltvai 2004). The identification and analysis of sub-graph, motif and modular composition of PIN-s is essential for understanding of their organizational and functional principles (Bader and Hogue 2003; Pereira-Leal, Enright, and Ouzounis 2004; Yeger-Lotem et al. 2004). This knowledge can be used in evaluation of the quality of experimental data on protein interactions and for assessing the reliability and accuracy of predicted protein associations (Goldberg and Roth 2003; Saito, Suzuki, and Hayashizaki 2003; Bader et al. 2004). In addition, protein interactions can be assessed by their correlations with gene expression data (Deane et al. 2002) since, as it has been previously described, proteins encoded by co-expressed genes are more likely to interact (Ge et al. 2001; Grigoriev 2001).

The network insight into protein interaction data also help integrating PIN-s with other types of bioinformatics information, including protein functional annotations (Deng, Chen, and Sun 2004; Yook, Oltvai, and Barabasi 2004), subcellular localizations (Yook, Oltvai, and Barabasi 2004), phenotypic effects (Walhout et al. 2002; Said et al. 2004) and expression levels (Ge et al. 2001; Grigoriev 2001; Ideker et al. 2002). For example, Ideker et al. (2002) demonstrated that overlaying of gene expression data onto networks of protein-protein and protein-DNA interactions allows identification of differentially expressed sub-networks that correspond to known yeast pathways.

As it has been discussed in greater details in the previous sections, there is a variety of other bioinformatics approaches that can help PIN reconstruction, visualization and data integration. In addition to already mentioned general tools such as *Cytoscape, VisANT, etc*, there are also several specialized applications such as *PathoLogic* that can automatically constructs metabolic pathways from annotated genomes (Karp, Paley, and Romero 2002) or the approach developed in work (Green and Karp 2004) that can identify missing components of metabolic pathways. The PathBLAST tool can compare PIN-s from different species to identify evolutionary conserved network elements (Kelley et al. 2004), while the TopNet identifies distinct sub-networks from protein expression levels, amino acid composition and solubility data (Yu et al. 2004b).

6. Modeling and Simulation Methods for Protein Interactions

The materials presented in the previous section illustrate that representation of protein-protein interactions in a form of two-dimensional networks can facilitate efficient integration of multiple sources of genomic and proteomic information. At the same time, to enable predicting complex behaviors of protein interactomes it is also necessary to place them into the context of cell models adequately reflecting spatial and temporal aspects of macromolecular interactions (Ideker and Lauffenburger 2003). Currently, there are several bioinformatics and cheminformatics approaches for simulating molecular interactions at macroscopic or microscopic levels that have been extensively discussed in a number of comprehensive review articles (de Jong 2002; Turner, Schnell, and Burrage 2004; Takahashi, Arjunan, and Tomita 2005; Lemerle, Ventura, and Serrano 2005). In the following section we will briefly describe the existing PIN-simulating approaches sorted by the increasing order of their realism and complexity.

6.1. Bayesian Network

The Bayesian Network approach (BN) represents a given gene as a node of the directed acyclic graph and interprets the corresponding gene expression levels through the joint conditional probabilities that, in turn, are described as connections between the interacting nodes (Friedman et al. 2000; de Jong 2002). Figure 18 illustrates that the BN can associate a gene i with a random variable X_i, describing the expression level of i (Friedman et al. 2000). Within the BN graph, this gene i is connected by its parents, denoting the regulators of its expression and linked to its children representing the effected genes. The BN specifies a conditional distribution, P(Xi|parents (Xi)) for each Xi, $1 \leq i \leq n$ and associates the gene expression levels with joint probability distribution P(Xi|parents (Xi) for all Xi (Friedman et al. 2000) (see Figure 18 for more details).

It has been demonstrated, that the Bayesian Networks can identify connections between genes more efficiently than widely used clustering methods that simply group them by the common expression patterns (Friedman et al. 2000). The approach has been successfully used for discovering numerous genetic regulatory interactions in *Saccharomyces cerevisiae* (Friedman et al. 2000; Hartemink et al. 2001; Imoto et al. 2003; Zou and Conzen 2005) and for modeling several cell signaling pathways involving focal adhesion kinase and extracellular signal-regulated kinase in humans (Sachs et al. 2002).

6.2. Boolean Networks

The Boolean Network (BN) approach considers a given gene as the two-state object associated with its active (1) and inactive (0) modes and represents an interaction between two genes as the Boolean function of their modes (Thomas 1973; Huang 1999; Kauffman et al. 2003). The method of Boolean Network has been successfully applied for reconstructing genetic regulatory mechanisms from experimental gene expression data (Huang and Ingber 2000; Kauffman et al. 2003; Kauffman et al. 2004; Mehra, Hu, and Karypis 2004; Serra, Villani, and Semeria 2004) and demonstrated good ability to accommodate the uncertainty of experimental results (Shmulevich et al. 2002; Pal et al. 2005). It has also been shown that the Boolean Networks can be useful for modeling gene activities under different cell conditions. Thus, by constructing the gene activity profiles (GAP) represented as strings of Boolean values (0|1), Huang (1999) efficiently modeled induced apoptosis cycles of tumor cells (Huang 1999).

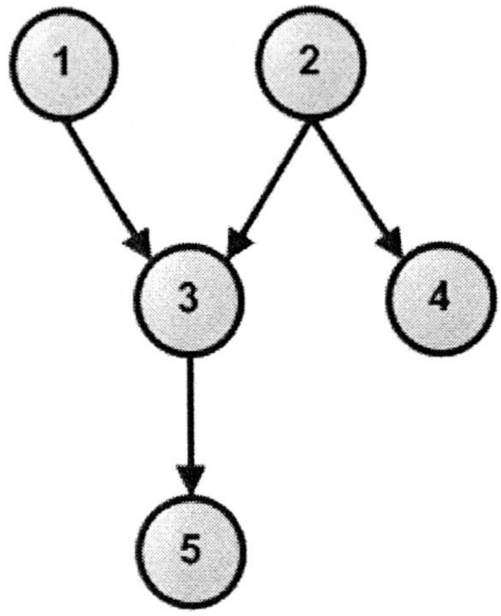

Figure 18. An example of a Bayesian network represents gene regulation (the example was taken from Friedman et al. 2000). Each node describes the gene expression level, Xi, of a gene i, and the graph follows the Markov Assumption: "each variable Xi is independent of its non-descendants, given its parents in the graph" (Friedman et al. 2000). Therefore several conditional independence statements are implied by the graph: $I(X_1;X_2,X_4)$, $I(X_2;X_1)$, $I(X_3; X_4|X_1, X_2)$, $I(X_4; X_1, X_3, X_5|X_2)$, $I(X_5; X_1, X_2, X_4|X_3)$, where I(Xi; Y | Z) indicates Xi is independent of Y given Z (Y and Z denoting sets of variables). The graph represents a joint probability distribution, $P(X_1, X_2, X_3, X_4, X_5) = P(X_1)P(X_2)P(X_3|X_1, X_2)P(X_4|X_2)P(X_5|X_3)$.

6.3. Ordinary Differential Equations

The cascades of individual cell chemical reactions can be modeled as a system of ordinary differential equations (ODE) operating on concentrations of participating chemicals (de Jong 2002). The ODE represents a macroscopic approach that does not distinguish between identical copies of the same molecule and operates solely on concentration variables that are fully determined by the past states of the system. For example, Bhalla utilized the ODE methodology of the 'well-stirred' cellular model to investigate the Michaelis-Menten kinetics of the cAMP pathway of the G-protein-coupled receptor (Bhalla 2002). Several simulating programs such as *Kinetikit* (Bhalla 2002), *GENESIS* (Bhalla 2002) and *Gepasi* (Mendes 1997) implemented the ODE methodology for quantification of other metabolic and cell signaling processes (Neves and Iyengar 2002) (Figure 19 features functionalities of the *Kinetikit* graphic interface). In particular, *Kinetikit* and *GENESIS* have been applied for modeling a wide range of biochemical signaling pathways (Sivakumaran et al. 2003; Vayttaden and Bhalla 2004) and complex neural networks (Nenadic, Ghosh, and Ulinski 2002; Kuriscak, Trojan, and Wunsch 2002; Cartwright et al. 2003).

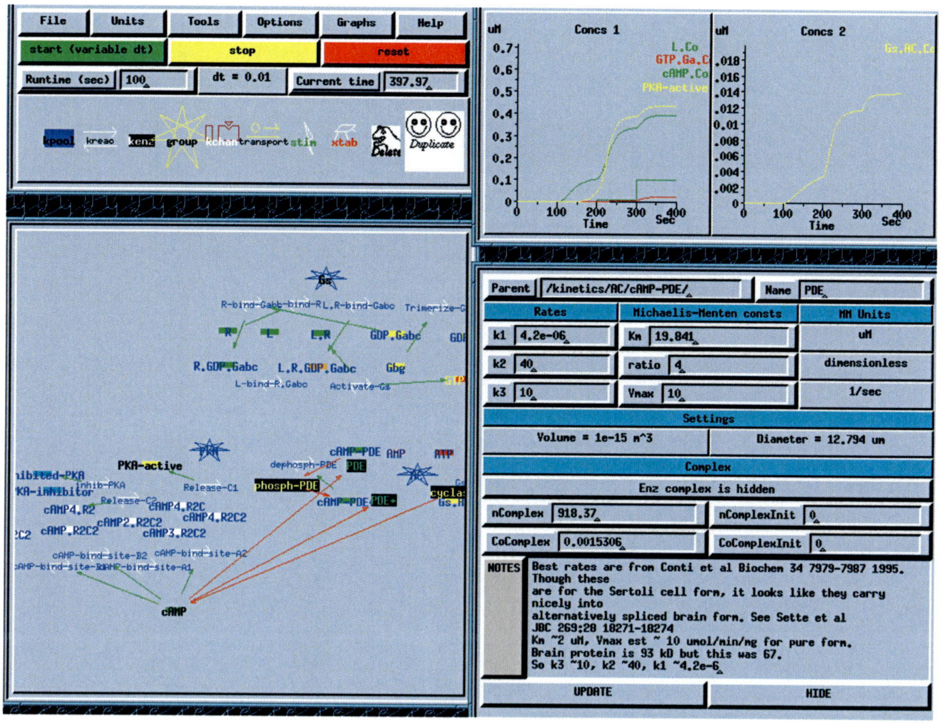

Figure 19. The Kinetikit graphic interface includes a menu bar (top-left), a interaction editor (bottom-left), a simulation display window to show concentration changes of molecules over time (top-right), and a parameter editor (bottom-right). The screenshot was reprinted with permission from the Kinetikit website.

6.4. Partial Differential Equations

The partial differential equations (PDE) can simulate the change of concentration of a given cellular component in both time- and space- dependent context (de Jong 2002; Takahashi, Arjunan, and Tomita 2005). In contrast to the ODE-based approaches considering cellular systems as homogenous spaces, the PDE-methods can adequately reflect compartmental cell organization and takes into account molecular diffusion between cellular compartments. The PDE-based tools can be viewed as macroscopic and deterministic (Takahashi, Arjunan, and Tomita 2005) and have been implemented within various simulation programs including the *Virtual Cell* (Loew and Schaff 2001). This technology can define the geometry of cellular compartments in one, two or three dimensional space and often produce simulation results directly comparable with experimental results (Loew and Schaff 2001). An example of *Virtual Cell* simulation outputs is featured on Figure 20 for the simulated diffusion and concentration changes of the RAN nuclear protein (see the figure captures for more details).

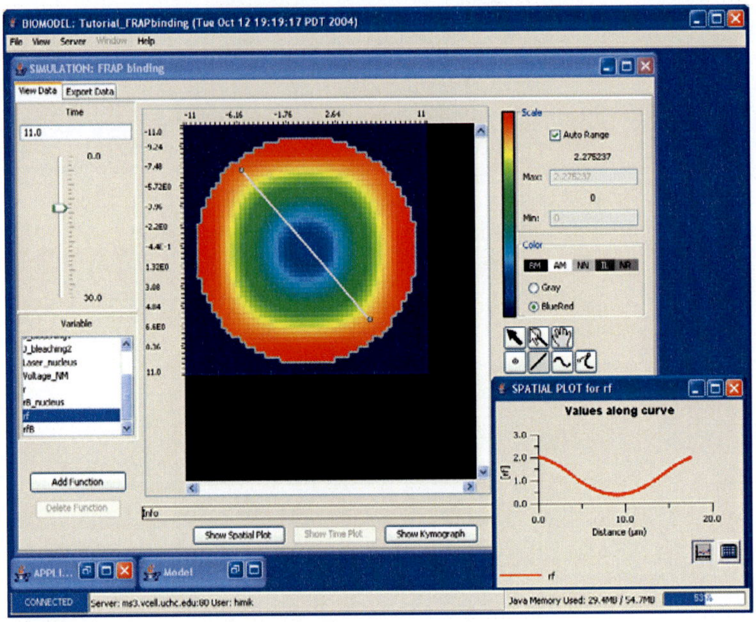

Figure 20. Virtual Cell simulates the change of concentration and diffusion of the nuclear protein RAN and its interacting proteins. The screenshot displays the concentration distribution of rf protein (RAN-FITC) in the nucleus of the cell as color gradient at 11.0 seconds (The center image represents the nucleus as circle and the exterior as cytosol). The bottom-right window plots the concentration of rf protein vs. distance alone the line as indicated in the nucleus diagram. Simulation results of different proteins at different time points can be displayed by selecting the variables (bottom-left) and adjusting the scrollbar for time (top-left). The screenshot was reprinted with permission from the Virtual Cell client program, using the example in the tutorial.

6.5. Gillespie Algorithm

Many cell signaling events involve a limited number of molecules subjected to random movements and collisions that probabilistically result in molecular interactions (Bhalla 2004). Such random behavior of certain cell components can be effectively captured by the stochastic approximation. The stochastic approach has originally been introduced by Gillespie (1976; 1977) for modeling biochemical systems involving irreversible, coupled autocatalytic or oscillatory reactions. Furthermore, the Gillespie algorithm has been extended to accommodate spatial simulations (Stundzia and Lumsden 1996; Elf, Doncic, and Ehrenberg 2003) and has been used for developing stochastic simulation programs such as *SmartCell* (Ander et al. 2004).

The SmartCell divides the intracellular space into a number of sub-volumes, where individual copies of constituent molecules can either interact or diffuse between the adjacent sub-volumes (see Figure 21 illustrating the SmartCell user interface for editing cell geometry and displaying simulation outputs). Thus, the SmartCell can adequately model localization and diffusion of molecules, transcription and translation, and chemical reactions that take place in different intracellular compartments in *E. coli* and other organisms (Ander et al. 2004).

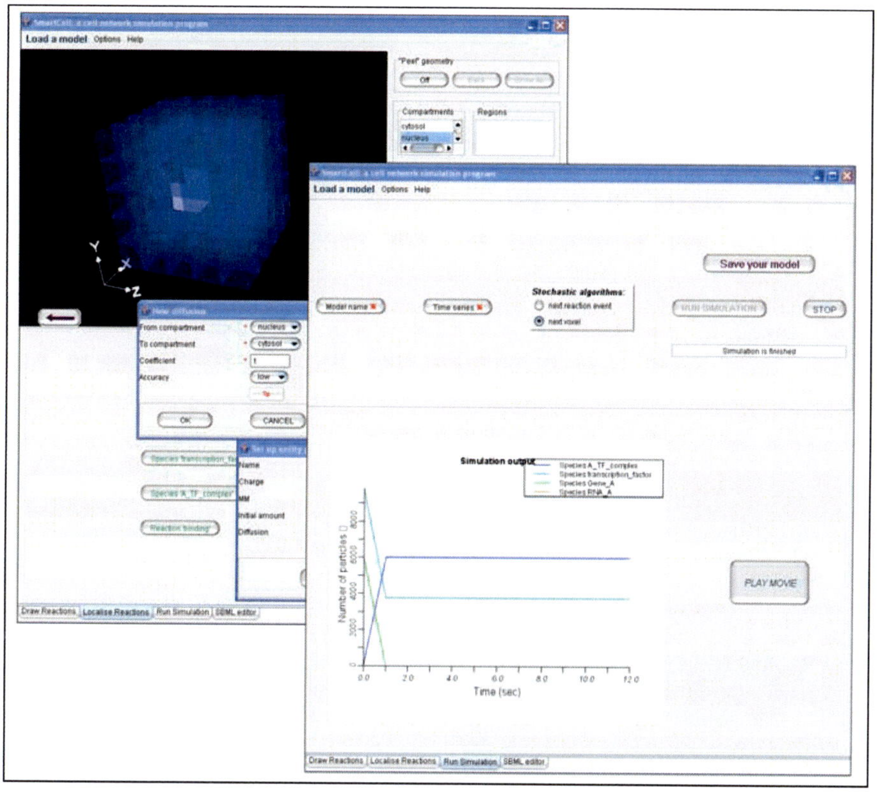

Figure 21. The top-left window allows users to define the geometry of a cell and create new compartments, and the bottom-right screen shows the simulation outputs as number of particles of molecular species over time. The screenshots were reprinted with permission from the SmartCell website.

6.6. Cellular Automata

The method of the *Cellular Automata* (CA) has been broadly used for modeling intracellular interactions at microscopic level as this approach not only allows distinguishing each instance of a molecule but it can quantify its stochastic behavior (Takahashi, Arjunan, and Tomita 2005). The *Cellular Automata* operates on the lattice of sites, occupied by molecules that can interact with the lattice neighbors (Weimar and Boon 1994). For example the *SimCell* approach, utilizing the CA methodology, models intracellular compartments by the two dimensional grid where each site contains one molecule such as a protein or a ligand (Wishart et al. 2004). At each simulation step, this molecule is allowed to randomly move to eight neighboring sites where it can interact with other molecule(s) according to the contact-dependent probability (an example of tricarboxylic acid cycle simulated by the *SimCell* technology is illustrated on Figure 22).

The Cellular Automata approach has been successfully used for simulating various enzymatic reactions (Weimar 2002) and other intracellular processes including intracellular diffusion, cellular metabolism and genetic regulation (Wishart et al. 2004).

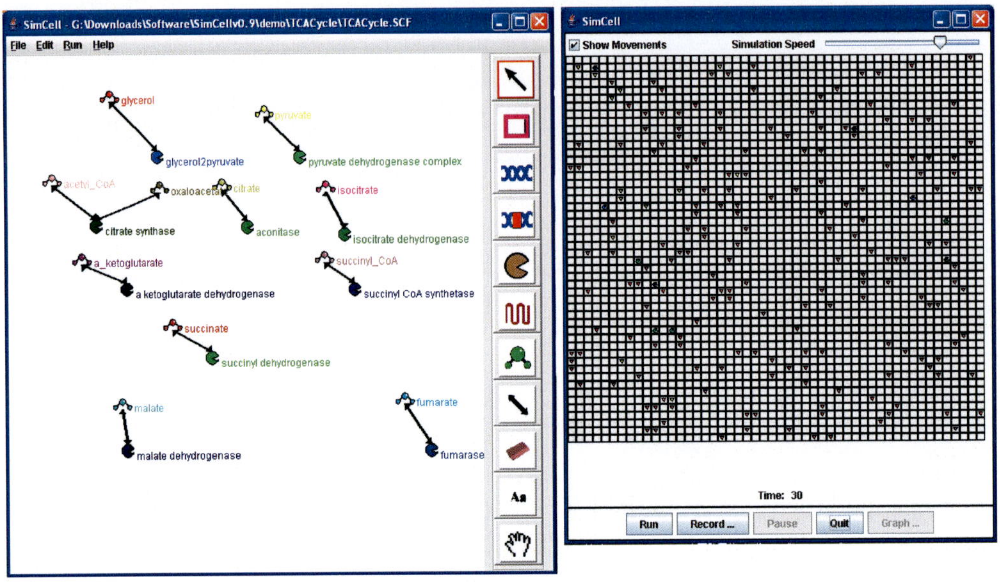

Figure 22. SimCell simulates interactions in tricarboxylic acid cycle by dynamic cellular automata. The left screen is the interaction editor, containing interactions among 10 types of enzymes and 11 types of small molecules. The right window displays the SimCell simulator, where each molecular instance is localized in one site of the grid. The "run" button performs the simulation and updates the grid image for every time step (each time interval represents 4 seconds in this example). The screenshots were reprinted with permission from the SimCell program.

6.7. Brownian Dynamics

The Brownian dynamics approximation operates in continuous time and space and allows conducting microscopic and stochastic simulation of molecular diffusion and collision using the Smoluchowski equation (Takahashi, Arjunan, and Tomita 2005). The Brownian dynamics has the advantage of effective modeling of molecular crowding effect that may have a major impact on simulation of protein diffusion rates and reaction kinetics (Schnell and Turner 2004; Minton 2005). Several simulation approaches such as *Smoldyn* (Andrews and Bray 2004), *MCell* (Stiles and Bartol 2000) and Green's function reaction dynamics (van Zon and ten Wolde 2005) utilize the Brownian dynamics methodology for modeling biochemical networks. For example, the *MCell* method adopts the Brownian movement of individual ligand(s) and applies Monte Carlo simulations for modeling interactions between free-diffusing ligands and stationary receptors (Stiles and Bartol 2000).

6.8. Molecular Dynamics

Among other conventional simulation techniques, the molecular dynamics (MD) approach is the most fundamental method that calculates molecular interactions by solving Newton's equations of motion (Takahashi, Arjunan, and Tomita 2005). Because of very high computational intensity, the MD simulations are currently used only for studying interactions involving small number of rather compact substances. Relatively recently, the MD-approaches have been adopted for detailed studying of protein-ligand interactions and demonstrated great promise for developing novel lead compounds (Baynes and Trout 2004; Friedel and Shea 2004). It is also expected that progress in computer hardware and algorithms will soon enhance the predictive ability of MD simulations and will extend applicability of contemporary MD packages such as AMBER (Pearlman et al. 1995), CHARMM (Brooks et al. 1983), GROMACS (Lindahl, Hess, and van der Spoel 2001), GROMOS (van Gunsteren et al. 1995), and NAMD (Kale et al. 1999) for prediction complex protein-protein interactions.

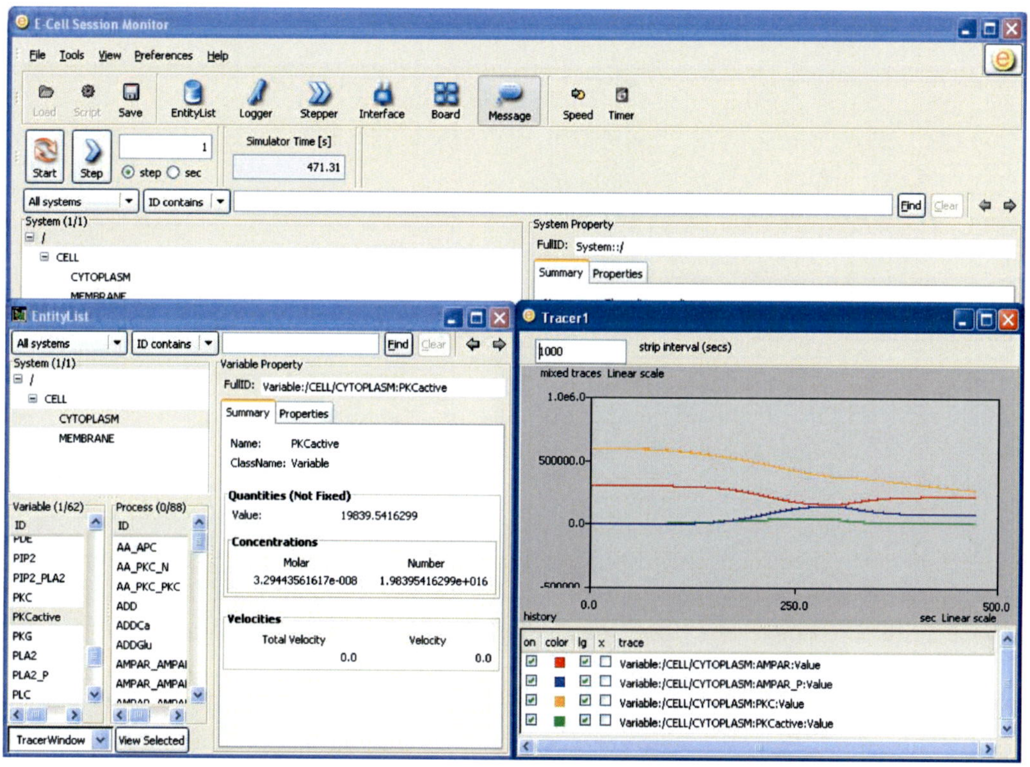

Figure 23. Pathways involved in cerebellar long term depression (LTD) are simulated by E-cell (the example was taken from the E-cell 3 tutorial). The interaction model file was loaded into the E-cell session monitor (top window), which starts and stops the simulation run. The EntityList window at bottom-left displays the variables (molecules) and processes in the interaction model. The Tracer window at bottom-right plots quantity of four selected variables (AMPAR, AMPAR_P, PKC, PKCactive) over time during the simulation. The screenshots were reprinted with permission from the E-cell program (version 3.1.103).

6.9. Other Simulation Approaches

Living cell represents a multi-scaled system of fast macroscopic non-covalent interactions and slow microscopic covalent reactions and, therefore, only combination of multiple micro- and macro-level simulation algorithms can result in plausible whole-cell simulations. The example of such integration has recently been reported – as the latest version of the *E-Cell Simulation Environment* has successfully combined both stochastic and deterministic approximations (Takahashi et al. 2004). In particular, the *E-cell* has been used to reconstruct the whole-cell models of *Mycoplasma genitalium* and human erythrocyte, where it successfully predicted the steady state and several metabolic responses (Tomita et al. 1999; Tomita 2001) (an example of E-cell simulation is shown in Figure 23).

Numerous other simulation approaches such as *MesoRD* (Hattne, Fange, and Elf. 2005), *CADLIVE simulator* (Kurata et al. 2005) and *StochSim* (Le Novere and Shimizu 2001) are also under development. Notably, these models tend to represent proteins as individual software objects composed of modification sites and simulate molecular interactions using compound concentrations and rate constants (Le Novere and Shimizu 2001). It is anticipated, that the object-oriented methodology will further facilitate protein-protein interaction modeling and will accelerate integration of multiple bioinformatics resources.

7. Semantic Networks for Modeling Protein Interactions

The recent developments in PIN detection methodologies, database technologies, methods of network analysis and computational simulations provide new unique opportunities for systemic investigation of complex protein interactions. On another hand, the advancement of genomic and proteomic technologies produced two major bioinformatics challenges.

The first challenge is to establish an adequate knowledge environment for representation and integration of biological data that is currently multi-leveled for genes, transcripts, proteins, intracellular compartments, cells and organisms. The conventional electronic tables or pathway maps of protein interactions might be sufficient for data storage and visual inspection but they have serious limitations when it comes to the knowledge integration. Thus, the table representations do not reflect the cause-effect characteristics of biological interactions while the conventional pathway diagrams miss their space and time-dependent content.

The adequate whole-cell simulations represent the second major challenge for the bioinformatics research (Tomita 2001; Ideker and Lauffenburger 2003; Uetz and Finley 2005). The transformation of individual molecular interactions into integrated cellular model requires representation of each molecular instance as a distinct programmable object while the programmable relationships between the objects should capture such diverse biochemical processes as gene transcription, protein translation, non-covalent binding, enzymatic reactions and translocation events among others.

All bioinformatics methodologies and tools described in the previous sections aim to resolve only some portions of these major bioinformatics challenges. There is a wide range of data integration and knowledge representation technologies available to date (Sowa 2000; Pollock 2004; Gardner 2005), including relational databases (Date 2005), object-oriented programming approaches (Budd 2001), description logics (Baader et al. 2003), data warehousing tools (Shah et al. 2005; Yona 2005) as well as the creation of expert -

(Giarratano and Riley 2004), frame - (Minsky 1975; Massar et al. 2005), and multiagent systems (Wooldridge 2002; Decker et al. 2002).

A powerful method that is common to many knowledge representation approaches is Semantic Networks (SN). The next part of the review will describe the method of semantic networks that in our opinion may offer possible solutions for adequate representation, integration and analysis of PIN-related bioinformatics data.

7.1. Theory and Background of Semantic Networks

Semantic Networks (SN) have been introduced as universal approach for generalizing any type of information in a form of edges and nodes (Richens 1956; Lehmann 1992). The nodes (semantic agents) represent abstract concepts, and the identity and behaviors of each agent is defined by its relationships (edges) with other semantic agents (Griffith 1982; Sowa 1991; Lehmann 1992). Notably, the topology of a semantic network can represent any type of knowledge describable by natural language (Lehmann 1992). One SN example is illustrated on Figure 24 where five semantic agents are connected by four pairs of semantic relationships: where agent [Protein A] has a relationship {instance of} with [Protein] agent that, in turn, has the opposite relationship {prototype of} with agent [Protein A]. Additionally, the objects [Protein A] and [Protein B] have relationships {composed of} with their constituent components: [Domain A] and [Domain B].

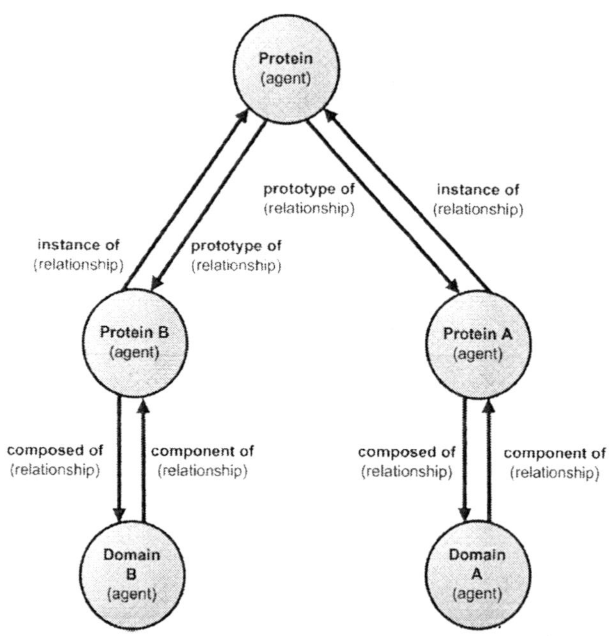

Figure 24. An example of a semantic network. Characteristics and behaviors of a semantic agent are defined by its relationships with other agents. Semantic agents are represented as nodes, and relationships are depicted as edges. This semantic network translates the information that Protein A and Protein B are instances of a Protein (a prototype), and they are composed of Domain A and Domain B respectively.

The example in Figure 24 illustrates two important types of relationships in the SN; the {prototype/instance} or often referred as IS-A relationship, and the {composed/component} or the PART-OF relationship. IS-A relationships define the hierarchy and inheritance between semantic agents, while the PART-OF relationships specify their composition. In addition to these two semantic relationships, other types of relationships can be created abstractly to represent information such as space and time for the occurrence of events within the SN.

Since the introduction of semantic networks in early 1950's, this methodology of knowledge representation influenced many areas of computer sciences, including artificial intelligence, relational database technology and object-oriented programming (Griffith 1982; Lehmann 1992). Recently, the method of semantic networks gained significant attention of biological community as powerful tool integrating diverse biological information (McCray and Nelson 1995). For instance, the semantic network approach has been adopted by the Unified Medical Language System (UMLS) (Lindberg, Humphreys, and McCray 1993). The UMLS stores over 900,000 biomedical concepts represented by distinct semantic agents and relationships that encode vast amount information on the patient care, health services, public health statistics and biomedical literature (Bodenreider 2004). The current bioinformatics projects focused on biological ontology also broadly use the SN; the recent cross-referencing of the Gene Ontology resource (Harris et al. 2004) with the UMLS platform enabled integrating numerous genomic concepts into the current biomedical knowledge (Lomax and McCray 2004). It is anticipated that the existing ontology and knowledge domains such as mammalian phenotypes (Smith, Goldsmith, and Eppig 2005), cell types (Bard, Rhee, and Ashburner 2005) and mouse anatomy (Hayamizu et al. 2005) will serve as important starting points for constructing comprehensive semantic networks of biological knowledge (Gardner 2005). In addition, the available ontology development and editing programs, including COBrA (Aitken et al. 2005), Protégé (Noy et al. 2003), Ontolingua Server (Farquhar, Fikes, and Rice 1997) and Chimaera Ontology Environment (McGuinness et al. 2000) will further facilitate data exchange within the SN environment.

The importance of semantic networks for the World Wide Web has also been recognized (Berners-Lee, Hendler, and Lassila 2001; Jenssen and Hovig. 2002; Hendler 2003). The concept of the "Semantic Web" has been developed to organize and consolidate diverse internet objects into individual semantic agents what would enable efficient communication between web services (Berners-Lee, Hendler, and Lassila 2001). Thus, the BioMOBY project applies the Semantic Web methodology for integrating most widely used bioinformatics tools and establishing effective communication between bioinformatics databases suffering from the lack of unified standards (Wilkinson and Links 2002). In addition, several contemporary Semantic Web applications such as FungalWeb (Shaban-Nejad et al. 2004), YeastHub (Cheung et al. 2005) and BioDASH (BioDASH 2005) integrate biological data using Semantic Web languages including RDF and OWL (World Wide Web Consortium 2004); the data are stored in RDF-compatible database systems, queried by the corresponding database query languages, and analyzed through automatic reasoners such as Racer (Haarslev and Moller 2003).

Other studies introduced the semantic approach into protein interaction modeling where molecules are viewed as "adaptive and logical agents", whose properties and behaviors emerge from the relationships with other semantic agents (Fisher, Paton, and Matsuno 1999; Fisher, Malcolm, and Paton 2000).

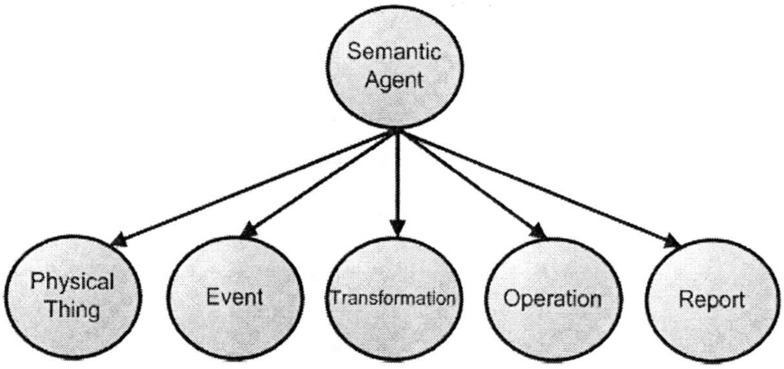

Figure 25. The basic classes of semantic agents in Visual Knowledge. Semantic agents in the Visual Knowledge environment are classified according to their common properties and functions (Visual Knowledge 2005). A semantic agent of the class "physical thing" models a physical object that has a shape and occupies space. An agent of the class "event" represents a phenomenon or a change that occurs on a physical object over a period of time. To enable application development such as a website, the VK contains application-specific agents including transformations, operations, and reports. A transformation agent creates other agents and modifies their relationships. An operation agent searches and collects other agents with certain properties, and a report agent displays the results in an application.

7.2. Semantic Networks for Application Development and Advanced Biological Modeling

In addition to knowledge representation and integration, semantic networks can be utilized for effective software implementation. For instance, the Visual Knowledge (VK) platform utilizes the SN in combination with other contemporary computer-science approaches including the set theory, frame system, object-oriented modeling and multi-agent system to successfully simulate the real-world applications, such as corporate enterprise environments, flight scheduling, hardware maintenance simulators and integrated currency exchange boards (Visual Knowledge 2005). The VK operates by several fundamental classes of semantic agents such as physical things, events and transformations, some of which are presented in Figure 25.

Within the VK environment, each semantic class contains its own computer-codes and a unique set of relationships that define the intrinsic behaviors of all its instances. Therefore, each semantic agent is reusable and contains instructions to act automatically when it is connected to other agents. In addition, new prototypical agents and relationships can be created by users to represent particular domain knowledge of interest. The corresponding graphic user-interface in VK allows users to conveniently create any semantic networks and computer applications without computer-code writing, but by simply dragging and dropping agents in-and-out of their relationships.

To address the current challenges in modeling biological systems, a specialized, biology-oriented VK-application package called BioCAD has been recently developed by the Visual Knowledge, Inc. and previously delegated to a bioinformatics company Upstream Biosciences, Inc. in collaboration with our group (Upstream Biosciences 2005). The BioCAD environment utilizes the semantic network methodology to integrate the gene expression, protein expression and protein interactions data by creating and connecting millions of biological agents including genes, proteins and cells defined from various bioinformatics resources. Figure 26 shows the graphic interface of a BioCAD client program, which enables creation and manipulation of biological agents stored at a central server.

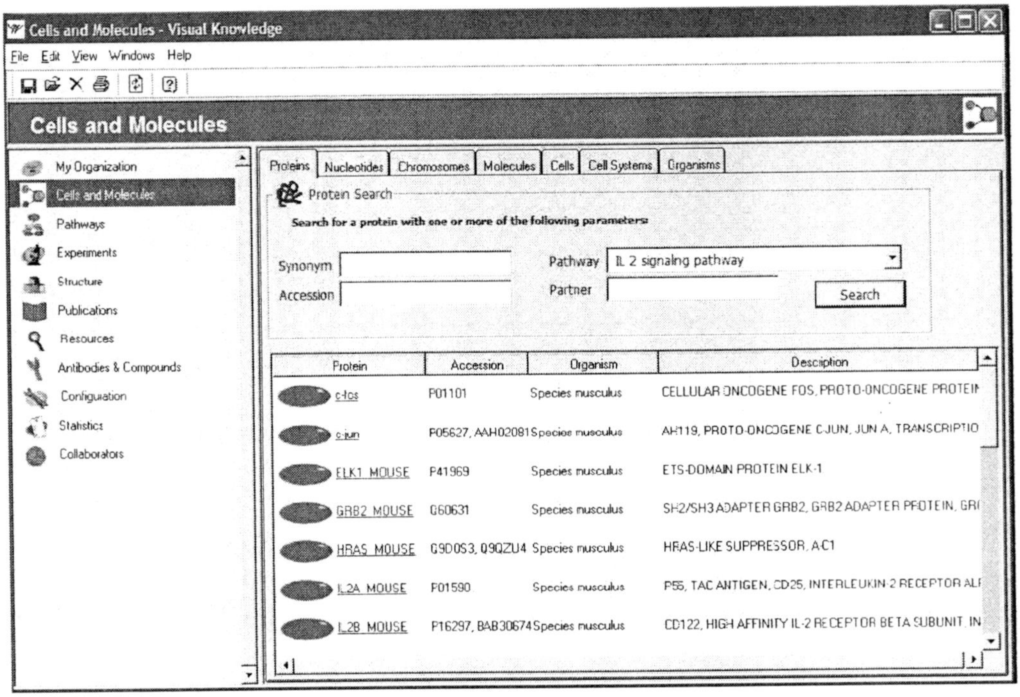

Figure 26. BioCAD - integrative environment for biological data. A BioCAD client program allows users to create and manipulate a variety of biological agents including genes, proteins, chromosomes, cells, cell systems and organisms. In addition, BioCAD enables pathway modeling and experimental data management by semantic networks. The BioCAD application was created from the Visual Knowledge SN-environment. The screenshot was reprinted with permission from the BioCAD website.

The BioCAD currently contains over 80,000 prototypical gene or protein agents from *Homo sapiens*, *Saccharomyces cerevisiae*, *Drosophila melanogaster*, *Mus musculus*, *Rattus norvegicus* and others. Within the BioCAD environment, a gene or a protein agent is connected to different annotation objects derived from the GenBank (Benson et al. 2005), RefSeq (Pruitt, Tatusova, and Maglott 2005), and Gene Ontology (Harris et al. 2004) (an example of the BioCAD annotation for the PIK3R1 protein agent is featured on Figure 27).

The experimental data on gene expression, protein expression or protein interactions obtained from other resources as tables, XML records, RDF or OWL files can be readily imported into the BioCAD database and converted into semantic agents and relationships.

For instance, protein-protein interactions determined from the yeast two-hybrid experiments are represented in the BioCAD as 'event' agents connected to the protein prototypes.

In addition, we have assigned the protein agents implemented in the BioCAD to protein domains defined by the *Pfam* (Bateman et al. 2004), *Prosite* (Sigrist 2002) and *InterPro* (Mulder 2005) classifications and converted into the individual SN agents. To enable the prediction of protein-protein interactions, the *InterDom* domain information (Ng et al. 2003) has also been imported into the BioCAD and the corresponding domain-domain interactions have been translated into the event agents.

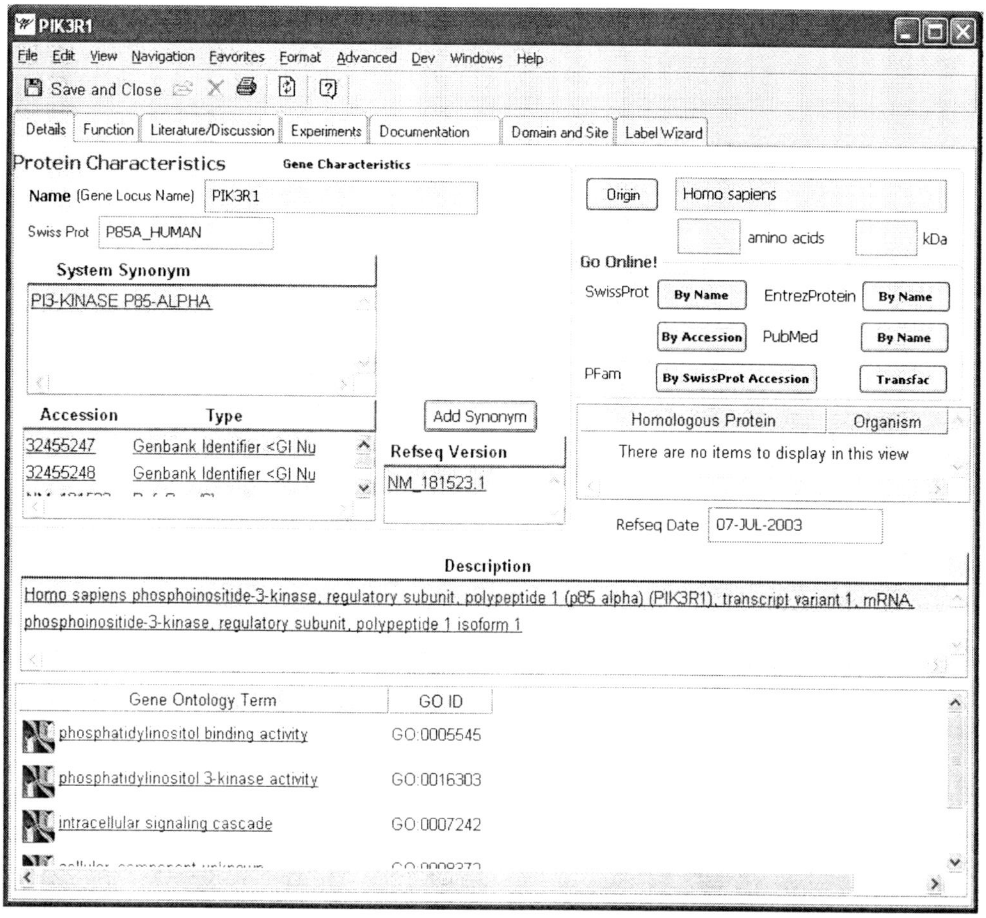

Figure 27. Protein annotations in BioCAD. The gene detail page for PIK3R1 shows the annotation integrated from various resources such as Genbank and Gene Ontology. Each piece of information such as a gene name, a description or an accession number is represented by a semantic agent in the database. The page is generated by collecting and displaying the relevant semantic agents through dynamic operations. This screenshot was reprinted with permission from the BioCAD program.

Hundreds of conventional pathways maps on metabolism and signal transduction have also been integrated into the BioCAD in a form of semantic agents and relationships and can be readily quarried by users. The BioCAD graphing capabilities allow visualizations of protein interaction networks enhanced with numerous layers of additional information – an example the BioCAD PIN construct can be found on Figure 28.

The existing biological agents and resources in BioCAD provide an excellent environment for modeling and simulating the mechanistic details of intracellular interactions. Using the BioCAD environment, we have developed and implemented a semantic network capable of representing and simulating complex molecular mechanisms such as PI3K regulation of cell signaling in human macrophages.

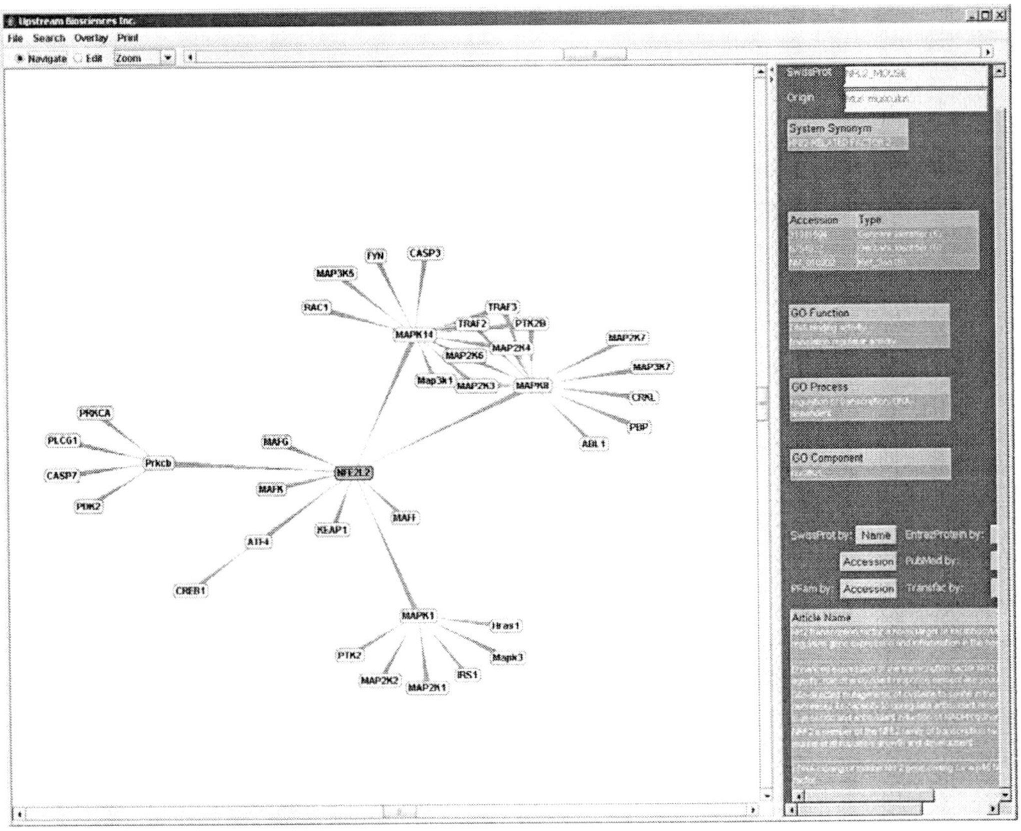

Figure 28. Visualization of protein interaction networks in BioCAD. The auto-graphic tool in BioCAD visualizes the underlying protein interaction data in a two-dimensional graph, where a node represents a protein agent and a line depicts a physical interaction event, connecting two proteins. This picture shows proteins that directly interact with a NFE2L2 protein (yellow node), and proteins that are one interaction away from NFE2L2. Clicking any node displays details about the protein in the right panel. This screenshot was reprinted with permission from the BioCAD's graphic tool.

7.3. A Case Study - Modeling and Simulating Molecular Interactions in Macrophages

The macrophages are the essential components of human immune system that engulf and digest pathogens by phagocytosis and phagosome maturation – the two cellular processes that are both regulated by the phosphoinositide 3-kinases (PI3Ks) (Tjelle, Lovdal, and Berg 2000; Stephens, Ellson, and Hawkins 2002; Cantley 2002; Wymann, Zvelebil, and Laffargue 2003). Previous experimental studies demonstrated that *Mycobacterium tuberculosis* (MTB) can interfere with the PI3K signaling pathways in order to survive within macrophages (Russell 2001; Fratti et al. 2001).

Based on the diverse roles of PI3Ks, we hypothesized that MTB's interference with the PI3K signaling could impact numerous other macrophage processes, far exceeding those that are currently studied.

We applied the semantic network approach to reconstruct the detailed picture of the PI3K signaling in human macrophages and used the resulting SN-model to predict additional cellular responses caused by the MTB interference (Hsing et al. 2004). Thus, we constructed the SN of human macrophage where cellular components have been modeled by SN agents of the [physical thing] class divided into six prototypes: [Cell], [Intracellular Compartment], [Macromolecule], [Domain and Site], [Small Molecule and Molecular Fragment], and [Atom]. Figure 29 illustrates that the spatial organization of the different types of structures is established through the {composition} relationships, linking the semantic agents.

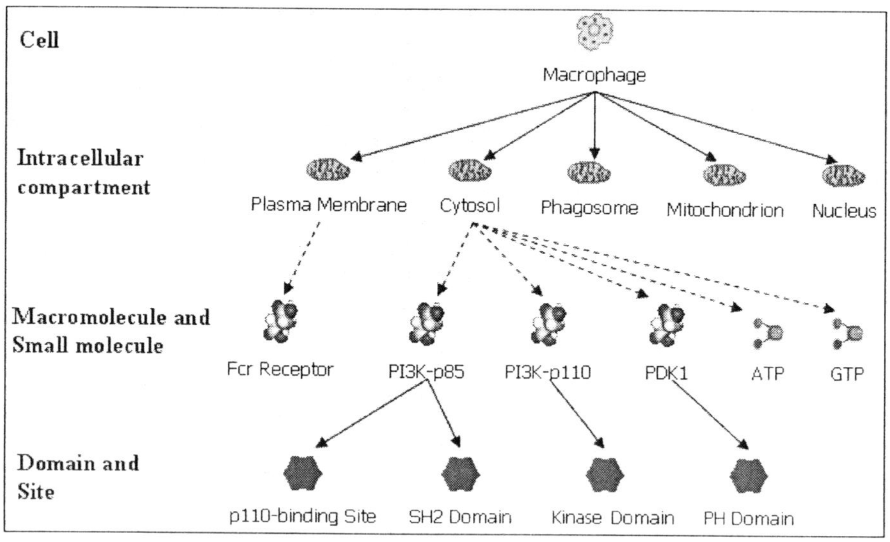

Figure 29. Spatial organization of intracellular structures in the semantic network. Biological structures are modeled by semantic agents (visualized as icons), which are related to their components by the composition relationships (shown as solid arrows). A human macrophage has been modeled as a semantic agent of the [Cell] prototype, and it is composed of various [Intracellular Compartment] agents, including plasma membrane, cytosol, nucleus and others. Each compartment such as cytosol has been linked to [Macromolecule] and [Small Molecule and Molecular Fragment] agents including proteins, ATP and GTP. A macromolecule such as a protein is further composed of [Domain and Site] agents. For simplicity, only one direction of the paired relationships is shown. We use dotted arrows to indicate that there are additional agents and relationships between two agents.

To represent molecular interactions inside the macrophage, we created six event prototypes: [localization], [translocation], [non-covalent interaction], [covalent interaction], [allosteric regulation] and [cellular response]. A biological example for each type of events is given in Table 1.

We have also modeled the mechanistic details and the cause-effect relationships through the following semantic events: non-covalent interactions, covalent interactions and allosteric regulations, each of which is composed of multiple semantic agents and relationships. For instance, a non-covalent event consists of two binding events, each connected to the corresponding binding domain, conformational states and binding states. An example of a covalent interaction, representing AKT1 phosphorylation by kinase PDK1 is illustrated in Figure 30.

Table 1. Classification of biological events in 6 prototypes in the semantic network. Six major event prototypes represent interactions among biological structures. The first column contains the six prototypes, and the second column contains biological examples of the correspond prototypes

Semantic Agent - Event	Biological Examples
Localization	A protein is located in the cytosol
Translocation	A protein moves from cytosol to plasma membrane.
Non-covalent Interaction	A ligand binds to a receptor.
Covalent Interaction	An enzyme catalyzes a chemical reaction where substrates are converted to products.
Allosteric Regulation	A ligand binding on site A of a protein causes a conformational change on site B of the protein.
Cellular Response	A qualitative cellular behavior such as cell survival, cell death, phagosome formation, and an increase of intracellular glucose level.

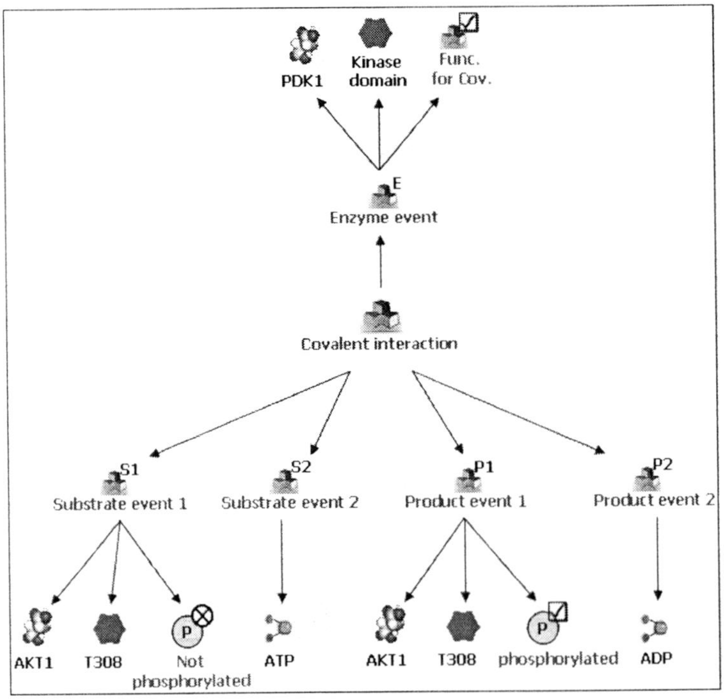

Figure 30. A covalent interaction event in the semantic network. This covalent interaction represents the phosphorylation of AKT1 by the enzyme PDK1. The event has been connected to the substrates (AKT1 and ATP), products (AKT1 and ADP) and the phosphorylation site on threonine 308 (T308).

An essential feature of signal transduction is the conformational and functional changes of proteins as a result of non-covalent binding with other molecules or chemical modification; however, such cause-effect information is poorly reflected by most interaction database and pathway maps. In the semantic network, we represented this information explicitly by the [allosteric regulation] event agents. Thus, a protein agent is modeled as a logical, integrating and adaptive device, where the functional responses of its domains are determined by a set of conditions from other domains.

Thus, the BioCAD SN-framework allowed us to translate information on PI3K signaling in literature and pathway database into semantic agents and relationships. As the result, we collected and placed into the SN-context the bioinformatics data from major public databases and 27 PI3K-specialized research articles covering 59 prototypical proteins and involving 46 non-covalent interactions, 17 covalent interactions, 27 allosteric regulations and 8 cellular responses. Figure 31 illustrate one of the interaction maps that summarize the different MTB interference scenarios in macrophages.

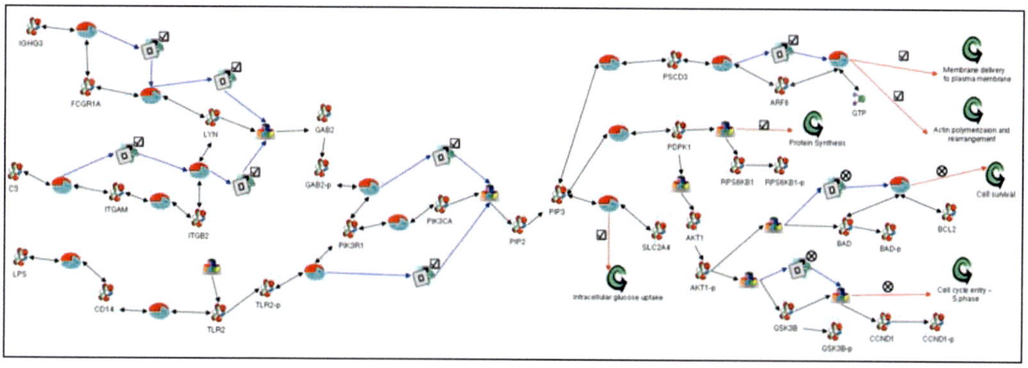

Figure 31. The interaction map was generated by traversing among the different molecules in the macrophage model. The starting points of this graph are the three MTB surface molecules IGHG3, C3 and LPS. All the arrows represent "derived relationships", which have consolidated several semantic agents and relationships. The black double-headed arrows are used to connect binding molecules to their non-covalent interaction (a double-headed arrow indicates the dual directionality in the interaction). The black single-headed arrows represent the connections among enzymes, substrates, and products. The blue arrows connect allosteric regulations to the molecular interactions. The map shows the cause-effect connections from the three surface molecules on MTB to the production of PIP3 in the macrophage. The map continues from the PIP3 downstream interactions to various cellular responses. The red arrow with a "check" sign represents "promoting" relationship (also a derived relationship). The red arrow with a "cross" sign represents "inhibitory" relationship. Icon definitions: Red/blue circle = non-covalent event; coloured cubes = covalent event; computer chip with "+" sign = positive allosteric regulation event; computer chip with "-" sign = negative allosteric regulation event; green arrow = cellular response event.

An occurrence of any event in a SN-simulation is represented by a semantic agent connecting the participating entities including 'molecule', 'location' and 'time' agents. Therefore, the SN-simulator provides a traceable "trajectory" of all the events that happened to every molecule (Hsing et al. 2004). Such an event history allows a detailed analysis of simulated molecular interactions. To illustrate this point, Figure 32 features an example of one macrophage simulation run, involving two instances of IGHG3, FCGR1A, LYN and

GAB2 molecules, each of which behaves according to its prototype in the model. As it can be seen on the chart, at certain time frame ('time 6') the SN-model identified the occurrence of 4 non-covalent interactions, 1 covalent interaction, 4 allosteric regulation and 11 translocation events. Similarly, by running the simulation on the SN-model for human macrophages we were able to traverse along signaling events starting from the MTB surface molecules through the macrophage cell receptors into the downstream interactions and leading to induced cellular responses. Such pathway reconstruction and simulation within the SN environment enabled us to predict several macrophage responses including increased macrophage cell survival and intracellular glucose uptake, some of which have not yet been appreciated in the literature.

Thus, the application of the SN methodology allowed not only effective integration of numerous layers of bioinformatics data such as gene expression, protein expression and protein interactions but also allowed reconstructing rather realistic cellular simulations conducted in space- and time-dependent manner. We also anticipate that future development of the SN methodology will enable simulating much larger numbers of interacting molecules and will lead to higher-resolution cellular model.

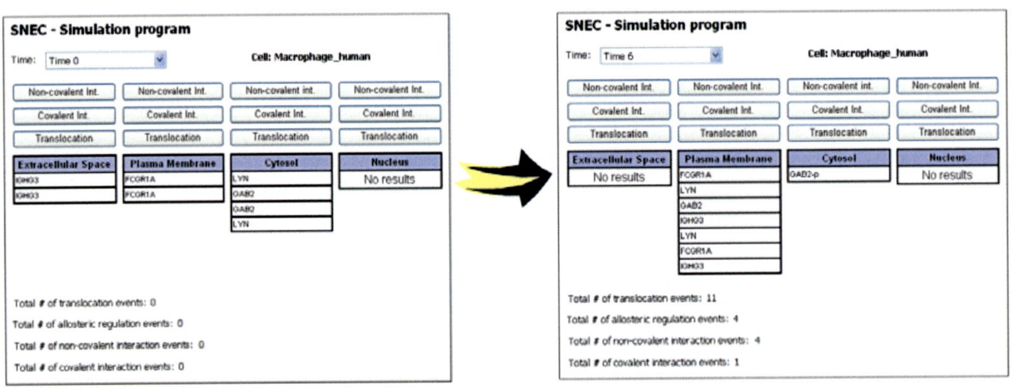

Figure 32. A SN-based simulator. An instance of a human macrophage cell is composed of four compartments; extracellular space, plasma membrane, cytosol and nucleus. There are three operations for each compartment, and they are activated by the three action buttons respectively: "Non-covalent int.", "Covalent int." and "Translocation." An operation such as "Non-covalent int." searches for a pair of molecules that have the ability to interact in the same location and creates an event agent accordingly. A translocation event randomly moves a molecule to an adjacent compartment. The combo boxes located at the top is used to increment the time. Before the simulation run (the screenshot at left), there was no event occurred as shown in the reports at the bottom of the screen. The screenshot at right shows the simulation outcomes at time 6. The molecules have changed their original locations, and many events have accumulated. There were 11 translocation events, 4 allosteric regulation events, 4 non-covalent interactions events, and 1 covalent interaction events. In particular, the allosteric events had switched the corresponding proteins to the active conformation that enabled different interactions in the following time steps. As a result, a GAB2 protein has been phosphorylated into GAB2-p by an activated LYN kinase. The phosphorylation event occurred at the plasma membrane, and GAB2-p moved from plasma membrane to cytosol through a subsequent translocation event.

8. Conclusion

The recent identification of organism-wide protein interactomes has shifted conventional thinking on the living cell from traditional protein-by-protein consideration to more coherent network perspectives. Although the rates of false predictions by the current high-throughput experiment techniques are still high, the integration of multiple experimental resources with other types of genomic information allows drawing rather adequate protein interactomes. Furthermore, the combination of experimental data with the predictive bioinformatics algorithms further facilitate the reconstruction of realistic protein networks and effective analyses of organizational principles of the living cells. It is also currently accepted that complete understanding of cellular behaviors requires not only static integration of biological data from multiple sources but demands for computational simulations considering molecular interactions in space and time and placing them in the contexts of stochastic and microscopic representations.

Among other current bioinformatics techniques for researching protein interaction systems, the method of semantic networks offers a broad range of capabilities for integrating and modeling large-scale biological data. This well-established method has already demonstrated its broad applicability in various fields of computer science including artificial intelligence, object-oriented programming, database technology and ontology-based projects. We anticipate further broader use of semantic networks in the field of bioinformatics and expect it to provide an efficient and general framework for both integration and simulation of bioinformatics data on protein-protein interactions.

Acknowledgements

We acknowledge Visual Knowledge, Inc. and Upstream Biosciences, Inc. for the software and database support during the macrophage case study. We thank Zakaria Hmama, Neil E. Reiner and Jimmy Lee (Division of Infectious Diseases, Department of Medicine, Faculty of Medicine, University of British Columbia) for their advice on modeling bacterial invasion processes. The research presented in the case study section of the book chapter was funded by the CIHR/MSFHR Strategic Training Program in Bioinformatics, Canadian Institutes of Health Research (CIHR), Michael Smith Foundation for Health Research (MSFHR), and Natural Sciences and Engineering Research Council of Canada (NSERC).

Appendix

Bioinformatics Internet Resources

Website	URL	Note
Protein interaction prediction tools		
iPfam	http://www.sanger.ac.uk/Software/Pfam/iPfam/	
InterDom	http://interdom.lit.org.sg/	domain-domain interaction
Scansite	http://scansite.mit.edu/	motif-domain interaction
InterPreTS	http://www.russell.embl.de/interprets/	tertiary Structure
PreBIND	http://www.blueprint.org/products/prebind/	text mining
MedScan	http://www.ariadnegenomics.com/products/medscan/	text mining
STRING	http://string.embl.de/	
Predictome	http://predictome.bu.edu/index.php	
Protein interaction databases		
BIND	http://bind.ca/	
DIP	http://dip.doe-mbi.ucla.edu/	
IntAct	http://www.ebi.ac.uk/intact/index.jsp	
GRID	http://biodata.mshri.on.ca/grid/servlet/Index	
MINT	http://mint.bio.uniroma2.it/mint/	
HPRD	http://www.hprd.org/	human protein interactions
Pathway databases		
KEGG	http://www.genome.jp/kegg/	
MetaCyc	http://metacyc.org/index.shtml	
STKE	http://stke.sciencemag.org/	
TRANSPATH	http://www.biobase.de/pages/products/transpath.html	
aMAZE	http://www.amaze.ulb.ac.be/	
Reactome	http://www.reactome.org/	
BioCarta	http://www.biocarta.com/	
Pathway Resource List	http://www.cbio.mskcc.org/prl/index.php	list of interaction and pathway resources on internet
Data standards		
PSI-MI	http://psidev.sourceforge.net/mi/xml/doc/user/	molecular interaction
BioPAX	http://www.biopax.org/	pathway
SBML	http://sbml.org/index.psp	simulation model
CellML	http://www.cellml.org/	simulation model
Visualization and analysis tools for interaction networks and pathways		
Cytoscape	http://www.cytoscape.org/index.php	
Osprey	http://biodata.mshri.on.ca/osprey/servlet/Index	
VisANT	http://visant.bu.edu/	
CADLIVE Editor	http://kurata21.bse.kyutech.ac.jp/cadlive/menu.htm	
CAVE	http://cave.ucalgary.ca/	
Pajek	http://vlado.fmf.uni-lj.si/pub/networks/pajek/	
BioLayout	http://www.biolayout.org/	
PATIKA	http://www.patika.org/	
NetBuilder	http://strc.herts.ac.uk/bio/maria/NetBuilder/	
Pathway Tools software	http://bioinformatics.ai.sri.com/ptools/	
PathwayAssist	http://www.ariadnegenomics.com/products/pathway.html	
PathBLAST	http://www.pathblast.org/bioc/pathblast/blastpathway.jsp	

Bioinformatics Internet Resources (Continued)

TopNet	http://networks.gersteinlab.org/genome/interactions/networks/core.html	
Simulation programs		
Kinetikit	http://www.ncbs.res.in/~bhalla/kkit/	Ordinary differential equations
GENESIS	http://www.ncbs.res.in/~bhalla/GENESIS/genesis.html	Ordinary differential equations
Gepasi	http://www.gepasi.org/	Ordinary differential equations
Virtual Cell	http://www.nrcam.uchc.edu/login/login.html	Partial differential equations
SmartCell	http://www.embl.de/ExternalInfo/serrano/smartcell/	Gillespie algorithm
SimCell	http://wishart.biology.ualberta.ca/SimCell/	Cellular automata
Smoldyn	http://sahara.lbl.gov/~sandrews/software.html	Brownian dynamics
MCell	http://www.mcell.cnl.salk.edu/	Brownian dynamics
AMBER	http://amber.scripps.edu/	Molecular dynamics
CHARMM	http://www.charmm.org/	Molecular dynamics
GROMACS	http://www.gromacs.org/	Molecular dynamics
GROMOS	http://www.igc.ethz.ch/gromos-docs/index.html	Molecular dynamics
NAMD	http://www.ks.uiuc.edu/Research/namd/	Molecular dynamics
E-Cell	http://ecell.sourceforge.net/	multi-algorithm and multi-timescale
MesoRD	http://mesord.sourceforge.net/	stochastic
StochSim	http://www.ebi.ac.uk/~lenov/stochsim.html	stochastic
CADLIVE simulator	http://kurata21.bsc.kyutech.ac.jp/cadlive/menu.htm	
Ontology projects		
Gene Ontology	http://www.geneontology.org/	
Open Biomedical Ontologies	http://obo.sourceforge.net/	List of current ontology projects
Ontology development software		
COBrA	http://www.xspan.org/cobra/index.html	
Protégé	http://protege.stanford.edu/	
Ontolingua Server	http://www.ksl.stanford.edu/software/ontolingua/	
Chimaera Ontology Environment	http://www.ksl.stanford.edu/software/chimaera/	
Semantic network development software		
Semantica	http://www.semanticresearch.com/products/	
Visual Knowledge	http://www.visualknowledge.com	
BioCAD	http://www.biocad.com	

REFERENCES

Agrawal, H. (2002). Extreme self-organization in networks constructed from gene expression data. *Phys Rev Lett, 89*, 268702.

Aitken, S., Korf, R., Webber, B., & Bard, J. (2005). COBrA: a bio-ontology editor. *Bioinformatics, 21*, 825-826.

Albert, R., Jeong, H., & Barabasi, A. L. (2000). Error and attack tolerance of complex networks. *Nature, 406*, 378-382.

Albert, S., Gaudan, S., Knigge, H., Raetsch, A., Delgado, A., Huhse, B., Kirsch, H., Albers, M., Rebholz-Schuhmann, D., & Koegl, M. (2003). Computer-assisted generation of a protein-interaction database for nuclear receptors. *Mol Endocrinol, 17*, 1555-1567.

Alfarano, C., Andrade, C. E., Anthony, K., Bahroos, N., Bajec, M., Bantoft, K., Betel, D., Bobechko, B., Boutilier, K., Burgess, E., Buzadzija, K., Cavero, R., D'Abreo, C., Donaldson, I., Dorairajoo, D., Dumontier, M. J., Dumontier, M. R., Earles, V., Farrall, R., Feldman, H., Garderman, E., Gong, Y., Gonzaga, R., Grytsan, V., Gryz, E., Gu, V., Haldorsen, E., Halupa, A., Haw, R., Hrvojic, A., Hurrell, L., Isserlin, R., Jack, F., Juma, F., Khan, A., Kon, T., Konopinsky, S., Le, V., Lee, E., Ling, S., Magidin, M., Moniakis, J., Montojo, J., Moore, S., Muskat, B., Ng, I., Paraiso, J. P., Parker, B., Pintilie, G., Pirone, R., Salama, J. J., Sgro, S., Shan, T., Shu, Y., Siew, J., Skinner, D., Snyder, K., Stasiuk, R., Strumpf, D., Tuekam, B., Tao, S., Wang, Z., White, M., Willis, R., Wolting, C., Wong, S., Wrong, A., Xin, C., Yao, R., Yates, B., Zhang, S., Zheng, K., Pawson, T., Ouellette, B. F., & Hogue, C. W. (2005). The Biomolecular Interaction Network Database and related tools 2005 update. *Nucleic Acids Res, 33*, D418-D424.

Aloy, P., & Russell, R. B. (2002). Interrogating protein interaction networks through structural biology. *Proc Natl Acad Sci U S A, 99*, 5896-5901.

Aloy, P., & Russell, R. B. (2003). InterPreTS: protein interaction prediction through tertiary structure. *Bioinformatics, 19*, 161-162.

Ander, M, Beltrao, P., Di Ventura, B., Ferkinghoff-Borg, J., Foglierini, M., Kaplan, A., Lemerle, C., Toma's-Oliveira, I., & Serrano, L. (2004). SmartCell, a framework to simulate cellular processes that combines stochastic approximation with diffusion and localisation: analysis of simple networks. *Syst Biol, 1*, 129-138.

Andrews, S. , & Bray, D. (2004). Stochastic simulation of chemical reactions with spatial resolution and single molecule detail. *Phys Biol, 1*, 137-151.

Baader, F., Calvanese, D., McGuinness, D., Nardi, D. , & Patel-Schneider, P., eds. (2003). The Description Logic Handbook : Theory, Implementation and Applications. Cambridge: Cambridge University Press.

Bader, G., & Cary, M. P. Pathway Resource List. 2005. Available from: URL: http://www.cbio.mskcc.org/prl

Bader, G. D., & Hogue, C. W. (2000). BIND--a data specification for storing and describing biomolecular interactions, molecular complexes and pathways. *Bioinformatics, 16*, 465-477.

Bader, G. D., & Hogue, C. W. (2002). Analyzing yeast protein-protein interaction data obtained from different sources. *Nat Biotechnol, 20*, 991-997.

Bader, G. D., & Hogue, C. W. (2003). An automated method for finding molecular complexes in large protein interaction networks. *BMC Bioinformatics, 4*, 2.

Bader, J. S., Chaudhuri, A., Rothberg, J. M., & Chant, J. (2004). Gaining confidence in high-throughput protein interaction networks. *Nat Biotechnol, 22*, 78-85.

Barabasi, A. L. (2003). Linked. Cambridge, MA: Plume.

Barabasi, A. L., & Oltvai, Z. N. (2004). Network biology: understanding the cell's functional organization. *Nat Rev Genet, 5*, 101-113.

Bard, J., Rhee, S. Y., & Ashburner, M. (2005). An ontology for cell types. *Genome Biol, 6*, R21.

Bartel, P. L., Roecklein, J. A., SenGupta, D., & Fields, S. (1996). A protein linkage map of Escherichia coli bacteriophage T7. *Nat Genet, 12*, 72-77.

Basalaj, W., & Eilbeck, K. (1999). Straight-Line drawings of protein interactions. In Kratochvil, J. (Eds.), *Graph Drawing* (pp.259-266). New York, NY: Springer-Verlag.

Batagelj, V, & Mrvar, A. (1998). Pajek - Program for large network analysis. *Connections, 21*, 47-57.

Bateman, A., Coin, L., Durbin, R., Finn, R. D., Hollich, V., Griffiths-Jones, S., Khanna, A., Marshall, M., Moxon, S., Sonnhammer, E. L., Studholme, D. J., Yeats, C., & Eddy, S. R. (2004). The Pfam protein families database. *Nucleic Acids Res, 32*, D138-D141.

Baynes, B. M., & Trout, B. L. (2004). Rational design of solution additives for the prevention of protein aggregation. *Biophys J, 87*, 1631-1639.

Benson, D. A., Karsch-Mizrachi, I., Lipman, D. J., Ostell, J., & Wheeler, D. L. (2005). GenBank. *Nucleic Acids Res, 33* Database Issue, D34-D38.

Berman, H. M., Westbrook, J., Feng, Z., Gilliland, G., Bhat, T. N., Weissig, H., Shindyalov, I. N., & Bourne, P. E. (2000). The Protein Data Bank. *Nucleic Acids Res, 28*, 235-242.

Berners-Lee, T., Hendler, J., & Lassila, O. (2001). The Semantic Web - A new form of Web content that is meaningful to computers will unleash a revolution of new possibilities. *Scientific American, 284*, 34-43.

Bhalla, U. S. (2002). Use of Kinetikit and GENESIS for modeling signaling pathways. *Methods Enzymol, 345*, 3-23.

Bhalla, U. S. (2004). (2004). Signaling in small subcellular volumes. I. Stochastic and diffusion effects on individual pathways. *Biophys J, 87*, 733-744.

BioCarta. BioCarta. 2005. Available from: URL: http://www.biocarta.com/

BioCyc. BioCyc. 2005. Available from: URL: http://biocyc.org/

BioDASH. BioDASH. 2005. Available from: URL: http://www.w3.org/2005/04/swls/BioDash/Demo/

BioPAX. BioPAX: Biological Pathways Exchange. 2005. Available from: URL: http://www.biopax.org/

Bock, J. R., & Gough, D. A. (2001). Predicting protein--protein interactions from primary structure. *Bioinformatics, 17*, 455-460.

Bodenreider, O. (2004). The Unified Medical Language System (UMLS): integrating biomedical terminology. *Nucleic Acids Res, 32*, D267-D270.

Bollobas, B. (1998). *Modern graph theory*. New York, NY: Springer-Verlag.

Borrebaeck, C. A., Ekstrom, S., Hager, A. C., Nilsson, J., Laurell, T., & Marko-Varga, G. (2001). Protein chips based on recombinant antibody fragments: a highly sensitive approach as detected by mass spectrometry. *Biotechniques, 30*, 1126-1132.

Bouwmeester, T., Bauch, A., Ruffner, H., Angrand, P. O., Bergamini, G., Croughton, K., Cruciat, C., Eberhard, D., Gagneur, J., Ghidelli, S., Hopf, C., Huhse, B., Mangano, R., Michon, A. M., Schirle, M., Schlegl, J., Schwab, M., Stein, M. A., Bauer, A., Casari, G., Drewes, G., Gavin, A. C., Jackson, D. B., Joberty, G., Neubauer, G., Rick, J., Kuster, B., & Superti-Furga, G. (2004). A physical and functional map of the human TNF-alpha/NF-kappa B signal transduction pathway. *Nat Cell Biol, 6*, 97-105.

Brajenovic, M., Joberty, G., Kuster, B., Bouwmeester, T., & Drewes, G. (2004). Comprehensive proteomic analysis of human Par protein complexes reveals an interconnected protein network. *J Biol Chem, 279*, 12804-12811.

Breitkreutz, B. J., Stark, C., & Tyers, M. (2003a). The GRID: the General Repository for Interaction Datasets. *Genome Biol, 4*, R23.

Breitkreutz, B. J., Stark, C., & Tyers, M. (2003b). Osprey: a network visualization system. *Genome Biol, 4*, R22.

Brooks, B. R., Bruccoleri, R. E., Olafson, B. D., States, D. J., Swaminathan, S., & Karplus, M. (1983). CHARMM: A Program for Macromolecular Energy, Minimization, and Dynamics Calculations. *J Comp Chem, 4*, 187-217.

Brown, C. T., Rust, A. G., Clarke, P. J., Pan, Z., Schilstra, M. J., De Buysscher, T., Griffin, G., Wold, B. J., Cameron, R. A., Davidson, E. H., & Bolouri, H. (2002). New computational approaches for analysis of cis-regulatory networks. *Dev Biol, 246*, 86-102.

Buck, M. J., & Lieb, J. D. (2004). ChIP-chip: considerations for the design, analysis, and application of genome-wide chromatin immunoprecipitation experiments. *Genomics, 83*, 349-360.

Budd, T. (2001). *An Introduction to Object-Oriented Programming*. Boston, MA: Addison Wesley.

Cahill, D. J. (2001). Protein and antibody arrays and their medical applications. *J Immunol Methods, 250*, 81-91.

Cantley, L. C. (2002). The phosphoinositide 3-kinase pathway. *Science, 296*, 1655-1657.

Cartwright, A. D., Gilchrist, D. P., Burgess, A. M., & Curthoys, I. S. (2003). A realistic neural-network simulation of both slow and quick phase components of the guinea pig VOR. *Exp Brain Res, 149*, 299-311.

Cary, M. P., Bader, G. D., & Sander, C. (2005). Pathway information for systems biology. *FEBS Lett, 579*, 1815-1820.

Cheung, K. H., Yip, K. Y., Smith, A., Deknikker, R., Masiar, A., & Gerstein, M. (2005). YeastHub: a semantic web use case for integrating data in the life sciences domain. *Bioinformatics, 21 Suppl 1*, i85-i96.

Choi, C., Krull, M., Kel, A., Kel-Margoulis, O., Pistor, S., Potapov, A., Voss, N., & Wingender, E. (2004b). TRANSPATH - a high quality database focused on signal transduction. *Comparative and Functional Genomics, 5*, 163-168.

Choi, C., Crass, T., Kel, A., Kel-Margoulis, O., Krull, M., Pistor, S., Potapov, A., Voss, N., & Wingender, E. (2004a). Consistent Re-Modeling of Signaling Pathways and Its Implementation in the TRANSPATH Database. *Genome Inform Ser Workshop Genome Inform, 15*, 244-254.

Christie, K. R., Weng, S., Balakrishnan, R., Costanzo, M. C., Dolinski, K., Dwight, S. S., Engel, S. R., Feierbach, B., Fisk, D. G., Hirschman, J. E., Hong, E. L., Issel-Tarver, L., Nash, R., Sethuraman, A., Starr, B., Theesfeld, C. L., Andrada, R., Binkley, G., Dong, Q., Lane, C., Schroeder, M., Botstein, D., & Cherry, J. M. (2004). Saccharomyces Genome Database (SGD) provides tools to identify and analyze sequences from Saccharomyces cerevisiae and related sequences from other organisms. *Nucleic Acids Res, 32*, D311-D314.

Cook, D. L., Farley, J. F., & Tapscott, S. J. (2001). A basis for a visual language for describing, archiving and analyzing functional models of complex biological systems. *Genome Biol, 2*, RESEARCH0012.

Crameri, R., & Kodzius, R. (2001). The powerful combination of phage surface display of cDNA libraries and high throughput screening. *Comb Chem High Throughput Screen, 4*, 145-155.

Dandekar, T., Snel, B., Huynen, M., & Bork, P. (1998). Conservation of gene order: a fingerprint of proteins that physically interact. *Trends Biochem Sci, 23*, 324-328.

Daraselia, N., Yuryev, A., Egorov, S., Novichkova, S., Nikitin, A., & Mazo, I. (2004). Extracting human protein interactions from MEDLINE using a full-sentence parser. *Bioinformatics, 20*, 604-611.

Das, P. M., Ramachandran, K., vanWert, J., & Singal, R. (2004). Chromatin immunoprecipitation assay. *Biotechniques, 37*, 961-969.

Date, C. J. (2005). *Database in depth: relational theory for practitioners*. Sebastopol, CA: O'Reilly.

de Jong, H. (2002). Modeling and simulation of genetic regulatory systems: a literature review. *J Comput Biol, 9*, 67-103.

Deane, C. M., Salwinski, L., Xenarios, I., & Eisenberg, D. (2002). Protein interactions: two methods for assessment of the reliability of high throughput observations. *Mol Cell Proteomics, 1*, 349-356.

Decker, K., Khan, S., Schmidt, C., Situ, G., Makkena, R., & Michaud, D. (2002). Biomas: A multi-agent system for genomic annotation. *International Journal of Cooperative Information Systems, 11*, 265-92.

Demir, E., Babur, O., Dogrusoz, U., Gursoy, A., Nisanci, G., Cetin-Atalay, R., & Ozturk, M. (2002). PATIKA: an integrated visual environment for collaborative construction and analysis of cellular pathways. *Bioinformatics, 18*, 996-1003.

Deng, M., Chen, T., & Sun, F. (2004). An integrated probabilistic model for functional prediction of proteins. *J Comput Biol, 11*, 463-475.

Deng, M., Mehta, S., Sun, F., & Chen, T. (2002). Inferring domain-domain interactions from protein-protein interactions. *Genome Res, 12*, 1540-1548.

Donaldson, I., Martin, J., de Bruijn, B., Wolting, C., Lay, V., Tuekam, B., Zhang, S., Baskin, B., Bader, G. D., Michalickova, K., Pawson, T., & Hogue, C. W. (2003). PreBIND and Textomy--mining the biomedical literature for protein-protein interactions using a support vector machine. *BMC Bioinformatics, 4*, 11.

Drees, B. L., Sundin, B., Brazeau, E., Caviston, J. P., Chen, G. C., Guo, W., Kozminski, K. G., Lau, M. W., Moskow, J. J., Tong, A., Schenkman, L. R., McKenzie, A., 3rd, Brennwald, P., Longtine, M., Bi, E., Chan, C., Novick, P., Boone, C., Pringle, J. R., Davis, T. N., Fields, S., & Drubin, D. G. (2001). A protein interaction map for cell polarity development. *J Cell Biol, 154*, 549-571.

Edwards, A. M., Kus, B., Jansen, R., Greenbaum, D., Greenblatt, J., & Gerstein, M. (2002). Bridging structural biology and genomics: assessing protein interaction data with known complexes. *Trends Genet, 18*, 529-536.

Elf, J. , Doncic, A. , & Ehrenberg, M. (2003). Mesoscopic reaction-diffusion in intracellular signaling. *Paper read at Proceedings of SPIE. 5110*, 114-124.

Enright, A. J., Iliopoulos, I., Kyrpides, N. C., & Ouzounis, C. A. (1999). Protein interaction maps for complete genomes based on gene fusion events. *Nature, 402*, 86-90.

Enright, A. J., & Ouzounis, C. A. (2001a). Functional associations of proteins in entire genomes by means of exhaustive detection of gene fusions. *Genome Biol, 2*, RESEARCH0034.

Enright, A. J., & Ouzounis, C. A. (2001b). BioLayout--an automatic graph layout algorithm for similarity visualization. *Bioinformatics, 17*, 853-854.

Farquhar, A. , Fikes, R. , & Rice, J. (1997). The Ontolingua Server: A tool for collaborative ontology construction. *International Journal of Human-computers Studies, 46*, 707-727.

Featherstone, D. E., & Broadie, K. (2002). Wrestling with pleiotropy: genomic and topological analysis of the yeast gene expression network. *Bioessays, 24*, 267-274.

Fields, S., & Song, O. (1989). A novel genetic system to detect protein-protein interactions. *Nature, 340*, 245-246.

Fields, S., & Sternglanz, R. (1994). The two-hybrid system: an assay for protein-protein interactions. *Trends Genet, 10*, 286-292.

Finn, R. D., Marshall, M., & Bateman, A. (2005). iPfam: visualization of protein-protein interactions in PDB at domain and amino acid resolutions. *Bioinformatics, 21*, 410-412.

Fisher, M. J., Malcolm, G., & Paton, R. C. (2000). Spatio-logical processes in intracellular signalling. *Biosystems, 55*, 83-92.

Fisher, M. J., Paton, R. C., & Matsuno, K. (1999). Intracellular signalling proteins as smart' agents in parallel distributed processes. *Biosystems, 50*, 159-171.

Fratti, R. A., Backer, J. M., Gruenberg, J., Corvera, S., & Deretic, V. (2001). Role of phosphatidylinositol 3-kinase and Rab5 effectors in phagosomal biogenesis and mycobacterial phagosome maturation arrest. *J Cell Biol, 154*, 631-644.

Friedel, M., & Shea, J. E. (2004). Self-assembly of peptides into a beta-barrel motif. *J Chem Phys, 120*, 5809-5823.

Friedman, N., Linial, M., Nachman, I., & Pe'er, D. (2000). Using Bayesian networks to analyze expression data. *J Comput Biol, 7*, 601-620.

Fromont-Racine, M., Mayes, A. E., Brunet-Simon, A., Rain, J. C., Colley, A., Dix, I., Decourty, L., Joly, N., Ricard, F., Beggs, J. D., & Legrain, P. (2000). Genome-wide protein interaction screens reveal functional networks involving Sm-like proteins. *Yeast, 17*, 95-110.

Fromont-Racine, M., Rain, J. C., & Legrain, P. (1997). Toward a functional analysis of the yeast genome through exhaustive two-hybrid screens. *Nat Genet, 16*, 277-282.

Fung, E. T., Thulasiraman, V., Weinberger, S. R., & Dalmasso, E. A. (2001). Protein biochips for differential profiling. *Curr Opin Biotechnol, 12*, 65-69.

Gabb, H. A., Jackson, R. M., & Sternberg, M. J. (1997). Modelling protein docking using shape complementarity, electrostatics and biochemical information. *J Mol Biol, 272*, 106-120.

Galarneau, A., Primeau, M., Trudeau, L. E., & Michnick, S. W. (2002). Beta-lactamase protein fragment complementation assays as in vivo and in vitro sensors of protein protein interactions. *Nat Biotechnol, 20*, 619-622.

Gardner, S. P. (2005). Ontologies and semantic data integration. *Drug Discov Today, 10*, 1001-1007.

Gavin, A. C., Bosche, M., Krause, R., Grandi, P., Marzioch, M., Bauer, A., Schultz, J., Rick, J. M., Michon, A. M., Cruciat, C. M., Remor, M., Hofert, C., Schelder, M., Brajenovic, M., Ruffner, H., Merino, A., Klein, K., Hudak, M., Dickson, D., Rudi, T., Gnau, V., Bauch, A., Bastuck, S., Huhse, B., Leutwein, C., Heurtier, M. A., Copley, R. R., Edelmann, A., Querfurth, E., Rybin, V., Drewes, G., Raida, M., Bouwmeester, T., Bork, P., Seraphin, B., Kuster, B., Neubauer, G., & Superti-Furga, G. (2002). Functional organization of the yeast proteome by systematic analysis of protein complexes. *Nature, 415*, 141-147.

Ge, H., Liu, Z., Church, G. M., & Vidal, M. (2001). Correlation between transcriptome and interactome mapping data from Saccharomyces cerevisiae. *Nat Genet, 29*, 482-486.

Giarratano, J. C., & Riley, G. D. (2004). *Expert Systems: Principles and Programming, Fourth* Edition: Principles and Programming. Boston, MA: Thomson Course Technology.

Gillespie, D. T. (1976). A general method for numerically simulating the stochastic time evolution of coupled chemical reactions. *Journal of Computational Physics, 22*, 403-434.

Gillespie, D. T. (1977). Exact stochastic simulation of coupled chemical reactions. *J Phys Chem, 81*, 2340-2361.

Giot, L., Bader, J. S., Brouwer, C., Chaudhuri, A., Kuang, B., Li, Y., Hao, Y. L., Ooi, C. E., Godwin, B., Vitols, E., Vijayadamodar, G., Pochart, P., Machineni, H., Welsh, M., Kong, Y., Zerhusen, B., Malcolm, R., Varrone, Z., Collis, A., Minto, M., Burgess, S., McDaniel, L., Stimpson, E., Spriggs, F., Williams, J., Neurath, K., Ioime, N., Agee, M., Voss, E., Furtak, K., Renzulli, R., Aanensen, N., Carrolla, S., Bickelhaupt, E., Lazovatsky, Y., DaSilva, A., Zhong, J., Stanyon, C. A., Finley, R. L., Jr., White, K. P., Braverman, M., Jarvie, T., Gold, S., Leach, M., Knight, J., Shimkets, R. A., McKenna, M. P., Chant, J., & Rothberg, J. M. (2003). A protein interaction map of Drosophila melanogaster. *Science, 302*, 1727-1736.

Gobel, U., Sander, C., Schneider, R., & Valencia, A. (1994). Correlated mutations and residue contacts in proteins. *Proteins, 18*, 309-317.

Goldberg, D. S., & Roth, F. P. (2003). Assessing experimentally derived interactions in a small world. *Proc Natl Acad Sci U S A, 100*, 4372-4376.

Gomez, S. M., & Rzhetsky, A. (2002). Towards the prediction of complete protein--protein interaction networks. *Pac Symp Biocomput*, 413-424.

Gough, N. R. (2002). Science's signal transduction knowledge environment: the connections maps database. *Ann N Y Acad Sci, 971*, 585-587.

Green, M. L., & Karp, P. D. (2004). A Bayesian method for identifying missing enzymes in predicted metabolic pathway databases. *BMC Bioinformatics, 5*, 76.

Griffith, R. L. (1982). Three Principles of Representation for Semantic Networks. ACM Trans. *Database Syst., 7*, 417-442.

Grigoriev, A. (2001). A relationship between gene expression and protein interactions on the proteome scale: analysis of the bacteriophage T7 and the yeast Saccharomyces cerevisiae. *Nucleic Acids Res, 29*, 3513-3519.

Haab, B. B., Dunham, M. J., & Brown, P. O. (2001). Protein microarrays for highly parallel detection and quantitation of specific proteins and antibodies in complex solutions. *Genome Biol, 2*, RESEARCH0004.

Haarslev, V. , & Moller, R. (2003). Racer: An OWL Reasoning Agent for the Semantic Web. Paper read at Proceedings of the International Workshop on Applications, Products and Services of Web-based Support Systems, in conjunction with the 2003 IEEE/WIC *International Conference on Web Intelligence, October 13*, at Halifax, Canada. 91-5.

Harris, M. A., Clark, J., Ireland, A., Lomax, J., Ashburner, M., Foulger, R., Eilbeck, K., Lewis, S., Marshall, B., Mungall, C., Richter, J., Rubin, G. M., Blake, J. A., Bult, C., Dolan, M., Drabkin, H., Eppig, J. T., Hill, D. P., Ni, L., Ringwald, M., Balakrishnan, R., Cherry, J. M., Christie, K. R., Costanzo, M. C., Dwight, S. S., Engel, S., Fisk, D. G., Hirschman, J. E., Hong, E. L., Nash, R. S., Sethuraman, A., Theesfeld, C. L., Botstein, D., Dolinski, K., Feierbach, B., Berardini, T., Mundodi, S., Rhee, S. Y., Apweiler, R., Barrell, D., Camon, E., Dimmer, E., Lee, V., Chisholm, R., Gaudet, P., Kibbe, W., Kishore, R., Schwarz, E. M., Sternberg, P., Gwinn, M., Hannick, L., Wortman, J., Berriman, M., Wood, V., de la Cruz, N., Tonellato, P., Jaiswal, P., Seigfried, T., & White, R. (2004). The Gene Ontology (GO) database and informatics resource. *Nucleic Acids Res,* 32, D258-D261.

Hartemink, A. J., Gifford, D. K., Jaakkola, T. S., & Young, R. A. (2001). Using graphical models and genomic expression data to statistically validate models of genetic regulatory networks. *Pac Symp Biocomput*, 422-433.

Hattne, J., Fange, D., & Elf, J. (2005). Stochastic reaction-diffusion simulation with MesoRD. *Bioinformatics, 21*, 2923-2924.

Hayamizu, T. F., Mangan, M., Corradi, J. P., Kadin, J. A., & Ringwald, M. (2005). The Adult Mouse Anatomical Dictionary: a tool for annotating and integrating data. *Genome Biol, 6*, R29.

Hendler, J. (2003). Communication. Science and the semantic web. *Science, 299*, 520-521.

Hermjakob, H., Montecchi-Palazzi, L., Bader, G., Wojcik, J., Salwinski, L., Ceol, A., Moore, S., Orchard, S., Sarkans, U., von Mering, C., Roechert, B., Poux, S., Jung, E., Mersch, H., Kersey, P., Lappe, M., Li, Y., Zeng, R., Rana, D., Nikolski, M., Husi, H., Brun, C., Shanker, K., Grant, S. G., Sander, C., Bork, P., Zhu, W., Pandey, A., Brazma, A., Jacq, B., Vidal, M., Sherman, D., Legrain, P., Cesareni, G., Xenarios, I., Eisenberg, D., Steipe, B., Hogue, C., & Apweiler, R. (2004a). The HUPO PSI's molecular interaction format--a community standard for the representation of protein interaction data. *Nat Biotechnol, 22*, 177-183.

Hermjakob, H., Montecchi-Palazzi, L., Lewington, C., Mudali, S., Kerrien, S., Orchard, S., Vingron, M., Roechert, B., Roepstorff, P., Valencia, A., Margalit, H., Armstrong, J., Bairoch, A., Cesareni, G., Sherman, D., & Apweiler, R. (2004b). IntAct: an open source molecular interaction database. *Nucleic Acids Res, 32* Database issue, D452-D455.

Ho, Y., Gruhler, A., Heilbut, A., Bader, G. D., Moore, L., Adams, S. L., Millar, A., Taylor, P., Bennett, K., Boutilier, K., Yang, L., Wolting, C., Donaldson, I., Schandorff, S., Shewnarane, J., Vo, M., Taggart, J., Goudreault, M., Muskat, B., Alfarano, C., Dewar, D., Lin, Z., Michalickova, K., Willems, A. R., Sassi, H., Nielsen, P. A., Rasmussen, K. J., Andersen, J. R., Johansen, L. E., Hansen, L. H., Jespersen, H., Podtelejnikov, A., Nielsen, E., Crawford, J., Poulsen, V., Sorensen, B. D., Matthiesen, J., Hendrickson, R. C., Gleeson, F., Pawson, T., Moran, M. F., Durocher, D., Mann, M., Hogue, C. W.,

Figeys, D., & Tyers, M. (2002). Systematic identification of protein complexes in Saccharomyces cerevisiae by mass spectrometry. *Nature, 415*, 180-183.

Hoffmann, R., Krallinger, M., Andres, E., Tamames, J., Blaschke, C., & Valencia, A. (2005). Text mining for metabolic pathways, signaling cascades, and protein networks. *Sci STKE*, 2005, pe21.

Holford, M., Li, N., Nadkarni, P., & Zhao, H. (2005). VitaPad: visualization tools for the analysis of pathway data. *Bioinformatics, 21*, 1596-1602.

Horak, C. E., Mahajan, M. C., Luscombe, N. M., Gerstein, M., Weissman, S. M., & Snyder, M. (2002). GATA-1 binding sites mapped in the beta-globin locus by using mammalian chIp-chip analysis. *Proc Natl Acad Sci U S A, 99*, 2924-2929.

Hsing, M., Bellenson, J. L., Shankey, C., & Cherkasov, A. (2004). Modeling of cell signaling pathways in macrophages by semantic networks. *BMC Bioinformatics, 5*, 156.

Hu, Z., Mellor, J., Wu, J., & DeLisi, C. (2004). VisANT: an online visualization and analysis tool for biological interaction data. *BMC Bioinformatics, 5*, 17.

Huang, S. (1999). Gene expression profiling, genetic networks, and cellular states: an integrating concept for tumorigenesis and drug discovery. *J Mol Med, 77*, 469-480.

Huang, S., & Ingber, D. E. (2000). Shape-dependent control of cell growth, differentiation, and apoptosis: switching between attractors in cell regulatory networks. *Exp Cell Res, 261*, 91-103.

Hucka, M., Finney, A., Sauro, H. M., Bolouri, H., Doyle, J. C., Kitano, H., Arkin, A. P., Bornstein, B. J., Bray, D., Cornish-Bowden, A., Cuellar, A. A., Dronov, S., Gilles, E. D., Ginkel, M., Gor, V., Goryanin, II, Hedley, W. J., Hodgman, T. C., Hofmeyr, J. H., Hunter, P. J., Juty, N. S., Kasberger, J. L., Kremling, A., Kummer, U., Le Novere, N., Loew, L. M., Lucio, D., Mendes, P., Minch, E., Mjolsness, E. D., Nakayama, Y., Nelson, M. R., Nielsen, P. F., Sakurada, T., Schaff, J. C., Shapiro, B. E., Shimizu, T. S., Spence, H. D., Stelling, J., Takahashi, K., Tomita, M., Wagner, J., & Wang, J. (2003). The systems biology markup language (SBML): a medium for representation and exchange of biochemical network models. *Bioinformatics, 19*, 524-531.

Ideker, T., & Lauffenburger, D. (2003). Building with a scaffold: emerging strategies for high- to low-level cellular modeling. *Trends Biotechnol, 21*, 255-262.

Ideker, T., Ozier, O., Schwikowski, B., & Siegel, A. F. (2002). Discovering regulatory and signalling circuits in molecular interaction networks. *Bioinformatics, 18 Suppl 1*, S233-S240.

Imoto, S., Kim, S., Goto, T., Miyano, S., Aburatani, S., Tashiro, K., & Kuhara, S. (2003). Bayesian network and nonparametric heteroscedastic regression for nonlinear modeling of genetic network. *J Bioinform Comput Biol, 1*, 231-252.

Ito, T., Chiba, T., Ozawa, R., Yoshida, M., Hattori, M., & Sakaki, Y. (2001). A comprehensive two-hybrid analysis to explore the yeast protein interactome. *Proc Natl Acad Sci U S A, 98*, 4569-4574.

Iyer, V. R., Horak, C. E., Scafe, C. S., Botstein, D., Snyder, M., & Brown, P. O. (2001). Genomic binding sites of the yeast cell-cycle transcription factors SBF and MBF. *Nature, 409*, 533-538.

Janin, J. (1995). Protein-protein recognition. *Prog Biophys Mol Biol, 64*, 145-166.

Janin, J. (2005). Assessing predictions of protein-protein interaction: the CAPRI experiment. *Protein Sci, 14*, 278-283.

Jansen, R., Greenbaum, D., & Gerstein, M. (2002). Relating whole-genome expression data with protein-protein interactions. *Genome Res, 12*, 37-46.

Jansen, R., Yu, H., Greenbaum, D., Kluger, Y., Krogan, N. J., Chung, S., Emili, A., Snyder, M., Greenblatt, J. F., & Gerstein, M. (2003). A Bayesian networks approach for predicting protein-protein interactions from genomic data. *Science, 302*, 449-453.

Jenssen, T. K., & Hovig, E. (2002). The semantic web and biology. *Drug Discov Today, 7*, 992.

Jeong, H., Mason, S. P., Barabasi, A. L., & Oltvai, Z. N. (2001). Lethality and centrality in protein networks. *Nature, 411*, 41-42.

Jeong, H., Tombor, B., Albert, R., Oltvai, Z. N., & Barabasi, A. L. (2000). The large-scale organization of metabolic networks. *Nature, 407*, 651-654.

Joshi-Tope, G., Gillespie, M., Vastrik, I., D'Eustachio, P., Schmidt, E., de Bono, B., Jassal, B., Gopinath, G. R., Wu, G. R., Matthews, L., Lewis, S., Birney, E., & Stein, L. (2005). Reactome: a knowledgebase of biological pathways. *Nucleic Acids Res, 33* Database Issue, D428-D432.

Kale, L., R., Skeel., Bhandarkar, M., Brunner, R., Gursoy, A., Krawetz, N., Phillips, J., Shinozaki, A., Varadarajan, K., & Schulten, K. (1999). Greater scalability for parallel molecular dynamics. *Journal of Computational Physics, 151*, 283-312.

Kanehisa, M., Goto, S., Kawashima, S., Okuno, Y., & Hattori, M. (2004). The KEGG resource for deciphering the genome. *Nucleic Acids Res, 32*, D277-D280.

Karp, P. D., Paley, S., & Romero, P. (2002). The Pathway Tools software. *Bioinformatics, 18* Suppl 1, S225-S232.

Kauffman, S., Peterson, C., Samuelsson, B., & Troein, C. (2003). Random Boolean network models and the yeast transcriptional network. *Proc Natl Acad Sci U S A, 100*, 14796-14799.

Kauffman, S., Peterson, C., Samuelsson, B., & Troein, C. (2004). Genetic networks with canalyzing Boolean rules are always stable. *Proc Natl Acad Sci U S A, 101*, 17102-17107.

Kelley, B. P., Yuan, B., Lewitter, F., Sharan, R., Stockwell, B. R., & Ideker, T. (2004). PathBLAST: a tool for alignment of protein interaction networks. *Nucleic Acids Res, 32*, W83-W88.

Kemmeren, P., van Berkum, N. L., Vilo, J., Bijma, T., Donders, R., Brazma, A., & Holstege, F. C. (2002). Protein interaction verification and functional annotation by integrated analysis of genome-scale data. *Mol Cell, 9*, 1133-1143.

Kitano, H. (2002). The standard graphical notation for biochemical networks. Paper read at ICSB-2002 workshop on SBML/SBW (Stockholm).

Kodadek, T. (2002). Development of protein-detecting microarrays and related devices. *Trends Biochem Sci, 27*, 295-300.

Kohn, K. W. (1999). Molecular interaction map of the mammalian cell cycle control and DNA repair systems. *Mol Biol Cell, 10*, 2703-2734.

Krieger, C. J., Zhang, P., Mueller, L. A., Wang, A., Paley, S., Arnaud, M., Pick, J., Rhee, S. Y., & Karp, P. D. (2004). MetaCyc: a multiorganism database of metabolic pathways and enzymes. *Nucleic Acids Res, 32* Database issue, D438-D442.

Krull, M., Voss, N., Choi, C., Pistor, S., Potapov, A., & Wingender, E. (2003). TRANSPATH: an integrated database on signal transduction and a tool for array analysis. *Nucleic Acids Res, 31*, 97-100.

Kurata, H., Masaki, K., Sumida, Y., & Iwasaki, R. (2005). CADLIVE dynamic simulator: direct link of biochemical networks to dynamic models. *Genome Res, 15*, 590-600.

Kurata, H., Matoba, N., & Shimizu, N. (2003). CADLIVE for constructing a large-scale biochemical network based on a simulation-directed notation and its application to yeast cell cycle. *Nucleic Acids Res, 31*, 4071-4084.

Kuriscak, E., Trojan, S., & Wunsch, Z. (2002). Model of spike propagation reliability along the myelinated axon corrupted by axonal intrinsic noise sources. Physiol Res, 51, 205-215.

Le Novere, N., & Shimizu, T. S. (2001). STOCHSIM: modelling of stochastic biomolecular processes. *Bioinformatics, 17*, 575-576.

Lee, T. I., Rinaldi, N. J., Robert, F., Odom, D. T., Bar-Joseph, Z., Gerber, G. K., Hannett, N. M., Harbison, C. T., Thompson, C. M., Simon, I., Zeitlinger, J., Jennings, E. G., Murray, H. L., Gordon, D. B., Ren, B., Wyrick, J. J., Tagne, J. B., Volkert, T. L., Fraenkel, E., Gifford, D. K., & Young, R. A. (2002). Transcriptional regulatory networks in Saccharomyces cerevisiae. *Science, 298*, 799-804.

Lehmann, F. (1992). Semantic networks. *Computers & Mathematics with Applications, 23*, 1-50.

Lehner, B., & Fraser, A. G. (2004). A first-draft human protein-interaction map. *Genome Biol, 5*, R63.

Lemer, C., Antezana, E., Couche, F., Fays, F., Santolaria, X., Janky, R., Deville, Y., Richelle, J., & Wodak, S. J. (2004). The aMAZE LightBench: a web interface to a relational database of cellular processes. *Nucleic Acids Res, 32* Database issue, D443-D448.

Lemerle, C., Di Ventura, B., & Serrano, L. (2005). Space as the final frontier in stochastic simulations of biological systems. *FEBS Lett, 579*, 1789-1794.

Li, S., Armstrong, C. M., Bertin, N., Ge, H., Milstein, S., Boxem, M., Vidalain, P. O., Han, J. D., Chesneau, A., Hao, T., Goldberg, D. S., Li, N., Martinez, M., Rual, J. F., Lamesch, P., Xu, L., Tewari, M., Wong, S. L., Zhang, L. V., Berriz, G. F., Jacotot, L., Vaglio, P., Reboul, J., Hirozane-Kishikawa, T., Li, Q., Gabel, H. W., Elewa, A., Baumgartner, B., Rose, D. J., Yu, H., Bosak, S., Sequerra, R., Fraser, A., Mango, S. E., Saxton, W. M., Strome, S., Van Den Heuvel, S., Piano, F., Vandenhaute, J., Sardet, C., Gerstein, M., Doucette-Stamm, L., Gunsalus, K. C., Harper, J. W., Cusick, M. E., Roth, F. P., Hill, D. E., & Vidal, M. (2004). A map of the interactome network of the metazoan C. elegans. *Science, 303*, 540-543.

Li, Z., Van Calcar, S., Qu, C., Cavenee, W. K., Zhang, M. Q., & Ren, B. (2003). A global transcriptional regulatory role for c-Myc in Burkitt's lymphoma cells. *Proc Natl Acad Sci U S A, 100*, 8164-8169.

Lieb, J. D., Liu, X., Botstein, D., & Brown, P. O. (2001). Promoter-specific binding of Rap1 revealed by genome-wide maps of protein-DNA association. *Nat Genet, 28*, 327-334.

Lindahl, E., Hess, B., & van der Spoel, D. (2001). GROMACS 3.0: a package for molecular simulation and trajectory analysis. *Journal of Molecular Modeling, 7*, 306-317.

Lindberg, D. A., Humphreys, B. L., & McCray, A. T. (1993). The Unified Medical Language System. *Methods Inf Med, 32*, 281-291.

Lloyd, C. M., Halstead, M. D., & Nielsen, P. F. (2004). CellML: its future, present and past. *Prog Biophys Mol Biol, 85*, 433-450.

Loew, L. M., & Schaff, J. C. (2001). The Virtual Cell: a software environment for computational cell biology. *Trends Biotechnol, 19*, 401-406.

Lomax, J., & McCray, A. T. (2004). Mapping the Gene Ontology into the Unified Medical Language System. *Comp Funct Genomics, 5*, 354-361.

Lopez, M. F., & Pluskal, M. G. (2003). Protein micro- and macroarrays: digitizing the proteome. *J Chromatogr B Analyt Technol Biomed Life Sci, 787*, 19-27.

Lu, L., Arakaki, A. K., Lu, H., & Skolnick, J. (2003). Multimeric threading-based prediction of protein-protein interactions on a genomic scale: application to the Saccharomyces cerevisiae proteome. *Genome Res, 13*, 1146-1154.

Lu, L., Lu, H., & Skolnick, J. (2002). MULTIPROSPECTOR: an algorithm for the prediction of protein-protein interactions by multimeric threading. *Proteins, 49*, 350-364.

MacBeath, G., & Schreiber, S. L. (2000). Printing proteins as microarrays for high-throughput function determination. *Science, 289*, 1760-1763.

Mann, M., Hendrickson, R. C., & Pandey, A. (2001). Analysis of proteins and proteomes by mass spectrometry. *Annu Rev Biochem, 70*, 437-473.

Marcotte, E. M., Pellegrini, M., Ng, H. L., Rice, D. W., Yeates, T. O., & Eisenberg, D. (1999). Detecting protein function and protein-protein interactions from genome sequences. *Science, 285*, 751-753.

Marcotte, E. M., Xenarios, I., & Eisenberg, D. (2001). Mining literature for protein-protein interactions. *Bioinformatics, 17*, 359-363.

Marcotte, E. M., Xenarios, I., van Der Bliek, A. M., & Eisenberg, D. (2000). Localizing proteins in the cell from their phylogenetic profiles. *Proc Natl Acad Sci U S A, 97*, 12115-12120.

Marsischky, G., & LaBaer, J. (2004). Many paths to many clones: a comparative look at high-throughput cloning methods. *Genome Res, 14*, 2020-2028.

Martin, S., Roe, D., & Faulon, J. L. (2005). Predicting protein-protein interactions using signature products. *Bioinformatics, 21*, 218-226.

Massar, J. P., Travers, M., Elhai, J., & Shrager, J. (2005). BioLingua: a programmable knowledge environment for biologists. *Bioinformatics, 21*, 199-207.

Matthews, L. R., Vaglio, P., Reboul, J., Ge, H., Davis, B. P., Garrels, J., Vincent, S., & Vidal, M. (2001). Identification of potential interaction networks using sequence-based searches for conserved protein-protein interactions or "interologs". *Genome Res, 11*, 2120-2126.

Matys, V., Fricke, E., Geffers, R., Gossling, E., Haubrock, M., Hehl, R., Hornischer, K., Karas, D., Kel, A. E., Kel-Margoulis, O. V., Kloos, D. U., Land, S., Lewicki-Potapov, B., Michael, H., Munch, R., Reuter, I., Rotert, S., Saxel, H., Scheer, M., Thiele, S., & Wingender, E. (2003). TRANSFAC: transcriptional regulation, from patterns to profiles. *Nucleic Acids Res, 31*, 374-378.

McCray, A. T., & Nelson, S. J. (1995). The representation of meaning in the UMLS. *Methods Inf Med, 34*, 193-201.

McGuinness, D. L., Fikes, R., Rice, J., & Wilder, S. (2000). The Chimaera Ontology Environment. *Paper read at the Seventeenth National Conference on Artificial Intelligence (AAAI 2000)*, July 30 - August 3, at Austin, Texas.

Mehra, S., Hu, W. S., & Karypis, G. (2004). A Boolean algorithm for reconstructing the structure of regulatory networks. *Metab Eng, 6*, 326-339.

Mellor, J. C., Yanai, I., Clodfelter, K. H., Mintseris, J., & DeLisi, C. (2002). Predictome: a database of putative functional links between proteins. *Nucleic Acids Res, 30*, 306-309.

Mendes, P. (1997). Biochemistry by numbers: simulation of biochemical pathways with Gepasi 3. *Trends Biochem Sci, 22*, 361-363.

Michnick, S. W. (2003). Protein fragment complementation strategies for biochemical network mapping. *Curr Opin Biotechnol, 14*, 610-617.

Milo, R., Shen-Orr, S., Itzkovitz, S., Kashtan, N., Chklovskii, D., & Alon, U. (2002). Network motifs: simple building blocks of complex networks. *Science, 298*, 824-827.

Minsky, M. (1975). A framework for representing knowledge. In Winston, P. H. (Eds.), *The Psychology of Computer Vision* (pp.211-277). New York, NY: McGraw-Hill.

Minton, A. P. (2005). Models for excluded volume interaction between an unfolded protein and rigid macromolecular cosolutes: macromolecular crowding and protein stability revisited. *Biophys J, 88*, 971-985.

Mulder, N. J., Apweiler, R., Attwood, T. K., Bairoch, A., Bateman, A., Binns, D., Bradley, P., Bork, P., Bucher, P., Cerutti, L., Copley, R., Courcelle, E., Das, U., Durbin, R., Fleischmann, W., Gough, J., Haft, D., Harte, N., Hulo, N., Kahn, D., Kanapin, A., Krestyaninova, M., Lonsdale, D., Lopez, R., Letunic, I., Madera, M., Maslen, J., McDowall, J., Mitchell, A., Nikolskaya, A. N., Orchard, S., Pagni, M., Ponting, C. P., Quevillon, E., Selengut, J., Sigrist, C. J., Silventoinen, V., Studholme, D. J., Vaughan, R., & Wu, C. H. (2005). InterPro, progress and status in 2005. *Nucleic Acids Res, 33* Database Issue, D201-D205.

Nenadic, Z., Ghosh, B. K., & Ulinski, P. S. (2002). Modeling and estimation problems in the turtle visual cortex. *IEEE Trans Biomed Eng, 49*, 753-762.

Neubauer, G., Gottschalk, A., Fabrizio, P., Seraphin, B., Luhrmann, R., & Mann, M. (1997). Identification of the proteins of the yeast U1 small nuclear ribonucleoprotein complex by mass spectrometry. *Proc Natl Acad Sci U S A, 94*, 385-390.

Neves, S. R., & Iyengar, R. (2002). Modeling of signaling networks. *Bioessays, 24*, 1110-1117.

Ng, S. K., Zhang, Z., & Tan, S. H. (2003). Integrative approach for computationally inferring protein domain interactions. *Bioinformatics, 19*, 923-929.

Ng, S. K., Zhang, Z., Tan, S. H., & Lin, K. (2003). InterDom: a database of putative interacting protein domains for validating predicted protein interactions and complexes. *Nucleic Acids Res, 31*, 251-254.

Nikitin, A., Egorov, S., Daraselia, N., & Mazo, I. (2003). Pathway studio--the analysis and navigation of molecular networks. *Bioinformatics, 19*, 2155-2157.

Noy, N. F., Crubezy, M., Fergerson, R. W., Knublauch, H., Tu, S. W., Vendetti, J., & Musen, M. A. (2003). Protege-2000: an open-source ontology-development and knowledge-acquisition environment. *AMIA Annu Symp Proc, 953.*

Obenauer, J. C., Cantley, L. C., & Yaffe, M. B. (2003). Scansite 2.0: Proteome-wide prediction of cell signaling interactions using short sequence motifs. *Nucleic Acids Res, 31*, 3635-3641.

Obenauer, J. C., & Yaffe, M. B. (2004). Computational prediction of protein-protein interactions. *Methods Mol Biol, 261*, 445-468.

Ouzounis, C., & Kyrpides, N. (1996). The emergence of major cellular processes in evolution. *FEBS Lett, 390*, 119-123.

Overbeek, R., Fonstein, M., D'Souza, M., Pusch, G. D., & Maltsev, N. (1999). The use of gene clusters to infer functional coupling. *Proc Natl Acad Sci U S A, 96*, 2896-2901.

Pal, R., Datta, A., Bittner, M. L., & Dougherty, E. R. (2005). Intervention in context-sensitive probabilistic Boolean networks. *Bioinformatics, 21*, 1211-1218.

Pandey, A., & Mann, M. (2000). Proteomics to study genes and genomes. *Nature, 405*, 837-846.

Pastor-Satorras, R., Smith, E., & Sole, R. V. (2003). Evolving protein interaction networks through gene duplication. *J Theor Biol, 222*, 199-210.

Paweletz, C. P., Charboneau, L., Bichsel, V. E., Simone, N. L., Chen, T., Gillespie, J. W., Emmert-Buck, M. R., Roth, M. J., Petricoin, Iii Ef, & Liotta, L. A. (2001). Reverse phase protein microarrays which capture disease progression show activation of pro-survival pathways at the cancer invasion front. *Oncogene, 20*, 1981-1989.

Pazos, F., Helmer-Citterich, M., Ausiello, G., & Valencia, A. (1997). Correlated mutations contain information about protein-protein interaction. *J Mol Biol, 271*, 511-523.

Pazos, F., & Valencia, A. (2002). In silico two-hybrid system for the selection of physically interacting protein pairs. *Proteins, 47*, 219-227.

Pearlman, D. A., Case, D. A., Caldwell, J. W., Ross, W. R., Cheatham III, T. E. , DeBolt, S., Ferguson, D., Seibel, G., & Kollman, P. (1995). AMBER, a computer program for applying molecular mechanics, normal mode analysis, molecular dynamics and free energy calculations to elucidate the structures and energies of molecules. *Comp Phys Commun, 91*, 1-41.

Pellegrini, M., Marcotte, E. M., Thompson, M. J., Eisenberg, D., & Yeates, T. O. (1999). Assigning protein functions by comparative genome analysis: protein phylogenetic profiles. *Proc Natl Acad Sci U S A, 96*, 4285-4288.

Pelletier, J. N., Remy, I., & Michnick, S. W. (1999). Protein-Fragment Complementation Assays: a General Strategy for the in vivo Detection of Protein-Protein Interactions. *J Biomol Tech, 10*, 32-39.

Pereira-Leal, J. B., Enright, A. J., & Ouzounis, C. A. (2004). Detection of functional modules from protein interaction networks. *Proteins, 54*, 49-57.

Phizicky, E. M., & Fields, S. (1995). Protein-protein interactions: methods for detection and analysis. *Microbiol Rev, 59*, 94-123.

Pollock, J. T., & Hodgson, R. (2004). *Adaptive information: improving business through semantic interoperability, grid computing, and enterprise integration.* Hoboken, NJ: John Wiley & Sons.

Poyatos, J. F., & Hurst, L. D. (2004). How biologically relevant are interaction-based modules in protein networks? *Genome Biol, 5*, R93.

Pruitt, K. D., Tatusova, T., & Maglott, D. R. (2005). NCBI Reference Sequence (RefSeq): a curated non-redundant sequence database of genomes, transcripts and proteins. *Nucleic Acids Res, 33* Database Issue, D501-D504.

Rain, J. C., Selig, L., De Reuse, H., Battaglia, V., Reverdy, C., Simon, S., Lenzen, G., Petel, F., Wojcik, J., Schachter, V., Chemama, Y., Labigne, A., & Legrain, P. (2001). The protein-protein interaction map of Helicobacter pylori. *Nature, 409*, 211-215.

Reiss, D. J., & Schwikowski, B. (2004). Predicting protein-peptide interactions via a network-based motif sampler. *Bioinformatics, 20 Suppl 1*, I274-I282.

Remy, I., & Michnick, S. W. (2001). Visualization of biochemical networks in living cells. *Proc Natl Acad Sci U S A, 98*, 7678-7683.

Ren, B., Robert, F., Wyrick, J. J., Aparicio, O., Jennings, E. G., Simon, I., Zeitlinger, J., Schreiber, J., Hannett, N., Kanin, E., Volkert, T. L., Wilson, C. J., Bell, S. P., & Young, R. A. (2000). Genome-wide location and function of DNA binding proteins. *Science, 290*, 2306-2309.

Richens, R. H. (1956). Preprogramming for mechanical translation. *Mechanical Translation, 3*, 20-25.

Rigaut, G., Shevchenko, A., Rutz, B., Wilm, M., Mann, M., & Seraphin, B. (1999). A generic protein purification method for protein complex characterization and proteome exploration. *Nat Biotechnol, 17*, 1030-1032.

Russell, D. G. (2001). Mycobacterium tuberculosis: here today, and here tomorrow. *Nat Rev Mol Cell Biol, 2*, 569-577.

Sachs, K., Gifford, D., Jaakkola, T., Sorger, P., & Lauffenburger, D. A. (2002). Bayesian network approach to cell signaling pathway modeling. *Sci STKE*, 2002, PE38.

Said, M. R., Begley, T. J., Oppenheim, A. V., Lauffenburger, D. A., & Samson, L. D. (2004). Global network analysis of phenotypic effects: protein networks and toxicity modulation in Saccharomyces cerevisiae. *Proc Natl Acad Sci U S A, 101*, 18006-18011.

Saito, R., Suzuki, H., & Hayashizaki, Y. (2003). Construction of reliable protein-protein interaction networks with a new interaction generality measure. *Bioinformatics, 19*, 756-763.

Salwinski, L., Miller, C. S., Smith, A. J., Pettit, F. K., Bowie, J. U., & Eisenberg, D. (2004). The Database of Interacting Proteins: 2004 update. *Nucleic Acids Res, 32*, D449-D451.

Schnell, S., & Turner, T. E. (2004). Reaction kinetics in intracellular environments with macromolecular crowding: simulations and rate laws. *Prog Biophys Mol Biol, 85*, 235-260.

Sensen, C. CAVE. 2005. Available from: URL: http://cave.ucalgary.ca/

Serra, R., Villani, M., & Semeria, A. (2004). Genetic network models and statistical properties of gene expression data in knock-out experiments. *J Theor Biol, 227*, 149-157.

Shaban-Nejad, A., Baker, C. J. O., Butler, G., & Haarslev, V. (2004). The FungalWeb Ontology: The core of a Semantic Web application for fungal genomics. Paper read at 1st Canadian Semantic Web Interest Group Meeting (SWIG' 04), at Montreal, Quebec, Canada.

Shah, S. P., Huang, Y., Xu, T., Yuen, M. M., Ling, J., & Ouellette, B. F. (2005). Atlas - a data warehouse for integrative bioinformatics. *BMC Bioinformatics, 6*, 34.

Shannon, P., Markiel, A., Ozier, O., Baliga, N. S., Wang, J. T., Ramage, D., Amin, N., Schwikowski, B., & Ideker, T. (2003). Cytoscape: a software environment for integrated models of biomolecular interaction networks. *Genome Res, 13*, 2498-2504.

Shmulevich, I., Dougherty, E. R., Kim, S., & Zhang, W. (2002). Probabilistic Boolean Networks: a rule-based uncertainty model for gene regulatory networks. *Bioinformatics, 18*, 261-274.

Sidhu, S. S., Fairbrother, W. J., & Deshayes, K. (2003). Exploring protein-protein interactions with phage display. *Chembiochem, 4*, 14-25.

Sigrist, C. J., Cerutti, L., Hulo, N., Gattiker, A., Falquet, L., Pagni, M., Bairoch, A., & Bucher, P. (2002). PROSITE: a documented database using patterns and profiles as motif descriptors. *Brief Bioinform, 3*, 265-274.

Sivakumaran, S., Hariharaputran, S., Mishra, J., & Bhalla, U. S. (2003). The Database of Quantitative Cellular Signaling: management and analysis of chemical kinetic models of signaling networks. *Bioinformatics, 19*, 408-415.

Smith, C. L., Goldsmith, C. A., & Eppig, J. T. (2005). The Mammalian Phenotype Ontology as a tool for annotating, analyzing and comparing phenotypic information. *Genome Biol, 6*, R7.

Sowa, J. F., ed. (1991). *Principles of Semantic Networks: Explorations in the Representation of Knowledge*. San Mateo, CA: Morgan Kaufmann Publishers.

Sowa, J. F. (2000). Knowledge Representation: Logical, Philosophical, and Computational Foundations. Pacific Grove, CA: Brooks/Cole Publishing Co.

Spirin, V., & Mirny, L. A. (2003). Protein complexes and functional modules in molecular networks. *Proc Natl Acad Sci U S A, 100*, 12123-12128.

Stanyon, C. A., Liu, G., Mangiola, B. A., Patel, N., Giot, L., Kuang, B., Zhang, H., Zhong, J., & Finley, R. L., Jr. (2004). A Drosophila protein-interaction map centered on cell-cycle regulators. *Genome Biol, 5*, R96.

Stephens, L., Ellson, C., & Hawkins, P. (2002). Roles of PI3Ks in leukocyte chemotaxis and phagocytosis. *Curr Opin Cell Biol, 14*, 203-213.

Stiles, J. R. , & Bartol, T. M. (2000). Monte Carlo Methods for Simulating Realistic Synaptic Microphysiology Using MCell. In DeSchutter, E. (Eds.), *Computational Neuroscience: Realistic Modeling for Experimentalists* (pp.87-128). Boca Raton, FL: CRC Press.

Stundzia, A. B., & Lumsden, C. J. (1996). Stochastic Simulation of Coupled Reaction - Diffusion Processes. *Journal of Computational Physics, 127*, 196-207.

Takahashi, K., Arjunan, S. N., & Tomita, M. (2005). Space in systems biology of signaling pathways--towards intracellular molecular crowding in silico. *FEBS Lett, 579*, 1783-1788.

Takahashi, K., Kaizu, K., Hu, B., & Tomita, M. (2004). A multi-algorithm, multi-timescale method for cell simulation. *Bioinformatics, 20*, 538-546.

Temkin, J. M., & Gilder, M. R. (2003). Extraction of protein interaction information from unstructured text using a context-free grammar. *Bioinformatics, 19*, 2046-2053.

Thomas, R. (1973). Boolean formalization of genetic control circuits. *J Theor Biol, 42*, 563-585.

Tjelle, T. E., Lovdal, T., & Berg, T. (2000). Phagosome dynamics and function. *Bioessays, 22*, 255-263.

Tomita, M. (2001). Whole-cell simulation: a grand challenge of the 21st century. *Trends Biotechnol, 19*, 205-210.

Tomita, M., Hashimoto, K., Takahashi, K., Shimizu, T. S., Matsuzaki, Y., Miyoshi, F., Saito, K., Tanida, S., Yugi, K., Venter, J. C., & Hutchison, C. A., 3rd. (1999). E-CELL: software environment for whole-cell simulation. *Bioinformatics, 15*, 72-84.

Turner, T. E., Schnell, S., & Burrage, K. (2004). Stochastic approaches for modelling in vivo reactions. *Comput Biol Chem, 28*, 165-178.

Uetz, P. (2002). Two-hybrid arrays. *Curr Opin Chem Biol, 6*, 57-62.

Uetz, P. , Ideker, T. , & Schwikowski, B. (2002). Visualization and integration of protein-protein interactions. In Golemis, E. (Eds.), *Protein-Protein Interactions - A Molecular Cloning Manual* (pp.623-646). Cold Spring Harbor Laboratory Press.

Uetz, P., & Finley, R. L., Jr. (2005). From protein networks to biological systems. *FEBS Lett, 579*, 1821-1827.

Uetz, P., Giot, L., Cagney, G., Mansfield, T. A., Judson, R. S., Knight, J. R., Lockshon, D., Narayan, V., Srinivasan, M., Pochart, P., Qureshi-Emili, A., Li, Y., Godwin, B., Conover, D., Kalbfleisch, T., Vijayadamodar, G., Yang, M., Johnston, M., Fields, S., & Rothberg, J. M. (2000). A comprehensive analysis of protein-protein interactions in Saccharomyces cerevisiae. *Nature, 403*, 623-627.

Uetz, P., Rajagopala, S. V., Dong, Y. A., & Haas, J. (2004). From ORFeomes to protein interaction maps in viruses. *Genome Res, 14*, 2029-2033.

Upstream Biosciences. BioCAD. 2005. Available from: URL: http://www.biocad.com

van Gunsteren, W. F., Hunenberger, P. H., Mark, A. E., Smith, P. E., & Tironi, I. G. (1995). Computer simulation of protein motion. *Computer Phys Communications, 91*, 305-319.

van Zon, J. S., & ten Wolde, P. R. (2005). Simulating biochemical networks at the particle level and in time and space: Green's function reaction dynamics. *Phys Rev Lett, 94*, 128103.

Vayttaden, S. J., & Bhalla, U. S. (2004). Developing complex signaling models using GENESIS/Kinetikit. *Sci STKE*, 2004, pl4.

Visual Knowledge. Visual Knowledge. 2005. Available from: URL: http://www.visualknowledge.com

von Mering, C., Jensen, L. J., Snel, B., Hooper, S. D., Krupp, M., Foglierini, M., Jouffre, N., Huynen, M. A., & Bork, P. (2005). STRING: known and predicted protein-protein associations, integrated and transferred across organisms. *Nucleic Acids Res, 33* Database Issue, D433-D437.

von Mering, C., Krause, R., Snel, B., Cornell, M., Oliver, S. G., Fields, S., & Bork, P. (2002). Comparative assessment of large-scale data sets of protein-protein interactions. *Nature, 417*, 399-403.

Wagner, A. (2001). The yeast protein interaction network evolves rapidly and contains few redundant duplicate genes. *Mol Biol Evol, 18*, 1283-1292.

Wagner, A., & Fell, D. A. (2001). The small world inside large metabolic networks. *Proc Biol Sci, 268*, 1803-1810.

Walhout, A. J., Reboul, J., Shtanko, O., Bertin, N., Vaglio, P., Ge, H., Lee, H., Doucette-Stamm, L., Gunsalus, K. C., Schetter, A. J., Morton, D. G., Kemphues, K. J., Reinke, V., Kim, S. K., Piano, F., & Vidal, M. (2002). Integrating interactome, phenome, and transcriptome mapping data for the C. elegans germline. *Curr Biol, 12*, 1952-1958.

Walhout, A. J., Sordella, R., Lu, X., Hartley, J. L., Temple, G. F., Brasch, M. A., Thierry-Mieg, N., & Vidal, M. (2000). Protein interaction mapping in C. elegans using proteins involved in vulval development. *Science, 287*, 116-122.

Walter, G., Konthur, Z., & Lehrach, H. (2001). High-throughput screening of surface displayed gene products. *Comb Chem High Throughput Screen, 4*, 193-205.

Weimar, J. R. (2002). Cellular Automata Approaches to Enzymatic Reaction Networks. *Paper read at 5th International Conference on Cellular Automata for Research and Industry*, ACRI, at Geneva, Switzerland. 294-303.

Weimar, J. R., & Boon, J. P. (1994). Class of cellular automata for reaction-diffusion systems. *Physical Review. E. Statistical Physics, Plasmas, Fluids, and Related Interdisciplinary Topics, 49*, 1749-1752.

Weinmann, A. S., Yan, P. S., Oberley, M. J., Huang, T. H., & Farnham, P. J. (2002). Isolating human transcription factor targets by coupling chromatin immunoprecipitation and CpG island microarray analysis. *Genes Dev, 16*, 235-244.

Wells, J., Yan, P. S., Cechvala, M., Huang, T., & Farnham, P. J. (2003). Identification of novel pRb binding sites using CpG microarrays suggests that E2F recruits pRb to specific genomic sites during S phase. *Oncogene, 22*, 1445-1460.

Wheeler, D. L., Chappey, C., Lash, A. E., Leipe, D. D., Madden, T. L., Schuler, G. D., Tatusova, T. A., & Rapp, B. A. (2000). Database resources of the National Center for Biotechnology Information. *Nucleic Acids Res, 28*, 10-14.

Wilkinson, M. D., & Links, M. (2002). BioMOBY: an open source biological web services proposal. *Brief Bioinform, 3*, 331-341.

Willats, W. G. (2002). Phage display: practicalities and prospects. *Plant Mol Biol, 50*, 837-854.

Williams, D. M., & Cole, P. A. (2001). Kinase chips hit the proteomics era. *Trends Biochem Sci*, 26, 271-273.

Wishart, D. S., Yang, R., Arndt, D., Tang, P., & Cruz, J. (2004). Dynamic cellular automata: an alternative approach to cellular simulation. *In Silico Biology, 5*, 0015.

Wooldridge, M. (2002). *Introduction to MultiAgent Systems*. Hoboken, NJ: John Wiley & Sons.

World Wide Web Consortium. OWL Web Ontology Language Overview. 2004. Available from: URL: http://www.w3.org/TR/owl-features/

Wuchty, S. (2004). Evolution and topology in the yeast protein interaction network. *Genome Res, 14*, 1310-1314.

Wuchty, S., Oltvai, Z. N., & Barabasi, A. L. (2003). Evolutionary conservation of motif constituents in the yeast protein interaction network. *Nat Genet, 35*, 176-179.

Wymann, M. P., Zvelebil, M., & Laffargue, M. (2003). Phosphoinositide 3-kinase signalling--which way to target? *Trends Pharmacol Sci, 24*, 366-376.

Yates, J. R., 3rd. (2004). Mass spectrometry as an emerging tool for systems biology. *Biotechniques, 36*, 917-919.

Yeger-Lotem, E., Sattath, S., Kashtan, N., Itzkovitz, S., Milo, R., Pinter, R. Y., Alon, U., & Margalit, H. (2004). Network motifs in integrated cellular networks of transcription-regulation and protein-protein interaction. *Proc Natl Acad Sci U S A, 101*, 5934-5939.

Yona, G. Biozon. 2005. Available from: URL: http://biozon.org

Yook, S. H., Oltvai, Z. N., & Barabasi, A. L. (2004). Functional and topological characterization of protein interaction networks. *Proteomics, 4*, 928-942.

Yu, H., Luscombe, N. M., Lu, H. X., Zhu, X., Xia, Y., Han, J. D., Bertin, N., Chung, S., Vidal, M., & Gerstein, M. (2004a). Annotation transfer between genomes: protein-protein interologs and protein-DNA regulogs. *Genome Res, 14*, 1107-1118.

Yu, H., Zhu, X., Greenbaum, D., Karro, J., & Gerstein, M. (2004b). TopNet: a tool for comparing biological sub-networks, correlating protein properties with topological statistics. *Nucleic Acids Res, 32*, 328-337.

Zanzoni, A., Montecchi-Palazzi, L., Quondam, M., Ausiello, G., Helmer-Citterich, M., & Cesareni, G. (2002). MINT: a Molecular INTeraction database. FEBS Lett, 513, 135-140.

Zhou, H., Roy, S., Schulman, H., & Natan, M. J. (2001). Solution and chip arrays in protein profiling. *Trends Biotechnol, 19*, S34-S39.

Zhu, H., Bilgin, M., Bangham, R., Hall, D., Casamayor, A., Bertone, P., Lan, N., Jansen, R., Bidlingmaier, S., Houfek, T., Mitchell, T., Miller, P., Dean, R. A., Gerstein, M., & Snyder, M. (2001). Global analysis of protein activities using proteome chips. *Science, 293*, 2101-2105.

Zou, M., & Conzen, S. D. (2005). A new dynamic Bayesian network (DBN) approach for identifying gene regulatory networks from time course microarray data. *Bioinformatics, 21*, 71-79.

In: Genome Research Advances
Editor: Alberto P. Bellini, pp. 69-164

ISBN: 978-1-60021-805-7

Chapter II

PIONEERING HUMAN MYOBLAST GENOME THERAPY AS A PLATFORM TECHNOLOGY OF REGENERATIVE MEDICINE

Peter K. Law[1,*], Danlin M. Law[1], Ping Lu[1], Eugene K. W. Sim[2], Khawja H. Haider[2], Lei Ye[2], Xun Li[3], Margarita N. Vakhromeeva[4], Ilia I. Berishvili[4], and Leo A. Bockeria[4]

[1]Cell Transplants Asia Ltd., Canada;
[2]National University Hospital, Singapore;
[3]Guangzhou Economic and Technological Development District Hospital, China;
[4]Bakoulev Scientific Center for Cardiovascular Surgery, RAMS, Russia.

ABSTRACT

Human Myoblast Genome Therapy (HMGT) is a platform technology of cell transplantation, nuclear transfer, and tissue engineering. Myoblasts are differentiated, immature cells destined to become muscles. Myoblasts cultured from muscle biopsy survive, develop and function to revitalize degenerative muscles upon transplantation. Transplant injury activates regeneration of host myofibers that fuse with the injected myoblasts, sharing their nuclei in a common gene pool of the syncytium. Thus, through nuclear transfer and complementation, human genome can be transferred into muscles of genetically-ill patients to achieve phenotype repair. Myoblasts are safe and efficient universal gene transfer vehicles endogenous to muscles that constitute 50% of the body. Myoblasts fuse among themselves to form new myofibers. Patients take only 2-month cyclosporine to immunosuppress allograft rejection because myofibers do not express MHC-1 antigens.

The first correction of human gene defect was published in the Lancet on July 14, 1990 when the therapeutic protein dystrophin was found in the myoblast-injected muscle

* Correspondence concerning this article should be addressed to Peter K Law, Cell Transplants Asia Ltd., Canada.

of a Duchenne muscular dystrophy (DMD) patient. Results over 280 HMGT procedures on MD subjects in the past 15 years demonstrated absolute safety. Myoblast-injected DMD muscles showed improved histology. Strength increase at 18 months post-operatively averaged 123%. FDA-approved clinical trials progressed unto Phase III in USA with direct cost recovery.

Heart muscle degeneration is the leading cause of human debilitation and death. The first human myoblast transfer into porcine myocardium in May 2000 revealed that it was safe to administer one billion myoblasts at 10^8/ml using 20 injections inside the left ventricle. Surgical HMGT into 50 infarcted porcine hearts demonstrated three mechanisms as proof of concept. Some myoblasts trans-differentiated to become cardiomyocytes. Others transferred their nuclei into host cardiomyocytes through cell fusion. As yet others formed skeletal myofibers with satellite cells. *De novo* production of contractile filaments augmented heart contractility. Myoblasts transduced with VEGF165 and Angiopoeitin-1 genes produced six times more mature capillaries in infarcted porcine myocardium than placebo. Xenograft rejection was not observed for up to 30 weeks despite cyclosporine discontinuation at 6 weeks.

Clinical trials on approximately 150 severe heart patients in 12 countries demonstrated $<10\%$ mortality since June 2000. Most subjects received autologous cells implanted epicardially after CABG or endomyocardially. Two subjects each receiving 1-billion myoblast allograft and 2-month cyclosporine survived over one year. Significant increases in left ventricular ejection fraction, wall thickness and motion have been reported, with reduction in perfusion defective areas, angina and shortness of breath. Two European trials are in Phase II.

HMGT on two Type II diabetic subjects aimed to correct the GLUT4 genomic defects, producing insulin receptors on myofibers to facilitate glucose uptake for metabolism. Each subject received 24- billion allogenic myoblasts and 2-month cyclosporine. It has been 10 months and HMGT appears to be safe for both subjects.

Other HMGT targets include cosmetic enhancement, bone degeneration, cancer, anemia, hemophilia, human growth deficiency, muscle trauma, pain and depression. Transduced myoblasts and/or their derivatives act as stable bioreactor(s) to deliver long-term supplies of therapeutic protein(s) at optimal levels.

INTRODUCTION

This chapter is about the landmark development of a biomedical platform technology called Human Myoblast Genome Therapy (HMGT), known previously as Myoblast Transfer Therapy (MTT). The procedure of HMGT/MTT involves cell transplantation, nuclear transfer and tissue engineering to correct genetic defects and to strengthen degenerative and weak muscles.

Originally developed as a treatment for human muscular dystrophies in Phase III clinical trial in the USA, HMGT/MTT is now in Phase II clinical trials for treating heart failure in Europe, Phase I clinical trial for treating Type II diabetes in China, and tested for anti-aging cosmetic enhancement in Singapore.

This is the most up-to-date review by the pioneer of the technology, with critical evaluation of both basic experimentation and clinical trials. The conceptual approach and technological development are discussed. State-of-the-art methodology and results of the clinical trials are described in details with future perspectives. The work traverses the fields of genetics, neurology, pediatrics, cardiology, endocrinology, immunology, surgery, and developmental biology. HMGT/MTT will emerge as a mainstream platform technology of Regenerative Medicine.

Development of Mammalian Skeletal Muscles

Myoblasts

Mammalian skeletal muscles are derived from the mesodermal germ layer in the embryo. In human, mesoderm first appears at 20 days after fertilization [1]. Concomitant is the appearance of somites that increase in number with time. The somites provide the microenvironment in which the muscular system develops.

Within the somites are uncommitted mitotic stem cells capable of giving rise to muscle, bone, cartilage, blood, lymphatic, fat and connective tissues (Figure 1). The commitment to being myogenic occurs early on since myoblasts, as these cells are called, are found in the limb buds at 26 days of gestation. Satellite cells which are myoblast reserves in adult muscles contain actin-like filaments in their cytoplasm. Such lineage determination is influenced by embryonic induction and irreversible gene expression [2].

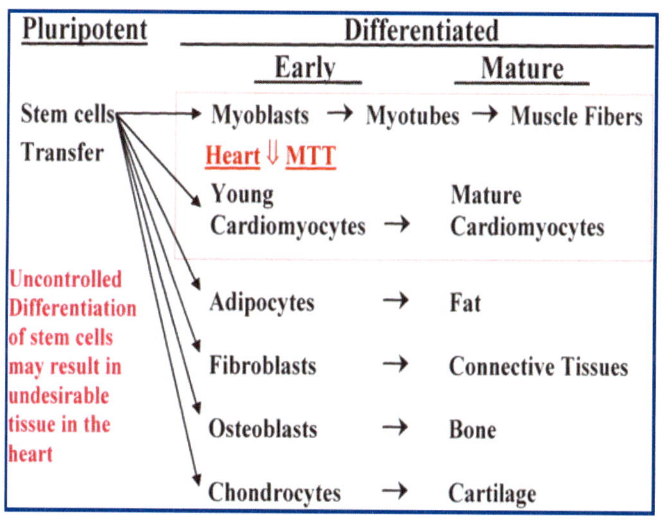

Figure 1. Pluripotent stem cells in the mesenchyme exhibit uncontrolled differentiation to become differentiated cells such as myoblasts, cardiomyocytes, fibroblasts, adipocytes, chondrocytes and osteoblasts. The benefits vs risk ratio of using myoblasts over stem cells in treating heart failure is greater. MTT, myoblast transfer therapy.

Lack of human fetal tissues for research in 1970s to 1980s has deemed myogenesis be studied in vitro. Figure 2 shows the different characteristics as the myoblasts develop. Beginning as small spheres of about 12 μm in diameter, myoblasts grow best on collagen attachment which somewhat resembles the extracellular matrix in vivo. Under proper culture condition, the transformation into spindle-shaped cells occurs within three days. These cells are characterized by their abilities to migrate, align, divide and fuse to form multinucleated myotubes that exhibit sarcomeres and immunostain positively for myosin. Equipped with these structural and biochemical bases of contraction, some myotubes can contract spontaneously in culture, and eventually pull from the culture surface. Without neural innervation the advance myotubes will undergo degeneration.

In the human fetus, close confinement within the somites ensures that the myoblasts are always at a state of confluence. The latter is a prerequisite for cell fusion. Cell division is an intrinsic property of myoblasts as evident by myoblast proliferation in serum-free culture media. However, proliferation in large quantity requires neurotrophic elements or factors [3,4]. It is known that myoblasts do not fuse when they are in the S, G2, M or even the early G1 phases of the cell cycle. Although it is not possible to study division synchronization in the human fetus, it would appear to be a less important factor in myoblast fusion, considering that myoblast doubling time in culture is approximately 20 hours, and that myoblasts within a somite are essentially replicating clones.

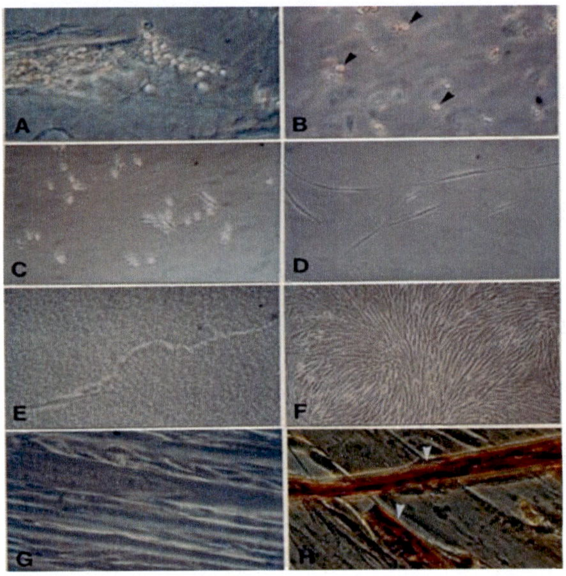

Figure 2. Human myogenic cell culture. (A) Satellite cells from human muscle biopsy. (B) Satellite cells 24 h after plating. (C) Satellite cells transforming to become spindle-shaped myoblasts at 48 h after plating. (D) Myoblasts at 96 h after plating. (E) Myoblast migrates. (F) Myoblasts align and confluent, ready for injection. (G) Higher magnification showing multi-nucleated myotubes with striations. (H) Myotubes immunostained with myosin antibody (arrowheads). One week after formation, spontaneously contracting myotubes detach from plate.

Natural Cell Fusion

Reduction in serum or chick embryo extract concentration precipitates cell fusion in culture. Cell fusion occurs only after the myoblasts have undergone considerable cycles of division and are withdrawn from the mitotic cycle [3]. The decision to fuse in vivo is genetically programmed, and appears to occur without any neural contact or influence. It involves structural gene transcription and RNA synthesis, especially in the formation of the receptors on the myoblast surface that are responsible for cell recognition, cell adhesion, and membrane restructuring [4, 5]. The end product are multinucleated myotubes each of which is formed from fusion of 200 or more myoblasts.

Myotubes

Myotubes, with the many nuclei within the syncytium, are differentiated cells capable of producing large quantities of contractile proteins and related molecules. These proteins include, among others, myosin, actin, tropomyosin, and myoglobin. The contractile proteins are structured into filaments as the myotube develops, and are packaged into basic contracting units called the sarcomeres. Hundreds of sarcomeres are aligned to form a myofibril many of which comprise a myofiber. Each sarcomere is made up of myosin and actin filaments linked together by cross-bridges. Upon excitation-contraction coupling, millions of cross-bridges are formed, sliding the actin filaments toward the center of each myosin filament, and results in force generation [6]. Contraction and relaxation require $Ca2+$ and ATP, and membrane excitation needs innervation from motoneurons.

Myotube Maturation

The myotubes will not develop into a myofiber unless innervated by a motor axon. With the arrival of the axonal terminal, an area of the myotube membrane which is adjacent to the axonal terminal becomes thickened and convoluted due to the formation of large amounts of acetylcholine receptors and ionic channels. The mature neuromuscular junction is a highly specialized relay station, transmitting the central command to effect contraction of the myofiber. Neuromuscular transmission is through the release of the neuromuscular transmitter acetylcholine which, upon combining with their postjunctional receptors, is metabolized by cholinesterase and produces significant membrane depolarization to effect excitation-contraction [6]. These junctions stain positively for acetylcholinesterase.

As the bones elongate with age, the passive stretch induces additional sarcomere production with subsequent increase in strength. Although the number of myofibers remain constant for individual muscles, the number of myofibrils within the myofiber can vary according to genetic and environmental differences. As more contractile proteins are deposited and the sarcoplasmic reticulum better developed, the centrally located nuclei immigrate peripherally. Numerous structural proteins and glycoproteins are synthesized and deposited during the transition from myotube to myofiber. Among these is dystrophin, a surface membrane protein which is not present in the myofibers of Duchenne muscular dystrophy subjects [7] as a result of the genetic defect [8,9]. A built-in regenerative measure resides with the satellite cells that are basically myoblast reserves found in adult skeletal muscle fibers. These are mononucleated cells located between the basal lamina and the plasma membrane.

Broad Perspective

Like murine dystrophy, Duchenne muscular dystrophy (DMD) serves as a point of reference for us to test the Human Myoblast Genome Therapy (HMGT) as a treatment for hereditary muscle degenerative diseases. In a broad sense, HMGT is being developed to repair abnormal cells and to replenish dead cells of all degenerative and weak muscles. This has important implication in the development of treatments for muscle conditions exhibited in muscular dystrophies, heart muscle degeneration, Type II diabetes, aging and many others.

Duchenne Muscular Dystrophy

The natural history of DMD depicts a continuous muscle degeneration and loss of strength, documented by subjective quantification [10-16], to begin at three years or younger, and continues throughout the course of the disease. Growth may outstrip the disease progress between four and five years of age, giving a false impression of remission or improvement; otherwise deterioration is continuous [11,17]. Degeneration is more severe in the proximal and anti-gravitational muscles than distal ones, and proximal muscle weakness in the lower body is responsible for the Gowers' sign used in physical diagnosis.

Debilitating and fatal, DMD affects 1 in 3300 live male births [18] and is the second most common lethal hereditary disease in human [19]. DMD individuals usually lose 50% of the strength in their leg muscles by age 9 due to muscle degeneration. They are wheelchair-bound by age 12, and three-quarters die before age 20. Pneumonia usually is the immediate cause of death, with underlying respiratory muscle degeneration, failure to inhale sufficient oxygen and to expel lung infection. Cardiomyopathic symptoms develop in mid adolescence [20] in about 10% of the DMD population. By age 18, all DMD individuals develop cardiomyopathy [21], but cardiac failure is seldom the primary cause of death [17].

The urgent need of a treatment for DMD prompted us to decide on using DMD in our initial clinical trials. We felt that if HMGT could benefit DMD individuals, it would likely benefit other muscle disease sufferers.

Murine Dystrophy Etiology-General Membrane Defect

The 129 ReJ-dydy [22] and the C57BL/6J dy2Jdy2J [23] dystrophic mice show autosomal recessive inheritance of dystrophy, with the dy mice having an earlier onset and a more rapid progression of the disease than the dy2J mice. Histopathology is similar to DMD, as is early death from respiratory failure due to muscle degeneration. Dystrophic symptoms appear at 3 to 4 weeks after birth and persist throughout the lives of these animals.

The *first* direct evidence of membrane abnormality in mammalian dystrophy was reported in 1972. The "cable" properties of normal and dy dystrophic mouse myofibers were compared using microelectrodes [24]. Subthreshold direct stimulation elicited a smaller response from the dystrophic fiber than from the normal one, indicating that the dystrophic fiber was "leaky" to the ionic current. Subsequently, the ability of dystrophic muscle fibers to generate action potentials in response to supramaximal nerve stimulation was studied with intracellular microelectrodes. About 55% of the fibers examined were abnormal in their ability to generate action potentials. Some of these fibers showed no detectable response to stimulation of the motor nerve, whereas in others, abortive spikes or localized end-plate potentials were recorded. The proportion of fibers showing very small or no electrical response increased when recordings were made away from the end-plate zone. The abortive spikes indicated a structural defect in the sarcolemma such that sodium conductance was diminished [25].

The abortive spikes did not propagate the whole length of the myofibers such that sarcomeres further away from the motor end-plates would not receive depolarization to contract and would be wasted. Dystrophic myofibers with abortive spikes generated localized contraction close to the endplates but the tensions developed were lower than normal. Furthermore, when these fibers were subjected to the passive stretch of the antagonistic muscle and the active pull of the localized contraction, breach in the plasma membrane [26] might occur at the region of non-activated sarcomeres.

With leakage of Ca2+ into the myofibers, there are mitochrondrial Ca2+ overload and localized hypercontraction [28]. Also, Ca2+ -activated neutral proteases and lysosomal activities are activated resulting in muscle necrosis. Some fibers showed positive acid phospatase reaction in muscle nuclei, mitochondria, sarcoplasmic reticulum and contractile myofilaments [27]. Membrane defect was found not only in the sarcolemma, but also in the sarcoplasmic reticulum and the nucleus. Eventually, the contractile elements are replaced with connective tissues. To conclude, the dystrophic gene is transcribed into a structural protein abnormality which directly or indirectly results in membrane "leakiness". The establishment of an abnormal ionic equilibrium across the membrane, especially that of Na+, constitutes the earliest detectable pathophysiology in mediating weakness and necrosis of dystrophic myofibers.

Using a normal/dystrophic parabiotic mice model, it was demonstrated that the dystrophic nervous system was normal in its ability to form new innervation, to induce fiber type differentiation, to bring to maturity, and to maintain the structure and function of a normal muscle [28-30]. Without such knowledge, it would be imprudent to attempt strengthening dystrophic muscles with normal myogenic cell transfer.

Muscle Transplantation and Regeneration

Normal Regenerates

Studies of mammalian muscle transplantation and regeneration in the seventies and earlier were confined to techniques of grafting whole muscles [31-35] and implanting muscle minces [36-39]. Grafting mature muscle suffers from slow reinnervation and revascularization of the inner myofibers that inevitably degenerated. Mincing traumatizes myofibers, producing pyknotic myonuclei, fragmented sarcolemma, denuded basement membranes, and degenerating cellular organelles [40-43]. Both procedures of muscle transplantation yield reduced function of the regenerates when compared to intact control muscles.

Satellite Cells

Satellite cells are the major source of myoblasts during regeneration [44,45]. The undifferentiated satellite cells are mobilized upon muscle injury to multiply and become myoblasts, that eventually fuse to form myotubes. It is estimated that 11% of all nuclei within the external lamina of normal muscle fibers in the gastrocnemius and soleus muscles of the rat belonged to satellite cells, and that only 6% of the nuclei remained viable 8 to 10 h after mincing. This 6% fraction would undergo mitosis before fusion [41,42]. When expressed as a percentage of the total number of myonuclei in a muscle fiber, satellite cell contribution becomes less with age [46].

Dystrophic Regenerates

The soleus of the C57BL/6J-dy2J dy2J dystrophic mouse is similar in number of muscle fibers to the soleus of a normal littermate at birth [47]. Muscle fiber loss begins at the second week after birth, and reaches a plateau of 50% loss at the fifteenth week. A similar percentage of fiber loss was observed from the third to the sixth postnatal week in 129/ReJ-dydy dystrophic mice [48,49]. Since there is no mechanism that can completely prevent muscle degeneration, and since dystrophic muscles exhibit poor regenerative capability [50,51], muscle weakness can best be reduced by replenishing the muscle fiber loss with genotypically normal cells.

Grafting mature muscles has not been successful. Because of the large size of a mature muscle graft, the deeper fibers in the muscle core are subjected to hypoxia. They often degenerate before becoming reinnnervated and revascularized [31-35]. Transplanting minced muscle fragments may overcome hypoxia and facilitate reinnervation [36]. However, mincing is such a traumatic procedure that the regenerated muscle always develops fewer fibers and contractile ability is greatly reduced [50,52].

Various factors contribute to account for the reduced capability of dystrophic muscles to regenerate. At the initiation of regeneration, myoblasts align themselves along the basement membrane and undergo mitosis [40,51]. Unlike crush lesions, which cause no disruption of the basement membrane, muscle minces cause basement membranes to fragment [41,42]. Such fragmentation destroys the scaffold for orderly cell replacement [54] and reduces the chances for complete regeneration. This explains why, even in normal animals, regenerated muscles from muscle minces cannot produce as much tension as unoperated muscles [52]. In the regeneration of dystrophic muscle minces, the problem is further complicated by a primary membrane defect such that the plasma membrane may not become properly regenerated after mincing. This deficiency would cause an infiltration of collagen and the deposition of connective tissues [40,53,55] that impedes innervation, thus aborting the maturation of the myotubes [51,55]. Failure to replace the damaged basement membrane may also result in a reduction in the number of regenerating myoblasts, because mitosis will stop as soon as the basement membrane is populated [40,53,55]. It was demonstrated that there was an increase in the number of satellite cells in muscles of patients with Duchenne muscular dystrophy compared to that in normal controls [56].

In regenerating dystrophic muscles, the regenerative capability is likely to be limited by the failure of basement and plasma membrane reformation, by the infiltration of collagen, and by the number of satellite cells that are present immediately after mincing. Another parameter that is important in the regeneration of muscle is muscle activity. Innervated young muscle fibers require constant active and passive muscle activities for normal differentiation and morphogenesis [40]. Such activities are not to be found in the regenerating dystrophic muscles, because neighboring muscles have already been undergoing degeneration, and the newly established motor endplates in the regenerate do not function as efficiently as normal.

Early muscle transplant studies in murine dystrophy research were designed to test whether muscular dystrophy is neurogenic or myogenic [57-61]. These studies always specified removal of a host muscle, which was then reimplanted or replaced with a foreign muscle. The procedure introduces two deficiencies. First, it replaces a functioning muscle with one which is nonfunctioning until it becomes reinnervated, thereby upsetting the balance in the activities of synergistic and antagonistic muscles of the host [62-65]. Such imbalance of muscle activity may eventually be reflected in the regenerative condition of the transplanted muscle [66-69]. Second, whenever the host nerve is cut, fewer than 50% of its fibers will survive to reinnervate [70-72]. This is especially significant in the dystrophic soleus nerve, which exhibits spontaneous reduction in the number of motor axons [73]. In designing a system to reduce muscle weakness, it is obviously advantageous not to remove the host muscle, so as to allow uninterrupted nerve and muscle function.

Gene Transfer

Dystrophyc Genes

The development of recombinant DNA, DNA sequencing and gene transfer techniques [74-77] hold the implicit promise to correct genetic diseases by correcting the abnormal gene itself. Through positional cloning [78], the dystrophin gene, whose absence or alteration results in Duchenne and Becker muscular dystrophies, has been located on chromosome Xp21 and sequenced [8,9,79,80]. Its 79 exons and introns span 2.5 megabases of genomic DNA and encode a 14 kilobase mRNA. The latter translates the missing protein dystrophin [7,80]. The gene responsible for myotonic dystrophy has also been isolated [81-83]. In addition, genes responsible for facioscapulohumeral dystrophy [84], limb-girdle dystrophy [85,86] and Emery Dreifuss muscular dystrophy [87] have been located on specific chromosomes through linkage analyses.

The above knowledge allows diagnosis at the molecular genetics level, which is especially significant in the differential diagnosis of Duchenne vs Becker muscular dystrophy [80,88]. It also offers advantages in prenatal counseling and carrier detection. Gene manipulation and transferal, being explored in animals, has sparked much hope for cure or treatment for the dystrophies. It is worthwhile to fully understand the conceptual approach, the evidence, the potentials and limitations of these approaches.

Dystrophin

Expression of the large dystrophin gene involves at least five promoters with very complex transcriptional and splicing control [89]. Dystrophin is a cytoskeletal membrane protein similar to β-spectrin and α-actinin [7,90]. It consists of 3685 amino acids with a molecular weight of 427 kilodaltons (kD), and folds into distinct structural domains: the amino (N) terminuus the "rod region" and the carboxy (C) terminus [89].

Dystrophin comprises about 0.001% of total protein in normal human muscles [7]. It is mainly localized on the inner surface of the plasma membrane. A different isoform is found in brain. Dystrophin is absent in DMD, and is reduced or altered in BMD [80,91]. Exceptions have been reported. Two BMD patients have been found without dystrophin [92], and a DMD patient shows a dystrophin phenotype indistinguishable from normal [93]. Overall, there is a correlation of deletion type (DMD vs. BMD), severity of the disease and the quantity and quality of dystrophin [89,93]. This is especially true if DMD and BMD are considered as one disease entity, the Xp21 myopathy.

Dystrophin is a large structural protein in close association to glycoproteins in the plasma membrane of normal myotubes or myofibers. The dystrophin/glycoproteins complex provides support to the cell membrane cytoskeleton.

Unlike the mdx mice, dystrophic mice (strain 129, Strain C57BL/6J-dy2Jdy2J) showed the presence of dystrophin in the plasma membrane of their skeletal muscles. The genetic defects of these mice are manifested in the absence of laminin, a membrane protein defect which is also found in Severe Childhood Autosomal Recessive Muscular Dystrophy (SCARMD). Their muscle degeneration parallels DMD in spite of the presence of dystrophin. The mdx mice exhibit essentially normal muscle function and lifespan in spite of the absence of dystrophin.

Whereas dystrophin serves as a good genetic/biochemical marker, more pertinent is the technical question of how to replace dystrophin. "The ultimate goal of rational therapy of Duchenne muscular dystrophy is clear: to replace dystrophin within muscle fibers" [94].

Dystrophin Up-Regulation

Up-regulation of dystrophin has been proposed as a therapeutic strategy for DMD. The incorporation of large structural protein(s) such as dystrophin into cell membranes has defied current technologies. Retroviral infection of mature myofibers with conjugated factor(s) to up-regulate dystrophin production seems inadequate in view of low efficiency, high risk and instability of transduction.

Contrary to common belief, dystrophin replacement constitutes only a part of the treatment. It has already been demonstrated using MTT in mdx mice [95-97] and in humans [98-105]. Since DMD pathology is one of muscle degeneration and weakness, structural and especially functional improvements have to be the primary concerns. Expression of a foreign gene requires appropriate integration and regulation involving numerous cofactors, many of which are transient during embryonic development. This is especially true in the deposit of structural proteins such as dystrophin, rather than in the secretion of enzymes or hormones.

Genes only transmit biological characteristics but are not primary functioning units like cells of the body. Like drug therapy, dystrophin up-regulation cannot replenish degenerated cells. At best, it can only repair degenerating cells. In a 12-year-old DMD boy in which over 70% of the muscle cells have been lost from the quadriceps, it is unlikely that dystrophin replacement in the remaining cells alone will bring forth a cure or a treatment, whatever the delivery system of dystrophin may be.

Plasmid Injection

When plasmid DNA of the dystrophin gene is injected intramuscularly into dystrophin-deficient mdx mice, dystrophin is found seven days later in about 1% of the myofibers [106-108]. Retroviral-mediated transfer increases the efficiency to 6% [109], whereas adenovirus-mediated transfer boosts it to 50% [110]. The authors conclude that there is a direct correlation between the level of muscle fibers expressing recombinant dystrophin and the level of muscle fibers with peripheral nuclei, indicating an improvement in muscle pathology [106,108].

Only a very small number of muscle fibers show dystrophin, and central nucleation is the only parameter used in monitoring normality in muscle pathology. Mature mdx mice are not totally devoid of dystrophin. Their muscles often contain over 1% of dystrophin-positive fibers. Since mdx mice do not exhibit muscle weakness and shortened life expectancy, the dystrophic dog (GRMD) is a better animal model [111] to demonstrate the efficacy of plasmid injection in which dystrophin production, cell structure and muscle function can be monitored. To date, plasmid delivery into dystrophic animals has not led to clinical trial of humanmuscular dystrophy.

Based on transgenic mice studies [112], clinical trials on cystic fibrosis (CF) and on severe combined immunodeficiency (SCID) [113] using viral vector single gene transduction were attempted and met with limited success. Unlike CF and SCID whose genetic defects are mediated through enzymatic deficiencies, DMD is caused by a deficit of a structural protein rather than a regulatory protein.

While optimism soars as transgenes were expressed in skeletal myofibers [114,115] and cardiac cells [116,117], the mechanism of uptake of plasmid DNA was not known [115], and the efficiency of transduction was low. Plasmid injection, if it ever becomes a treatment for DMD or heart failure, would have to be administered before any major cell degeneration occurs.

Transfected Myoblast Transfer

This was the original approach proposed in gene manipulation toward DMD treatment [118]. The procedure consists of a) obtaining a muscle biopsy from the DMD patient, b) transducing a "seed" amount of satellite cells with the dystrophin gene, c) confirming the myogenicity of the transfected cells, d) proliferating the transduced myoblasts to an amount enough to produce beneficial effect and e) injecting the myoblasts into the patient.

Retroviral vectors have been used to transfer genes into rat and dog myoblasts in primary cultures under conditions that permit the transduced myoblasts to differentiate into myotubes and expressing the transferred genes [119]. Furthermore, mice injected with murine myoblasts that are transfected with human growth hormone (hGH) show significant levels of hGH in both muscle and serum that are stable for 3 months [120,121]. The transduced myoblast transfer for DMD is similar in approach to the study of adenosine deaminase (ADA) deficiency. In the latter, T cells from the patient were transduced with functional ADA genes and returned to the patient after expansion in the number of the transduced cells through cell culture [122].

How realistic is such an approach for DMD? When the recombinant human dystrophin gene is integrated stably into mouse myoblasts in culture, these cells express the gene and synthesize recombinant dystrophin protein of the correct size and appropriate antibody reactivity [123]. However, the precise mechanisms of gene integration, regulation and expression are not known (Table 1). Imagine a large DNA-retroviral conjugate traversing the myoblast cell membrane, its cytoskeleton and possibly the nuclear membrane. The myoblast has to remain physically undamaged. Oncogenes should not be turned on to produce tumors. A sufficient amount, e.g. 10^5, of such "seed" myoblasts will have to be produced with molecular manipulation, and then be cloned to produce 50 X 10^9 cells for the myoblast injection treatment, without any loss of myogenicity or any acquisition of tumorgenicity.

Table 1. Gene manipulation versus myoblast transfer in repairing DMD muscle fibers

Procedure	Gene manipulation	Myoblast transfer
Identify defective gene	achieved	bypass
Synthesize normal gene	achieved	bypass
Target gene to tissue	injection	injection
Target gene into cells	viruses	natural cell fusion
Integrate normal gene	rare, random	natural
Regulate normal gene	not achieved	natural
Express normal gene	rare, inconsistent	natural

(Reproduced with permission from Law et al; Cell Transplantation 1:235, 1992).

Unfortunately, delivering transduced autologous myoblasts to treat DMD muscles has encountered low efficiency and mutation in transduction [119-121], and will yield insufficient myogenic cells to provide for a whole body MTT [124]. The down-sized dystrophin gene is still at least five times too big for efficient viral vector transduction. It cannot perform exactly like the full-size gene. Viral vectors may revert to pathologic states and cause acute pneumonia as encounted in a clinical trial of CF using adenovirus mediated vector. Although the myoblast clones are histocompatible with the recipient, the dystrophin recombinant, being derived from a foreign gene, is foreign to the host immune system. This is a complicated, uncertain, and a very expensive way to obtain histocompatible normal myoblasts, and does not offer any advantage over normal MTT.

With this approach, myoblast transfer is still part of the therapy. Adenovirus vector in high titre [10^{10} pfu/ml or above] can efficiently transduce immature muscle cells but not mature nuscle fibers in vivo [125]. Every step of the therapy will have to be custom tailored for each patient for histocompatibility. Even if such a procedure does become available many years from now, the technology will be too late to help many.

Arterial delivery of genetically transformed myoblasts into rat skeletal muscles has been reported [126]. It is difficult to reconcile the survival of myoblasts in the blood stream where macrophages and leukocytes abound. If these myoblasts indeed survive, they will undoubtedly divide in the rich milieu of the blood stream and fuse when they reach confluence in the capillaries. The latter will introduce risks of thrombi formation, possibly leading to pulmonary embolism. The recombinant DNA research has not provided a cure or treatment for DMD [124,127].

The success in transgenic mouse production, though rare, does indicate that gene manipulation can be effective in phenotype transformation when applied in embryos [128]. It is through manipulating the embryo that the normal gene may possibly transfect all cells of the adult to achieve a cure. However, such approach in humans would seem impractical for ethical, religious and technical reasons.

Dystrophin-Associated Glycoproteins (DAGs)

Dystrophin exists in a large oligomeric complex with the transmembrane glycoproteins of 35 kD (35-DAG), 43 kD (43-DAG) and 50 kD (50-DAG), an extracellular glycoprotein of 156 kD (156-DAG) and a cytoskeletal protein of 59 kD (59-DAG) [129-132]. The complex links the subsarcolemmal cytoskeleton to the extracellular matrix through the binding of 156-DAG to laminin. The DAGs are reduced in muscles of DMD [133], Fukuyama congenital muscular dystrophy [134], mdx [135] and the cardiomyopathic hamster [136]. Absence of DAGs can cause degeneration by themselves [133].

Hypothetically, the absence of dystrophin causes a disruption of the sarcolemmal linkage, with the loss of DAGs, especially during muscle contraction, and results in cell necrosis and muscle weakness in DMD [133]. As more myogenic transcription factors are discovered and their functions defined [137,138], it is hoped that those regulating the expression of dystrophin and DAGs will soon be unveiled.

Perspectives

Human gene therapy is a procedure that was used in attempts to treat genetic diseases such as cystic fibrosis (CF) and severe combined immunodeficiency (SCID). Human genetic engineering raises unique safety, social and ethical concerns [139]. Unlike CF and SCID whose defects through life are single enzymatic deficiencies, DMD symptoms are caused by a deficit of several structural proteins rather than a single regulatory protein. This latter difference plays an important role in deciding whether single gene manipulation is the direction we should take in developing treatment for DMD and the other muscular dystrophies.

As mentioned earlier, expression of a foreign gene requires appropriate integration and regulation involving polygenomic interaction and numerous cofactors, many of which are transient and unknown during embryonic development. This is especially true in the deposit of structural proteins such as dystrophin than in the secretion of enzymes such as adenosine deaminase (ADA). Since muscle cells of DMD boys lack not only dystrophin but dystrophin-related proteins and glycoproteins [133,140], and since the absence of these latter proteins can cause dystrophy by themselves [133,140], single gene replacement will not be effective in repairing the defective cells except for the very young patients in whom cell degeneration has not begun. There is no better method than MTT to replace all of these proteins through natural integration of the normal genome. Thus, MTT is not a stop-gap but conceptually the best possible treatment for the current DMD population.

Whereas single gene manipulation still harbors many unknowns [141], its success toward treatment of muscular dystrophy has been very limited. Even when fully developed, the procedure can only repair degenerating cells but cannot replenish degenerated cells. Gene therapists need to be cognizant of the fact that multiple gene integration will be necessary to repair myofibers of the current DMD population, and that these clinical trials are still many years away. The sooner researchers acknowledge the limitations of gene manipulation [124,126,141,142], the sooner will research efforts be focused toward developing myoblast transfer as the therapy for muscular dystrophy.

Myoblast Transfer

The terms "Myoblast Transfer" and "Myoblast Transfer Therapy" (MTT) were coined in a workshop held in Perth, Western Australia on February 20-22, 1989 [143]. Within four months, the First International Conference on Myoblast Transfer Therapy was held in New York City on June 10-12, 1989 [144]. In both meetings, international experts concluded that MTT was the most logical approach in the development of muscle disease treatment. In addition to continued animal research on MTT, small-scale clinical trials were initiated to determine the safety and efficacy of the procedure in humans. MTT thus appeared as a breakthrough with a quantum leap from animal experimentation into human study. Until now, the term "MTT" has been used in the scientific literature interchangeably for animal and human studies. However, in human studies, a more appropriate term is "Human Myoblast Genome Therapy" or "HMGT" because of the concomitant cell therapy and genome therapy procedure as explained below.

What is MTT or HMGT?

MTT or HMGT is a platform technology of cell transplantation, nuclear transfer, and tissue engineering developed to treat diseases associated with muscle degeneration and weakness. It is also used to enhance normal or near-normal physical conditions.

Myoblasts are differentiated, immature cells destined to become muscles. Myoblasts cultured from muscle biopsy survive, develop and function to revitalize degenerative muscles after transplantation.

The myoblast transplant procedure incorporates cell therapy and genome therapy concomitantly. Transplant injury activates regeneration of the host myofibers that fuse with the injected myoblasts, sharing their nuclei in a common gene pool of the myofiber synctium. Through nuclear transfer and complementation, the normal human genome can be transferred into muscle cells of genetically-ill patients to achieve phenotype *repair*. Myoblasts are safe and efficient gene transfer vehicles endogenous to muscles. Since muscles constitute 50% of the human body, extremely precise dosing of secreted biologics can be achieved from transduced myoblasts implanted. In addition, myoblasts fuse among themselves to form new myofibres to *replenish* those that underwent degeneration. The procedure leads to the formation of a genetically mosaic muscle (Figure 3).

Only 2-month cyclosporine immunosuppression is sufficient to prevent allograft rejection because myofibres do not express MHC-1 antigens. Only allografts instead of autografts can be used in the treatment of genetic disorders because myoblasts of the genetically-ill patients carry the same genetic defect(s) (Table 2). Allografts also have advantage over autografts when transduced myoblasts are targeted to secret foreign therapeutic protein(s).

The procedure involves isolation and culturing of satellite cells derived from muscle biopsies of genetically normal donors. It is customary to refer to myoblast reserves in adult muscles as satellite cells, and satellite cells in culture as myoblasts. Cell culture mass-produces the myoblasts and removes the antigenic leukocytes, adipocytes and fibroblasts. These donor myoblasts are then injected or surgically implanted into a number of foci in the degenerative muscles where host satellite cells and regenerating myofibers are abound.

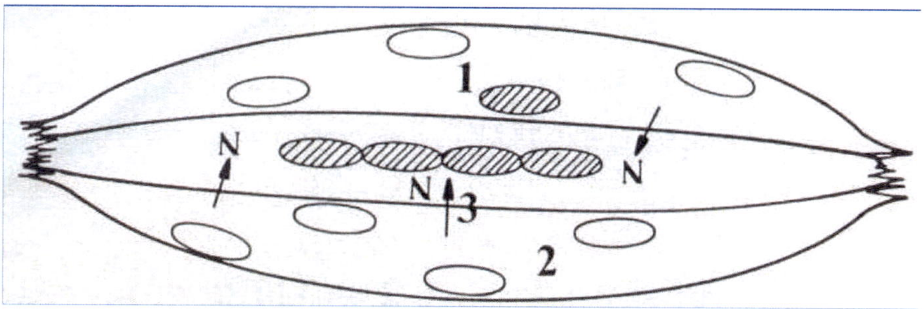

Figure 3. Injection of normal myoblasts into the dystrophic mouse muscle resulted in the formation of a mosaic muscle with normal, dystrophic, and mosaic myofibers. Two genetic complementation mechanisms were shown to alleviate muscle weakness: (1) normal gene product, N, in the mosaic fibers suppresses dystrophy expression; (2) developing normal fibers progressively replaced degenerative fibers. A third possibility (3), normal fibers supplying a diffusible "corrective factor" N to supplement neighboring dystrophic fibers, was not examined. Open ovals, normal nuclei; shaded ovals, dystrophic nuclei. (Reproduced with permission from Law et al. 1988.)

Table 2. Autograft versus allograft for the regenerative heart

	AUTOGRAFT	ALLOGRAFT	
PROS	No immunosuppression	2-mo immunosuppression	CONS
CONS	Cell culture takes 1-mo	Good cells readily available	PROS
	Graft delay ⟶ scar	Grafting within 12 hr ↛ scar	
	Elderly cells less regenerative	Young cells rejuvenate	
	High QA/QC costs for each individual graft	Low QA/QC costs for 50 grafts at a time	
	Host limitation	Unlimited supply	
	Infected/mutated host tissues may contaminate system	Patients with infections and genetic diseases can be treated	

The idea of using myoblasts as donor cells in MTT/HMGT is to provide the genetic blueprint to be activated through the regeneration process. Although mammalian skeletal muscles are capable of regeneration, there are not enough satellite cells to fully compensate for major muscle damage in the body. Furthermore, muscle regeneration is abortive in DMD. Treatment or cure involving single gene manipulation will have to be conducted on young patients or even fetuses rather than on older individuals, whose muscle cells have significantly degenerated. This differs from MTT/HMGT which *replenishes* lost cells and *repairs* degenerating cells.

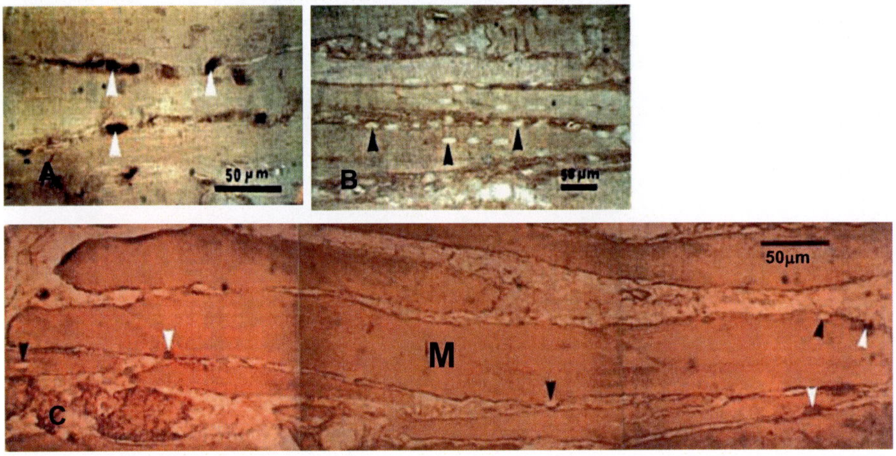

Figure 4. Immunocytochemical localization of donor (stained, white arrowheads) and host (unstained, dark arrowheads) nuclei in longitudinal muscle sections. A and B are normal and dystrophic controls, respectively. C is from a dystrophic muscle 18 months after normal myoblast injection. A mosaic fiber (M) is demonstrated by the presence of both stained and unstained nuclei. (Reproduced with permission from Law et al. 1990.)

How MTT/HMGT Works?

Dystrophic cells degenerate because they lack a normal gene. Such degeneration can be prevented by incorporation of full complements of normal genes through cell fusion with donor myoblasts [145-148]. Cell fusion is a spontaneous occurrence during myogenesis [149] and muscle regeneration [150,151]. When normal myoblasts are transferred into a regenerating dystrophic muscle as in MTT/HMGT, the donor cells fuse spontaneously with the host satellite cells and/or the host myofibers themselves to form mosaic fibers, sharing their normal genes to produce normal cell phenotype (Figure 4). Furthermore, donor myoblasts fuse among themselves to form myotubes that develop into normal myofibers (Figures 5, 6C, 6D). These normal myofibers progressively replace the degenerative host fibers. MTT/HMGT thus repairs degenerating cells and replenishes lost cells.

It does not matter which gene is abnormal or which protein is missing; MTT/HMGT has the potential to prevent genetically abnormal muscle cells from degenerating, regardless of the genetic defect. Through genetic complementation, the inserted normal genes encode one or more missing protein(s) (or factors) to supplement the metabolism or development of the heterokaryons, thereby sustaining the normal integrity of these genetically transformed cells. In this respect, MTT/HMGT differs from bone marrow grafting which is based strictly on the replacement strategy where cell fusion and genetic complementation do not occur. MTT/HMGT is the first therapy that has demonstrated significant improvement in cell genetics and phenotypes of live mammals [145-148].

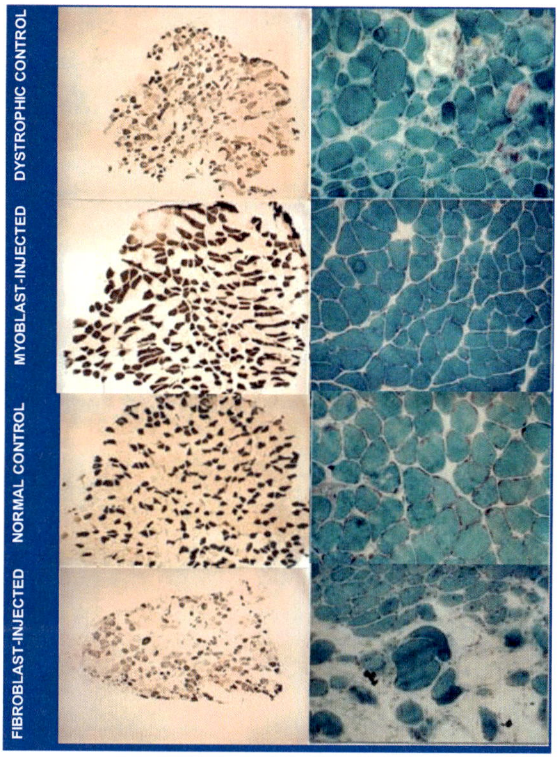

Figure 5. When compared with contralateral controls 6 months postoperatively, myoblast-injected dystrophic solei exhibited greater cross-sectional area and total fiber number. They contained more normal appearing and less abnormal-appearing fibers. Injection of about 10^6 fibroblasts as a control for cell type resulted in less absolute number of fibers, higher percentage of abnormal fibers, and more connective tissues in test dystrophic solei. (Reproduced with permission from Law et al. 1988).

The MTT/HMGT concept is based on previous work on muscle development, regeneration, transplantation, reinnervation, and genetic complementation. Because skeletal muscles of all mammals behave similarly in these situations, MTT/HMGT should have application to all mammalian muscle diseases including those of humans. Augmentation in muscle size, shape, consistency and force are not only therapeutic for muscle diseases, but should also be useful in disease prevention and enhancement of physical conditions.

Origin of MTT/HMGT

It is not possible to turn an already dystrophic individual into a totally normal one. The best alternative is to transform the dystrophic individual to become similar to the X-linked carrier in order to achieve normal or near-normal function. According to the Lyon hypothesis [152], random inactivation of one or the other X-chromosome can occur in any nucleus throughout the body of a Duchenne female carrier; individual muscle fibers would therefore contain mixtures of normal and dystrophic nuclei similar to the mosaic fibers produced by injecting normal myoblasts into dy2Jdy2J muscles (Figure 4). The mosaic fibers exhibit normal phenotype, presumably due to metabolic or developmental complementation as a result of sharing the normal nuclei and therefore the normal genes. In addition, muscles of chimeric mice containing fibers of normal, dystrophic and mosaic genotypes exhibit normal function [153]. It thus appears that the logical and practical approach to reverse the expression of dystrophy in muscles of adult animals is to induce genetic mosaicism through the incorporation of the normal nuclei. This can be accomplished by transplantation of myogenic cells.

Neural Hypothesis of Muscular Dystrophy

A serious consideration in these muscle transplant studies is whether the nervous system of dystrophic individuals will support normal muscle development. The approach of muscle transplant would not be effective if the dystrophic nervous system induces degenerative changes in the regenerate. In the parabiotic mice study of Law et al [28], in which a "fast" dystrophic nerve was transposed onto a "slow" normal muscle, it was demonstrated that the dystrophic nerve is normal in its ability to form synapses, to induce fiber-type differentiation, to bring fibers to maturity, and to maintain the structural and functional integrity of normal muscle fibers. This parabiotic mice study also provides strong evidence against a neural etiology of murine dystrophy.

A study often cited to support the neural hypothesis is the chimeric mouse study of Peterson11 who concludes that "extramuscular factors must be implicated in muscular dystrophy". This conclusion is based on Peterson's argument that a small population of normal nuclei could not have prevented dystrophic pathology in the chimeric muscles that showed normal histology and isometric twitch tensions. However, such argument contradicts Peterson's later claim that "only a very small proportion of genotypically normal myonuclei were required for expression of an apparently normal phenotype [154]. In view of the parabiotic mice studies [28, 29, 30] it would appear that muscle degeneration in the dystrophic mice is not primarily caused by neurotrophic deficiency or by a lack of myelin as previously claimed by others [155-158]. Whereas both nerve and muscle cells are susceptible to the dystrophic gene action, muscle cells are more vulnerable to degeneration because of their contractile activity [27,30].

Animal Models

The dystrophic mice (129 ReJ-dydy and C57BL/6J-dy2Jdy2J strains) are excellent models of hereditary muscle degeneration and weakening [22,23,159] and are most studied. The autosomal recessive inheritance is manifested as progressive cell degeneration mainly in muscle and, to a minor extent, in nerve [148,73]. The crippling pathology, associated with significant muscle weakness, is similar to most forms of human dystrophies. In addition, these mice bear a close resemblance to DMD boys and children with SCARMD in that premature death occurs as a result of respiratory failure. Another animal model is the mdx mouse which does not show functional deficit or premature death [160-163]. Its muscle pathology does not show fiber splitting, phagocytic necrosis, and infiltration of fat and connective tissues, all of which are diagnostic features of dystrophy. Muscle pathology of mdx includes a brief period of central nucleation which is often interpreted as necrosis but may also represent delayed maturation or regeneration of the myofibers. Nevertheless, many consider mdx as a better genetic model of DMD because both diseases are X-linked recessive and both lack dystrophin.

The more ideal model for DMD is the dystrophic dog which shows a similar mode of inheritance, severe progressive muscle degeneration and weakness, and a lack of dystrophin [164,165]. The dystrophic dog offers advantage over the dy2Jdy2J mouse for MTT/HMGT research because, as in human, its larger muscles provide a more realistic challenge to the culturing and placement of donor myoblasts. In addition, dystrophin can be used as a biochemical marker to demonstrate successful engrafting in the dystrophic dog but not in the dy2Jdy2J mouse, since muscles of the mouse produce dystrophin [166].

Cell Therapy VS Gene Therapy

It is erroneous to believe that MTT/HMGT stems from the molecular studies of dystrophic genes or the proteins they encoded. The original testing of the MTT/HMGT idea was published in 1978 [150], and the complete cloning of the DMD gene did not appear until 1987 [9]. MTT/HMGT is developed according to the knowledge of genetic complementation which circumvents the time-consuming and expensive processes of abnormal gene identification, therapeutic gene synthesis, tissue targeting, gene integration, regulation, and expression that are not fully understood. The spontaneous cell fusion process inherent in myogenesis and muscle regeneration is utilized to incorporate the normal nuclei into dystrophic muscle cells. Since the fusion process is a natural occurrence, there should not be any problem with specificities of integration, complementation, regulation, and expression of the normal genes inserted. It is not necessary to know which gene(s) is responsible for the defect. Furthermore, the injection of normal myoblasts directly into the dystrophic muscle eliminates the uncertainty of tissue targeting with gene therapy. By virtue of incorporation of full complements of normal genes into genetically abnormal cells to achieve repair, MTT/HMGT is a gene therapy, or better still, a genome therapy.

The essence of MTT/HMGT is that it does not matter which gene or protein is missing. Through the incorporation of donor myonuclei, dystrophic cells are replenished with full complements of normal genes that are allowed to express and interact spontaneously. This differs from gene therapy in which a single gene is manipulated into the abnormal cell where cofactors essential for the regulation and expression of the gene may no longer be present at the developmental stage of the cell.

MTT/HMGT is the first cell therapy in which genetic engineering has demonstrated significant success in mammals. The ability to produce genotypic and phenotypic improvement through incorporation of the normal genes opens new avenues for other genetic diseases. Since many genetic diseases are fatal, the less time-consuming genetic complementation approach should be the method of choice in the development of treatments.

Development of MTT/HMGT

The original idea of MTT/HMGT, and experiments testing MTT/HMGT, were first published by Law in 1978 [50]. A deliberate attempt was made in adult dystrophic mice to produce mosaic muscles containing normal, dystrophic and mosaic myofibers from the regenerates of normal and dystrophic minced muscle mixes. The mince and mix procedure was designed to provide an environment for the intermingling and fusion of normal and dystrophic satellite cells. These muscle regenerates produced greater twitch and tetanus tensions and more normal appearing cell structure than control regenerates of minced dystrophic muscles. It was concluded that the introduction of cell contents of normal genotype into dystrophic muscles could improve the function of the latter.

The result contradicts the study of Partridge and Sloper who concluded, in transplanting normal minced muscles into normal hosts, that little or none of the regenerates was of donor origin [167]. These latter authors suggested the possibility of introducing, by a systemic route, normal muscle nuclei into diseased muscle fibers. Attractive though it seemed, this latter idea would not have been in agreement with knowledge of myoblast characteristics. Apart from the risks of having the circulatory system congested with rapidly dividing myoblasts, capillaries ruptured by growing myotubes, and severe host immune response against myoblast surface antigens, it is doubtful if the myoblasts are small enough to exit the wall of the capillaries. In the Anglo-French Cellular Therapy Meeting held in Cambridge, England on April 2-3, 1990, Kornegay reported the absence of a beneficial effect after intraarterial injection of donor myoblasts into two dystrophic golden retriever dogs. There was no evidence that any component of the donor myoblasts entered the skeletal muscles.

Concurrent with Law's 1978 publication on normal/dystrophic transplant was that of Partridge, Grounds, and Sloper, which described fusion between host and donor myogenic cells of normal genotypes using skeletal muscle grafts [168]. Contrary to what these authors previously published, they now claimed satellite cells did survive and develop in normal host muscles. Although the study did not involve dystrophic animals, the author inferred that MTT was a distinct development with potential applicability to hereditary myopathies.

Law's subsequent experiments were designed to bypass the trauma introduced by mincing. Interruption of the connections of nerve, blood, and tendon had to be avoided. A new muscle transplant method had to be developed. Knowing that the dystrophic nerve would support normal muscle development [28,30], Law grafted intact muscles of newborn normal mice into recipient soleus muscles of dsytrophic mice at the first clinical sign of dystrophy [169]. Results obtained 6 months after the grafting indicated that muscle fibers of the grafts survived, developed, and functioned in the dystrophic environment The regenerates had larger cross-sectional areas and more muscle fibers than the contralateral dystrophic solei. *That was the first study in muscular dystrophy research to have successfully improved the twitch tension of an adult dystrophic muscle to that of a normal.*

Grafting newborn muscle in animal is an innovative method of muscle transplant. The method eliminates tendon damage, overcomes hypoxia, and facilitates reinnervation and revascularization of the grafted muscle fibers, thus promoting the survival and development of the characteristics of the donor muscle [169]. The result achieved is superior to that obtained from mature muscle grafts or from minced muscle transplants. In addition to treatment of recessive myopathies of autosomal dominant inheritance, where rescue relies more on replenishing lost cells rather than on nuclear complementation after mosaic fiber formation. *A corollary to MTT/HMGT is to allow donor myoblasts to fuse in culture and to transplant the resulting myotubes that form the major component in newborn muscles.* By 1979, the concept of replenishing lost cells and repairing degenerative cells through the production of muscle mosaicism using MTT/HMGT was firmly established [169]. In the same year, it was established that myoblasts cultured from muscle biopsies of adult normal rats could survive and develop in the original donor after implantation [170].

Nevertheless, grafting newborn muscles has limited applicability in humans, because human newborn muscles are too large for rapid reinnervation and revascularization. There are also the ethical considerations. It is necessary to develop the myoblast transplant techniques.

A convenient way to obtain myoblasts in mice is through dissection of *limb-bud mesenchyme* of embryos. The first successful implantation of normal myoblasts into dystrophic mouse muscles was published by Law in 1982 [171]. Mesenchymal tissue rich in myoblasts (about 80% myoblasts, 20% fibreblasts) was dissected from limb-buds of day-12 normal mouse embryos. It was surgically implanted into the right solei of 20-day-old normal or dystrophic C56BL6J-dy2Jdy2J mice. Host and donors were histocompatible. Unoperated left solei served as controls. Sham control solei receiving similar surgical treatment but no mesenchyme transplant did not differ from contralateral, unoperated solei. This argues against the possibility that trauma to dystrophic muscles or their innervation could induce long-term improvement in muscle structure and function as claimed by Ontell [172].

Six to seven months postoperatively, the myoblast-implanted solei exhibited greater cross-sectional area, total fiber number, and twitch and tetanus tensions than their contralateral controls. Test dystrophic solei contained more normal-appearing and less abnormal-appearing fibers than their controls. The results indicate that such transplantation improves the structure and function of the dystrophic muscles [184]. In this study, the donor cells were not cultured before implantation, and recipient muscles were not subjected to denervation.

Meanwhile, the laboratory of Sloper and Partridge continued to explore factors that affected the incorporation and fusion of allogeneic muscle precursor cells in vivo [173]. Their studies involved only normal but not dystrophic mice. The implants consisted of either minced muscle mixes or newborn muscles disaggregated by trypsin and pargestin [174-176]. In both cases, they were implanted for two months or less in muscle compartments where the host muscles were removed. Such experimental design introduced denervation, devascularization and tendon detachment. It did not allow the muscle regenerates enough time to develop good structure or function. The authors demonstrated the survival and development of donor cells in the host muscles, using electrophoretic analyses of glucose phosphate isomerases (GPI), the genetic markers to identify hosts vs. donor cells.

Partridge's laboratory did not report implantation into abnormal muscles until 1988, when they employed the ICR/IAn mice as hosts [176]. These mice exhibited a mild glycogen-storage myopathy associated with a genetic lack of muscle phosphorylase kinase (PhK), but no significant structural or functional abnormality was apparent [177]. Enzymic disaggregation of neonatal muscle was employed to produce muscle precursor cells, resulting in only a small proportion of myogenic cells in the injected suspension. Although there have been frequent claims of supplying normal muscle precursor cells to alleviate hereditary myopathies by these authors, no evidence of any structural or functional improvement after transplantation was presented. In fact, these authors have not reported working with any dystrophic muscles having structural and functional deficits at all.

1988 witnessed the explosive development of MTT/HMGT following two significant publications in June by Law et al. [145,146]. In the first study, primary myoblast cultures from limb-bud explants of normal mouse embryos were injected into the soleus muscles of histocompatible dystrophic hosts [145]. In the second study, clones of normal myoblasts were injected into the leg and intercostal muscles on both sides of histoincompatible hosts with cyclosporine-A (CsA) as host immunosuppressant [146]. In both cases, donor myoblasts fused among themselves, developing into normal myofibers that progressively replaced the degenerated fibers. They also fused with dystrophic host myogenic cells to form mosaic myofibers of normal phenotype [145-148]. These two mechanisms of genetic complementation were shown to be responsible for significant improvement in muscle genetics, structure, function and animal behavior of the test dystrophic mice. In later reports, prolongation of the life-spans of the myoblast-injected dystrophic mice was demonstrated [147,148].

Morgan et al. reported the synthesis of trace amounts of PhK in about 5% of the myoblast-injected muscles of the PhK-deficient mice [176]. Seven months later, Partridge et al. reported the conversion of mdx myofibers from dystrophin-negative to -positive 20 to 50 days after injections of normal myoblasts [96]. The authors claimed rescue of the host fibers from their biochemical defect. The study demonstrates biochemical improvement in the mdx mouse model, an additional evidence to confirm the efficacy of MTT/HMGT as previously reported by Law et al. who had demonstrated improved muscle cell genetics and structure, strength, mobility, general well-being, and longevity of the dy2Jdy2J mice [50,145,148,169,171]. Karpati et al. provided further confirmation that dystrophin is expressed in mdx myofibers after MTT [97].

Myoblast Transfer Improves Muscle Genetics/Biochemistry/Structure/Function and Normalizes the Behavior and Life Span of Dystrophic Mice

We shall now examine the experimental results from animal studies that lead to MTT/HMGT clinical trials in humans. All of the developmental work was published by essentially two research teams whose approaches were disparate but complementary. While Law and associates were demonstrating the safety and efficacy of transferring normal myogenic cells into the dy2Jdy2J dystrophic mice, Partridge and associates were examining the developmental fate of donor cells in normal mice.

This was at a time when neither the golden retriever muscular dystrophy (GRMD) nor the xmd canine dystrophy was widely known or available [164,165]. Indeed, most of what can be done in the dog can be done in half the time and at a fraction of the cost in the mouse. Although GRMD and xmd are excellent models for testing potential treatments, they are available to a few chosen laboratories that, because of technical deficiencies, have not produced any convincing experimental results with MTT [178].

It was not until 1989 that study of MTT on mdx mice was first published [96,97]. The majority of evidence in support of MTT/HMGT efficacy is derived from previous studies using the dy2Jdy2J mice [50,145-148,169,171].

Central to MTT/HMGT is the correlation of genetic and phenotypic improvement at the cellular and at the whole muscle levels. Such correlation has to be derived from demonstrations that in fact there are genotypic and phenotypic improvements. These studies play an essential role in the elucidation of the mechanisms by which MTT/HMGT exerts its beneficial effects [50,145-148,169,171].

Genetic Improvement

Isozymes of glucose-6-phosphate isomerase (GPI) have been used as genetic markers [179-181] to quantify myonuclear genotypes of mice. Initially the procedure was used to determine the genotypic constitution in muscles and other tissues of chimeric [182,153] mice. Eventually it was adopted for myoblast transfer studies because of the persistence of the GPI isozymes within individual cells throughout the whole life of the animal [145-148,183].

Each cell expresses the isoform(s) which is specific to the genes encoding GPI, regardless of whether the latter is of donor or host origin. For example, Watt et al. implanted myogenic cells of the 129/ReJ mouse, which is homozygous for the GPI-1a allele, into muscles of either C57BL/6J or ICR/IAn hosts that are homozygous for the GPI-1b allele [183]. Grafts of the muscle precursor cells were found to have survived and developed in the host muscles as demonstrated by the presence of the parental isozymes GPI-1AA, GPI-1BB in gel electrophoresis of whole muscles. In addition, the presence of the hybrid isozyme GPI-1AB indicated fusion between host and donor cells in about 10% of the host muscles examined. Because the host mice selected exhibit normal genotypes and normal motor function, it is impossible to determine if there was any phenotypic improvement. The study does demonstrate that myogenic cell implantation allows the introduction of normal genome into normal muscles at the cellular and at the whole muscle level. This study provides support and confirmation to the mechanisms responsible for the structural and functional improvement observed after myogenic cell transfer into the dy2Jdy2J mice [50,169,171].

Genetic improvement of dystrophic muscles appeared in publication in 1988, when Law et al. reported that normal myoblast injections provided genetic treatment for murine (dy2Jdy2J) dystrophy. Cultured myoblasts (10^6) from genetically normal mouse embryos were injected into the right soleus of 20-day-old normal or dystrophic mice. Hosts and donors were histocompatible but exhibited different genotype markers. Donor cells produced GPI-1CC. Host cells produced GPI-1BB.

The GPI-1c allele codes for an isozyme of greater cathodal electrophoretic mobility than the GPI-1b allele [181]. The presence of GPI-1BC is a demonstration of mosaicism of single myofibers. GPI-1BC appears as a hybrid band with intermediate electrophoretic mobility between its parental bands GPI-1BB and GPI-1CC.

The survival and development of donor cells in host muscles were demonstrated with agarose gel electrophoresis of the strain-specific isozymes of GPI at 6 months after MTT. Analyses of six test normal and eight test dystrophic solei showed that all contained both parental isozymes, but only half of each group exhibited GPI-1BC.

The presence of GPI-1CC in test dystrophic solei implied the survival and development of donor cells in host muscles. The normal myoblasts fused among themselves to form normal fibers that functioned within the dystrophic muscles. Furthermore, the presence of GPI-1BC in test solei substantiated intracellular mosaicism through fusion of host and donor cells. This initial evidence substantiates the mechanisms responsible for muscle improvement after MTT/HMGT (Figure 3). The demonstration of GPI-1CC and GPI-1BC, both of normal genotypic origin, in dystrophic host muscles provides the first evidence of genetic improvement. The latter is well-correlated with structural and functional improvement observed after MTT [145-148].

It is known that gene products encoded by local myonuclei are uniformly distributed throughout the myofiber syncytium [184]. Furthermore, heterokaryons containing nuclei from mouse muscle cells and human nonmuscle cells can be induced to produce proteins typical of human muscle [185]. Thus, in the MTT study it is likely that the normal donor myoblasts fused with host myofibers, sharing normal genes and restoring normal structure and function to the now mosaic myofibers. This latter contention, though partially supported by the presence of GPI-1BC, did require direct confirmation at the single cell level.

Whereas gel electrophoresis of muscle isozymes may be used to demonstrate genetic mosaicism of the myoblast-injected muscles, it does not allow correlation of genotypic constitution with phenotypic expression in single cells. The latter correlation requires the use of immunocytochemistry or nuclear labeling to localize, in mosaic fibers, donor myonuclei that contain the normal genome.

For immunocytochemical localization of host and donor nuclei within the injected muscles, polyclonal antibodies highly specific for GPI-1CC but not for GPI-1BB were produced [186-188]. These antibodies, when conjugated with nuclear stain and horseradish peroxidase (HRP), stained the nuclei of normal control muscles but not the dystrophic control muscles (Figure 4). Using this immunocytochemical procedure, Law et al. observed normal appearing mosaic fibers that contained host and donor nuclei 18 months after myoblast injection (Figure 4). Normal nuclei of donor origin were stained. Dystrophic nuclei of the host were not stained [147,148]. Whereas neighboring dystrophic fibers of the host were at different stages of degeneration, the mosaic fiber appeared histologically normal. The study provides strong direct evidence of gene therapy at the single cell level. Cell genetics of the dystrophic fiber is improved through incorporation of the normal nuclei with all of the normal genes. The dystrophic cell is replenished with normal copies of the defective gene. This genetic/structural correlation of the presence of donor nuclei of normal genotype in mosaic fibers of normal histology (Figure 4) provides the second evidence in support of the MTT/HMGT concept (Figure 3).

The mosaic muscles resulted from MTT also contained normal and dystrophic muscle fibers. The presence of genetically normal fibers demonstrates genetic improvement at the whole muscle level (Figure 6 C). This latter mechanism is better illustrated using gel electrophoresis showing the GPI-1CC band (Figure 6D).

Thus, with the use of isozymes of GPI as genotype markers for host and donor cells, both electrophoretography and immunocytochemistry indicated the survival and development of donor myoblasts in the host muscles. Donor myoblasts fused among themselves to form normal fibers. They also fused with host cells, sharing their normal gene pool to effect repairing in the genetically mosaic cell. These mechanisms have been explored as early as 1978 [50,96,145-148,183].

Analysis of sections of the mosaic fibers will allow the determination of how many genetically normal nuclei are necessary to convert the dystrophic muscle cell back to normal phenotype. In determining the minimal ratio of normal/dystrophic genotype mosaicism necessary to produce normal phenotype, we have developed, in addition to the antibodies specific for GPI-1CC [213-215], a vital dye muscle nuclear stain to label donor nuclei [189].

This method corroborates well with the GPI-1CC immunocytochemistry and is more reliable and convenient.

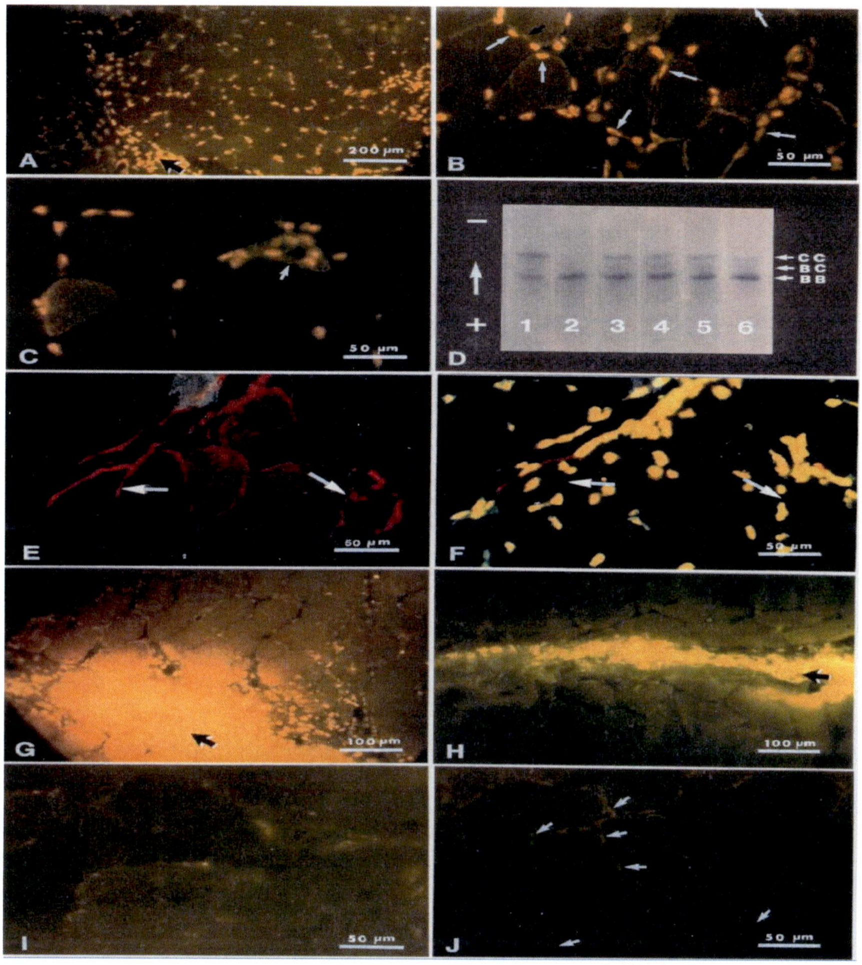

Figure 6. A. Fluorescent microscopy showing the wide and even distribution of F-G labeled donor myoblasts (yellow nuclei) in normal host muscle cross section. Arrow shows the site of injection. B. Host muscle showing mosaic muscle fibers as early as 7 days after myoblast injection. White arrows, donor nuclei; black arrows, host nuclei. C. Injected host muscle showing the fusion of donor myoblasts to form normal myotube (Arrow). D. All of the injected muscles show three different muscle fiber enzymes (GPI-1BB, BC, CC) in agarose gel electrophoresis. Lane 1, controls. E,F. Human myoblast transfer into mdx mouse muscle monitored by FG myoblast labeling, Hoechst 33342 nuclear counterstain, and dystrophin immunbocytochemistry. (F) Fluorescent microscopy showing the distribution of FG-labeled donor myoblast nuclei (yellow) and host nuclei (green) counterstained with Hoechst 33342. (E) Fluorescent microscopy of the same muscle section as (F), respectively, in which dystrophin (red) is observed with a Texas red filter. Arrows show fibers with distinct correlation between dystrophin immunocytochemistry and the presence of normal donor nuclei. G. (Control 1) Poor distribution as a result of poor injection technique. Arrow shows the site of injection. H.(Control 2) The distribution of FG-labeled dead myoblasts in the host muscle 7 days after injection. I. (Control 3) Host muscle injected with 0.01% F-G saline 7 days after injection. J. (Control 4) Auto fluorescence of host myonuclei. (Reproduced with permission from Fang et al.1991.)

The usage of GPI as genotype marker to follow the fate of donor myonuclei in host muscles is necessary for long-term studies, which in the case reported above, lasted up to 18 months after MTT [146-148]. However, the production of antibodies is extremely difficult and time consuming when the antibodies have to be specific enough to differentiate between isozymes [186-188]. For studies that involve shorter experimental periods, labeling the donor myonuclei with the vital dye Fluoro-Gold (FG) is more appropriate [189].

In vitro incubation with 0.01% FG (Fluorochrome Inc. Englewood, CO) for 16 h resulted in 100% nuclei labeling. Intensive fluorescence persisted following 9 days of subculture when labeled human myoblasts were injected into the quadriceps of mouse recipients immunosuppressed with cyclosporine A (CsA) [189].

FG was used as a donor cell marker to observe the status of donor cell nuclei in the host muscle. Donor myoblasts were found widely distributed in the injected host muscle. Myoblast fusion began one week after injection. Mosaic muscle fibers could be found in sections of muscles which had received myoblast injections. The incorporation of normal nuclei from cultured human myoblasts into host muscles of normal mice suggested that a similar procedure can be used to genetically improve dystrophic muscles. FG labeling was not observed in control muscles injected with an equal volume of FG-labeled dead myoblasts, 0.01% FG medium, or phosphate-buffered saline.

Aside from donor cell survival in an immunologically hostile host, cell fusion is the key to strengthening dystrophic muscles with MTT. To improve the fusion rate between host and donor cells, various injection methods aimed at wide dissemination of donor myoblasts were tested and compared. The goal was to achieve maximum cell fusion with the least number of injections.

The results indicate that delivery of myoblasts is best conducted by diagonal placement of needle into the host muscle with ejaculation of the myoblasts as the needle is withdrawn. This method of myoblast injection yields an even and wide distribution of donor myoblasts with a high rate of cell fusion. Myoblasts injected perpendicular to myofiber orientation are partially distributed. Myoblasts injected longitudinally through the core of the muscles and parallel to the myofibers are poorly distributed. Thus myoblast injection method regulates cell distribution and fusion [195].

Biochemical Improvement

To further examine the fate of the donor myoblasts and to further demonstrate genetic and biochemical improvement after MTT/HMGT, we developed a new technique which enables correlation of dystrophin expression with the location of donor versus host nuclei in the same sections of mdx mouse muscle injected with normal myoblasts.25 Myoblasts from C57BL/6J mice or from humans were labeled with 0.01% FG in Dulbecco's Modified Eagles Medium (DMEM) for 16 h at 37^0C before myoblast transfer. About 3 x 10^4 myoblasts were injected into the quadriceps muscles of mdx mice immunosuppressed with CsA. At 11, 21, or 25 days after myoblast transfer, injected muscles were dissected out and sectioned. These muscle sections were processed for dystrophin and then labeled with a fluorescent nucleus counterstain, 5 μg% Hoechst 33342 in phosphate-buffered saline (PBS), for 10 min at room temperature. FG labeling corresponding with Hoechst 33342 staining indicated survival of normal nuclei in dystrophic muscle. Dystrophin was found in the sarcolemma of myofibers containing FG-labeled nuclei but not of myofibers containing only Hoechst 33342 labeled nuclei (Figures 6E, F). Control muscle samples showed neither FG labeling nor dystrophin. The study demonstrates that the donor human and mouse myoblasts survived and developed in host mouse muscles after myoblast transfer, and that the localization of their normal nuclei correlates with dystrophin expression in muscle fibers of immunosuppressed mdx host mice. This genetic/biochemical correlation provides the third evidence in support of the MTT/HMGT concept (Figure 3), and confirms previous contention of genetic improvement as demonstrated using GPI gel electrophoresis and immunocytochemistry.

Structural Improvement

The first demonstration that cultured myoblasts can survive and develop in vivo in a dystrophic muscle after transplantation was reported by Law et al. in 1988 [145]. Genotypically normal (+/+) myoblasts were derived from limb bud explants of day 14 mouse embryos of the gpi-1c/c mice. The muscle primordia yielded samples consisting of 80% spindle-shaped myoblasts and 20% polygonal fibroblasts after proliferation in culture. Without purification, about 10^6 donor cells were injected into the right soleus of 20-day old normal or dystrophic mice.

Host and donors were histocompatible but exhibited different genotype markers. Donor cells produced GPI-1CC. Host cells produced GPI-1BB.

When compared with contralateral controls 6 months postoperatively, test dystrophic solei exhibited greater cross-sectional area and total fiber number (Figure 5). They contained more normal appearing and less abnormal-appearing fibers [145].

Improvement in histology correlated well with genetic improvement which was shown using GPI gel electrophoresis. Presence of GPI-1CC with or without hybrid isozyme GPI-1BC in these muscles implied the survival and development of donor myoblasts into normal myofibers, and fusion of normal myoblasts with dystrophic satellite cells to form genetically mosaic myofibers.

Injection of about 10^6 fibroblasts as a control for cell type resulted in less absolute number of fibers, higher percentage of abnormal fibers, and more connective tissues (Figure 5) in both test dystrophic and test normal solei [145]. This indicates that the structural improvement was indeed mediated through myoblasts and not through fibroblasts.

It should be noted that this study did not involve the use of CsA because of the histocompatibility between hosts and donors. This would indicate that the observed structural improvement was not caused by CsA.

Aside from the use of inbred mice that afford histocompatible MTT, the reality is that myoblasts exhibit MHC-1 surface antigens that provoke immunological rejection [190]. Furthermore, fully matched human donors and dystrophic recipients are rarely available [105,191]. Current technology would thus necessitate the inclusion of host immunosuppression to facilitate myoblast survival after transfer. Cyclosporine is the most widely documented immunosuppressant in transplantation studies [192]. FK 506 provides another alternative [193].

The first histoincompatible myoblast transfer into dystrophic muscles was reported in 1988 [146]. Donor myoblasts were clones of G8 cell line (ATCC) originally derived from limb muscles of the Swiss Webster mice. Host mice, either normal or dystrophic, belonged to the C57BL/6J-dy2Jdy2J strain. These mice were primed one week with CsA injected subcutaneously everyday at 50mg/kg body weight before receiving myoblasts. The same CsA treatment continued for 6 months after MTT. About 10^6 donor myoblasts were injected into each of the following muscle groups on both sides of the hosts: the quadriceps femoris, hamstrings, abductors, extensors, flexors, peroneal, and the external intercostal muscles. Control mice, both normal and dystrophic, received similar myoblast treatment but no CsA.

Histologically, there were no indications of the presence of donor cells in the myoblast injected normal muscles without CsA treatment. These muscle preparations, showing polygonal myofibers with peripheral nuclei and minimal intercellular connective tissue, were as normal as any intact normal controls. Similarly, the dystrophic muscles receiving myoblast injections but no CsA treatment did not differ from the intact dystrophic controls. However, both CsA-treated normal and dystrophic muscles showed immature and developing myogenic cells that were not observed in no-CsA-treated preparations and were thus likely to be donor in origin. Two months was not long enough for all of the donor cells to mature. Nonetheless, there was a significant improvement in muscle structure in the CsA-treated dystrophic mice as compared to the no-CsA-treated dystrophic ones, both receiving normal myoblasts. Dystrophic characteristics such as muscle fiber splitting, central nucleation, phagocytic necrosis, variation in fiber shape and size, and increase in intercellular connective tissues were rarely present in the CsA-treated dystrophic muscle receiving normal myoblasts. It should be noted that the action of CsA is one of an immunosuppressant, and that CsA does not improve muscle cell structure by itself [194]. The latter is especially obvious considering the positive results of histocompatible MTT with using CsA.

The survival and development of donor cells in host muscles were also demonstrated in another series of experiments in which only the right legs received myoblast injections, with the left legs serving as controls. The injected leg showed muscle enlargement which was not observable in the contralateral leg. Such muscle enlargement was present in the CsA-treated hosts but not in those without CsA treatment, regardless of whether the host was normal or dystrophic. These results were obtained from 12 mice from each of the four groups 2 months after myoblast injection [146].

Functional Improvement

Histocompatible Transfer

After 6 months, injection of 10^6 histocompatible cultured myoblasts into the 20-day-old dy2Jdy2J mouse soleus produced significantly greater muscle twitch tension, tetanus tension and wet weight than the contralateral unoperated control (Figure 7) [145]. Injection of fibroblasts instead of myoblasts caused significant reduction in twitch and tetanus tensions in the test dystrophic solei, although the test normal solei were unaffected. Results were obtained from 10 normal and 6 dystrophic mice.

The study demonstrated for the *first* time that cultured myoblasts could survive, develop in vivo, and function in a disease environment. It established a link between the cell culture and the in vivo systems. In vivo genetic complementation study on mammalian skeletal muscle had not been reported. These results indicate that histocompatible MTT with cultured myoblast is safe and effective in improving mammalian dystrophic muscle function.

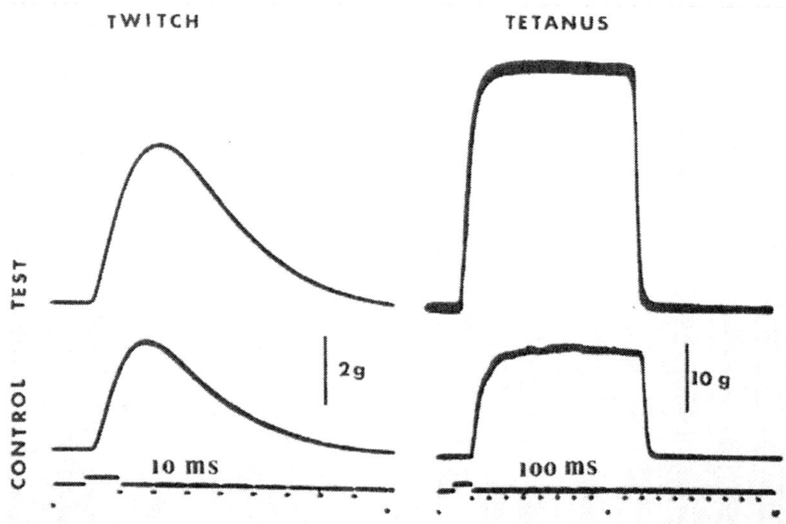

Figure 7. Mechnophysiology in vivo at 37°C showing significantly increased twitch and tetanus tensions for the myoblast-injected dystrophic soleus as compared to the contralateral uninjected dystrophic control.

Histoincompatible Transfer

The ideal situation was to develop healthy clone lines of normal myoblasts in reserve for injection. This was already available through the G8-1 cell line which was histoincompatible with the C57BL/6J hosts. G8-1 cloned myoblasts were purchased from ATCC. The initial injection study using these clones occasionally showed tumorigencity. Subsequently, we developed and used myoblast clones from Swiss Webster mouse embryos from which the original G8-1 clones were derived.

Immunosuppression was induced by daily subcutaneous injection of CsA at 50 mg/kg body weight. CsA was withdrawn after 6 months of treatment. The survival of cloned myoblasts in muscles of CsA-treated and non-treated C57BL/6J normal or dystrophic mice was examined.10-12 Donor myoblasts that produced GPI-1AA were injected multifocally via a 30 gauge needle into the major muscle groups of the calf in both legs: tibialis anterior and extensor digitorum longus representing the extensors, the triceps surae representing the flexors, the quadriceps, and the hamstrings. About 10^6 myoblasts were injected into each muscle group of the 20 to 30-day-old mice. The external intercostal muscles on both sides also received 10^6 normal myoblasts each as previously described. In vivo mechanophysiology with supramaximal nerve stimulation showed significant increase in tetanus tensions in myoblast injected dystrophic muscle groups of CsA-treated mice. The latter study included 12 normal and 19 dystrophic mice. The increase in tetanus tensions indicated that histoincompatible MTT is safe and effective in strengthening mammalian dystrophic muscles [146-148].

Behavioral Improvement

From 2 to 18 months after myoblast injection, eleven dystrophic mice showed such behavioral improvement that their locomotive patterns were indistinguishable from those of the unoperated normal mice [146-148]. Sporadic flexion associated with myotonia and flaccid extension of the hindlimbs was not seen. They were able to use their hindlimbs and toes. Their hindlimb muscles were strong enough to support them and to allow them to balance themselves on a glass rod and to climb out of a cage. Occasionally they would still walk on duck feet, but the mice could now run. Muscle bulk was increased in both legs and in the chest.

Five mice did not respond at all. Three other mice showed intermediate recovery 2 months after myoblast injection indicating that there was some functional improvement of the muscles. However, when they were tested on the glass rod, their hind-limbs were not strong enough to hold onto the glass rod. Sporadic flexion and flaccid extension of the hind-limbs could still be demonstrated.

One year after myoblast injection, the mice that showed significant behavioral improvement continued to be able to support themselves on their hind-limbs and walked normally. They could grasp the glass rod with their toes, and were able to balance themselves and moved along the glass rod.

The mice that showed intermediate improvement could now walk on the glass rod without showing sporadic flexion and flaccid extension of the hind-limbs. They continued to walk with duck feet.

Normal littermates treated similarly were hyperactive, showed enlarged intercostal and leg muscles, but were otherwise normal.

Before myoblast injection, four-month old dystrophic mice could not run the treadmill at 2 meters/minute. After myoblast injection, they could run at 2.8 meters/minute. The one that showed most improvement actually outran that speed and could jump out of the treadmill. Young dystrophic mice injected around one month of age could run at a speed of 7.1 meters/minute and could easily climb out of the treadmill, if the speed was slowed down to 2.8 meters/minute [146-148].

Life Prolongation

This study was remarkable in that because of the improvement of cell genetics, biochemistry, muscle function and animal behavior, the life span of the responding dystrophic mice (normally 8-9 months) was increased to nineteen months. This was likely due to strengthening of the intercostal and locomotive muscles, allowing the dystrophic mice to obtain enough oxygen, nutrients and exercise [202-204].

Among the five injected dystrophic mice that did not respond behaviorally, only one survived to a year after injection. Nevertheless, its life span was extended.

Perspectives

Two mechanisms (Figure 3) were responsible for these phenotype improvements. First, donor myoblasts fused among themselves, developed into normal fibers and replenished lost dystrophic myofibers. Second, normal myoblasts fused with dystrophic myogenic cells, sharing the normal genome to effect genetic repair at the cellular level.

These two mechanisms of genetic complementation were demonstrated using GPIs as genotype markers (Figure 6D). The presence of parental isozymes in the myoblast injected muscles indicated the survival and development of donor cells in host muscles. Furthermore, the presence of hybrid isozymes GPI-1BC indicated fusion between host and donor cells.

The immunocytochemical localization of host and donor nuclei within myofibers of the injected muscles using GPI polyclonal antibodies further substantiated the mechanism. These antibodies stained the nuclei of normal control muscles but not the dystrophic control muscles. Normal-appearing mosaic fibers containing host and donor nuclei could be observed eighteen months after myoblast injection (Figure 4). This genetic/structural correlation provides the second evidence in support of the MTT/HMGT concept. Analysis of sections of the mosaic fibers will allow the determination of how many genetically normal nuclei are necessary to convert the dystrophic muscle cell back to normal phenotype. The mosaic muscles also contained normal and dystrophic muscle fibers (Figure 5).

The third and final proof of concept utilizes genetic/biochemical correlation in which dystrophin induction was demonstrated in the sarcolemma of myofibers containing FG-labeled normal nuclei but not in myofibers containing only Hoechst 33342 labeled dystrophic nuclei (Figure 6).

Animal experimentation from 1975 to 1990 demonstrated that MTT/HMGT was safe and effective. The beneficial effect was not due to the use of CsA, fibroblasts or needle trauma. Normal myoblast injections provide long-term genetic complementation treatment for muscle degeneration in murine dystrophy. The procedure repairs degenerating cells and replenishes degenerated cells, thus normalizing cell genetics, biochemistry, structure, muscle function, animal behavior and life span.

Since the treatment design is based on development processes universal to all mammals, it has potential for clinical application to treat all types of muscle diseases. This animal experimentation lays the scientific groundwork for the development of a clinical treatment, with the promise of bringing restoration of function and quality of life to patients who suffer muscular dystrophy. Such treatment may potentially prevent or remedy muscle weaknesses in locomotive and respiratory muscles, thereby improving the quality of life and prolonging the life span of the patients.

Myoblast Transfer in DMD Subjects

Single Muscle Treatment

Phase I MTT/HMGT clinical trial began on February 15, 1990, after lengthy reviews despite the initial protocol approval on April 7, 1989 [196]. *It was the first human gene therapy clinical trial* [197]. Cultured myoblasts were used as vehicles to deliver all of the normal genes into DMD myofibers to repair genetic defects. As a cell therapy, MTT/HMGT was to replenish the degenerated myofibers also.

The study, a single muscle treatment (SMT), was based on the safety and efficacy of MTT/HMGT previously demonstrated in the dydy, dy2Jdy2J and mdx mice. The randomized, double-blind study was designed to determine the survival, development and functioning of donor myoblasts in dystrophic muscles of DMD boys. We hypothesize that intramuscular injection of normal myoblasts can significantly improve the biochemistry, structure and function of dystrophic muscles. Our goal is to demonstrate the safety and efficacy of MTT/HMGT in humans.

The experimental design and the results of the SMT study had previously been published [98-101,198].

Subjects and Methods

Approval

This study protocol received approval from 1) the Institutional Review Board of the University of Tennessee Memphis (UTM); 2) the Scientific Advisory and Administrative Committees of the Clinical Research Center in UTM; 3) the LeBonheur Children's Medical Center Institutional Review Board, and 4) the UTM Data and Safety Monitoring Board.

Subjects

Eleven 6 to 10-year-old boys previously diagnosed in different MDA clinics of having DMD on the basis of physical examination, muscle biopsy, serum CPK, electromyography and/or absence dystrophin were selected. They were in good general health and were not on any medication. They showed no evidence of renal or liver disease as judged by normal levels of BUN, creatinine, SGOT, SGPT and alkaline phosphatase.

Muscle Preparation

Before admission into the project, the DMD boys demonstrated at least 25% reduction in isometric twitch tension in their extensor digitorum brevis (EDB) muscles as compared to age-matched normal boys. The EDB was chosen because of its early involvement in dystrophy [199,200], its easy access for myoblast injection, and its distinct innervation as the last muscle innervated by the common peroneal nerve so that its mechanophysiological properties could be objectively measured in vivo without movement artifacts. Its small size necessitated fewer injections of myoblasts to effect improved muscle structure and function. Confined to a small muscle, donor cells could easily be located during biopsy three months after myoblast injection. Its relative insignificance protected the subjects from the loss of use of an important muscle, should adverse reaction occur which might necessitate the removal of the whole muscle. Its unique location on top of the foot protected the developing muscle fibers and neuromuscular junctions from undue stretches as a result of the contracture of the antagonistic muscles.

Donor Myoblasts

These were derived from cell culture of muscle biopsies from normal fathers, male guardians or brothers of the subjects. Female family members were not used as muscle donors since they might be carriers of DMD. All donors were previously screened for human immunodeficiency virus (HIV) and hepatitis B infection.

Under local anesthesia, about 1 g of muscle was surgically removed from the rectus femoris muscle of each donor using an open biopsy technique. The muscle biopsy was immediately dissociated with 0.1% collagenase (Sigma type II) and 0.2% crude trypsin (Sigma) in phosphate buffered saline, pH 7.3. The latter contained 5.4mM KCl; 144 mM NaCl; 5 mM Na phosphate; 25 mM glucose; 14 mM sucrose; 50 units/ml penicillin; 10 μg/ml streptomycin; and 1 μg/ml phenol red. Complete dissociation of 1 g of fresh skeletal muscle occurred in 45 minutes of stirring with three changes of the enzyme solution, alternated with three changes of neutralizing medium. The latter was composed of 100 parts of Dulbecco's Modified Eagle's Medium (DMEM-Gibco) containing 0.37% NaHCO3 and 4 mM glutamine, 10 parts horse serum (Gibco) and 1% antibiotic-antimycotic (Gibco).

Cells were then cultured in the neutralizing medium supplemented with 2 parts of chick embryo extract (Gibco). Myoblasts were fed with fresh medium every two days and were incubated in 7% CO_2.

Before myoblast transfer into human recipients, myoblasts were tested for their ability to migrate, align, fuse and form myotubes and immature myofibers that could contract spontaneously in vitro. Myotubes were immunocytochemically stained for myosin. Fusion medium consisted of 98% DMEM and 2% horse serum. In addition, about 106 myoblasts from each donor were injected into a leg muscle of a normal C57BL/6J mouse to determine if the myoblasts were tumorigenic. The mice were immunosuppressed with daily subcutaneous injections of Cyclosporine (Cy) (50 mg/kg body weight) and were killed four weeks after myoblast transfer. Injection sites were examined for tumors.

Myoblast Transfer

Myoblasts were rinsed in Dulbecco phosphate-buffered saline (DPBS) which was used as the carrier solution for myoblast injection and for sham injections. Injections of 10^6 cells each were delivered at eight foci into each test EDB. A total volume of 400 μl was injected. Four foci were injected per puncture of the skin. Injection of an equal volume of DPBS into the control EDB was administered in a similar manner. Myoblast-injected EDBs were randomly selected by the injecting physician. All other participants in the study including the patients were unaware as to which side received the myoblasts.

Immunosuppression

All donors and recipients were HLA tested. Results from HLA testings were used retrospectively to evaluate the significance of histocompatibility differences in determining the likelihood of engraftment in this initial series of patients. Forty-eight hours prior to myoblast injection, patients began receiving Cy. Sandimmune® (Cy) was administered orally at 5 to 7 mg/kg [201,202] divided in two doses per day (Sandoz, 100 mg/cc dissolved in an olive oil-Labrafil base to be further diluted by milk, chocolate milk, or orange juice). The dosage was varied to reach a whole blood trough Cy level of 100 to 200 ng/ml, as measured by the fluorescence polarization method (Abbott TDX). Cy concentration initially was monitored weekly for 3 weeks and then monthly. Cy treatment continued for three months. Routine blood and urine tests were performed weekly and then monthly to evaluate renal and liver function. Tests included BUN, creatinine, SGOT, SGPT and blood pressure determinations.

Effect of Cy on Myoblasts

The goal of Cy therapy was to maintain a serum trough concentration of 100 to 200 ng/ml. [145,146] To determine if cell proliferation and fusion of donor myoblasts might be affected by such Cy concentration, a study was conducted to examine the cell count, protein content and the fusion rate of clone myoblasts when exposed to 0, 50, 100, 150, 200, 300 or

500 ng/ml of Cy in vitro. Myoblasts were cultured for twelve days. Cell counts and total protein contents were determined seven days after plating. The fusion rate was determined five days after changing to the fusion medium, and in the presence of 0, 200, 500 ng/ml of Cy only.

Functional Monitor

Three days prior to myoblast injection and three months after, the isometric twitch and maximum voluntary contraction of the left and right EDBs were measured. The twitch tensions were elicited through supramaximal stimulation of the peroneal nerve at the ankle, such that the extensor hallucis longus (EHL) could not be stimulated to contract. Even if the EHL contributed to the maximum voluntary contraction of the big toe, this contribution should cancel out when muscle contractions before and after myoblast transfer were compared. The maximum voluntary contraction yielded registers of tetani tensions. The tensions were measured in vivo with an Omega LCL-20 force transducer connected via a stiff wire looping around the big toe. Six readings of each measurement were taken and averaged. Movement artifact was minimized by securing the foot to a frame. The transduced electric signals were transmitted through a WPI Transbridge, a MacLab and displayed on an Apple Macintosh SE screen. Tension measurements were confirmed from signal traces copied with an Apple Macintosh Laser-writer Plus.

Structural Monitor

Muscle biopsies (5x3x3 mm3) were obtained by open biopsy method three months after myoblast injection. Sections were cut at 6 μm in a cryostat and stained with Modified Gomori Trichrome [145].

Biochemical Monitor

Dystrophin is a cytoskeletal protein absent in DMD muscle cell membrane. If donor myoblasts survive and develop in DMD host muscles, dystrophin immunocytochemistry can be used to monitor biochemical success as shown in mice [95,96].

The immunocytochemical staining for dystrophin followed the methods of Hoffman et al.15 and Bonilla et al.16 Muscle biopsies ($30mm^3$) of normal adults and of DMD boys were frozen in 2-methylbutane chilled in liquid nitrogen. Six-micron thick transverse sections were cut in a cryostat maintained at -20^0 C. Muscle sections were thawed on 0.5% gelatin-coated slides. They were then fixed for one minute at room temperature with acetone.

The sections were incubated for two hours with pre-centrifuged antidystrophin serum diluted 1:1,000 with 95% PBS and 5% horse serum. They were then washed three times in PBS. They were further incubated for 60 minutes in biotinylated donkey anti-sheep IgG (Amersham) diluted 1:100 with 95% PBS and 5% horse serum. After being washed three times in PBS, they were mounted with 50% glycerol in PBS and examined. Control sections were incubated with non-immune serum and were processed on the same slides in the same manner.

Results

Myoblast Culture

After seven days in culture, the satellite cells released from the donor biopsy took on the spindle-shape characteristics of the myoblasts. Healthy myoblasts were selected for cloning. The myoblasts had the ability to migrate, align, undergo mitotic division and form myoblast clones. Myoblasts were harvested for injection between 20 to 30 days after culturing. When left alone in culture, the myoblasts fused together to form multinucleated myotubes. These myotubes possessed striations or sarcomeres. They contracted spontaneously and eventually pulled themselves away from the bottom of the culture flask. Immunocytochemical staining indicated that myotubes formed from fusion of these myoblasts contained the contractile protein myosin (Figure 2). Myoblasts injected into immunosuppressed mice survived and developed without any sign of tumorigenicity.

Due to Professor Law's move from the University of Tennessee Memphis to the Cell Therapy Research Foundation, complete data on muscle function, structure, and biochemistry were obtained from only 3 out of the 11 subjects that received SMT.

Functional Improvement

The EDB of 9-year-old DMD boys developed only about half of the amplitudes in twitch and tetanus tensions of the age-matched normal controls. Three months after myoblast injection, the myoblast-injected EDBs showed significant increases in tensions whereas sham-injected EDBs showed reductions. Functional improvement in maximum voluntary contraction ranged from 13% with myoblasts derived from a ward (patient 1), to 30% (patient 2) and 80% (patient 3) from brothers [98-100,198].

Structural Improvement

Myoblast-injected EDBs showed gross anatomical improvement over the sham-injected EDBs. In the sham-injected muscle biopsies, dystrophic characteristics such as muscle fiber splitting, central nucleation, variation in fiber shape and size, infiltration of fat and connective tissue and phagocytic necrosis were significant. The myoblast-injected muscle biopsy contained larger fibers that were relatively uniform in fiber shape and size. There was very little intercellular connective tissue. Muscle fibers were polygonal in cross-section. Nuclei were peripherally located. Dystrophic characteristics, while present, were less apparent [98-100,198].

Biochemical Improvement

Both immunocytochemical staining and immunoblot revealed dystrophin in the myoblast-injected EDBs but not in the sham-injected EDBs. Dystrophin was immunocytochemically localized at the sarcolemma. This was not seen in the sham-injected muscle. Many muscle fibers with dystrophin in the myoblast-injected muscle showed normal appearances. They were polygonal in shape, and nuclei were peripherally located.

Clinical Tests

At no time during the 92 days after myoblast injection of the 11 subjects was there any sign of erythema, swelling or tenderness at the injection sites. Serial laboratory evaluation, including electrolytes, creatinine and urea, did not reveal any significant changes before or after MTT/HMGT. There was no clinical evidence of an adverse reaction to MTT/HMGT or to Cy.

Discussion

This is the *first* study to have provided corroborative data on biochemical, structural and functional improvement of DMD muscles after MTT/HMGT. By virtue of its absence in DMD muscles, dystrophin can be used a biochemical marker to monitor MTT/HMGT success. The presence of dystrophin in myoblast-injected but not in the sham-injected muscles provided unequivocal evidence of the survival and development of donor myoblasts in the myoblast-injected muscles. Although correction of the primary genetic defect is demonstrated, such evidence cannot be extrapolated to indicate structural and functional improvement, which have to be demonstrated by themselves as in this study. MTT/HMGT is a safe and effective procedure to alleviate some of the biochemical, structural and functional deficits inherent in small muscles of DMD.

SMT began on February 15, 1990. It was the *first human gene therapy* clinical trial [98,197]. Cultured myoblasts were used to deliver all of the normal genes into DMD myofibers to repair genetic defects. Correction of the DMD genetic defect was published in July 1990, when dystrophin was found in the myoblast-injected DMD muscle [98]. Improvement in muscle structure and function after MTT/HMGT were also announced [205]. Two months later, Anderson announced the beginning of the single gene manipulation therapy. T cells from an ADA deficient SCID patient were transduced with the functional ADA gene and returned to the patient after expansion through culture [206,207]. Correction of ADA deficiency in the SCID patient [122,139] was published eight months after dystrophin replenishment was reported in myofibers of a DMD patient [197]. Anderson's success has sparked much hope for gene therapy.

Different HMGT techniques that accounted for success or failure, were employed in different laboratories. At the initiation of SMT, hardly any group besides Law's had published significant animal experimentation with MTT/HMGT. Like Karpati [208,209] in Montreal, Miller [102,210] in San Francisco used myoblasts cultured in commercialized laboratories away from the myoblast transfer suites. Donor myoblasts might have been subjected to changes of nutrients, temperature, pH, and sterility in transit. Myoblast purity was doubtful especially in the early studies. Karpati transferred 55 million cells with 55 injections into the biceps brachii. Miller transferred 100 million cells with 100 injections into the tibialis anterior. In both cases, cells were transferred at lower concentrations in which cells reportedly leaked from the sites of injection. In Law's later study, 312 million myoblasts were transferred through three injections into the tibialis anterior, at a concentration of 55.6×10^6 cells/mL.

Apart from fewer cells delivered at a lower concentration, the much larger numbers of injections spaced indiscriminately 1 cm apart undoubtedly introduced a higher risk of embolism and much more damage to the recipients' blood vessels, nerves, and muscles which were known to lack the normal ability to regenerate [211]. A traumatized nerve takes 6 to 8 weeks to regenerate. Myotubes formed from myoblast fusion within the first 4 weeks of MTT/HMGT required nerve innervation to survive. The trauma of injection could further mobilize macrophages to remove not only the degenerate but also the donor myoblasts. It is possible that repeated injections over several months would increase the possibility of triggering a general immune reaction to the donor cells and that they put the subject at greater risk. Such technique was used by Mendell in Columbus,Ohio.

Lower Body Treatment

With the SMT results partially confirmed [102,104,105,] the feasibility, safety, and efficacy of HMGT were assessed in an experimental lower body treatment (LBT) involving 32 DMD boys aged 6-14 yr, half of whom were nonambulatory [141,191]. The LBT protocol received approval from the Essex Institutional Review Board Inc in Lebanon, NJ, which is in compliance with the regulations of the Food and Drug Administration. The protocol was also approved by the Patient Participation Committee of the Baptist Memorial Hospital Medical Center in Memphis, TN.

Through 48 injections, five billion (55.6 x 10^6/ mL) normal myoblasts were injected into 22 major muscles in both lower limbs, in 10 minutes with the subject under general anesthesia. Ten subjects received myoblasts cultured from satellite cells derived from 1-g fresh muscle biopsies of normal males aged 9-21 yr. Donor myoblasts for the remaining 22 boys were subcultured from reserves frozen 1 mo to 1.5 yr ago.

Only four donors were known to have identical histocompatibility with their recipients. All subjects took oral doses of the immunosuppressant cyclosporine (Cy), beginning at 2 days before HMGT and lasting for 6 mo after HMGT to facilitate donor cell survival. There was no evidence of an adverse reaction to HMGT or Cy as determined by serial laboratory evaluations including electrolytes, creatinine, and urea.

Objective functional tests using the KinCom Robotic Dynamometer measured the maximum isometric contractile forces of the ankle plantar flexors (AF), knee flexors (KF), and knee extensors (KE) before HMGT and at 3, 6, and 9 mo after HMGT. The AF, being distal muscles and less degenerative than the KE and the KF, showed no decrease in mean contractile force 3 mo after HMGT, and progressive increases in force at 6 and 9 mo after HMGT.

At 9 mo after HMGT, 60% of the 60 AF examined showed a mean increase of 50% in force ; 28% showed no change; and only 12% showed a mean decrease in force of 29% when compared to the function of the same muscles before HMGT. The KF, being proximal muscles and more degenerative, showed no change in function at 9 mo after HMGT. The KE, being proximal and anti-gravitational, were most degenerative before HMGT. They showed no statistically significant change in force at 3 mo after HMGT but showed decreases at 6 and 9 mo after HMGT.

At 9 mo after HMGT, 23% of the 60 KE examined showed a mean increase of 65% in force; 22% showed no change; and 55% showed a mean decrease of 24% in force. When results of all muscle groups (AF, KF, KE) were pooled, there was no change in force at 3, 6, or 9 mo after HMGT versus before HMGT according to the Wilcoxon signed rank test.

The ambulatory subjects showed more muscle improvement than the non-ambulatory ones at various times after HMGT. Statistically significant progressive increase in force in the AF and arrest of weakening in the KF and KE were observed in the ambulatory subjects as early as 3 months and continued up to 9 months after HMGT.

The results indicate that 1) HMGT is safe; 2) HMGT improves muscle function in DMD: 88% of the AF, 49% of the KF, and 45% of the KE showed either increase in strength or didnot show continuous loss of strength 9 mo after HMGT; 3) LBT significantly strengthens the lower bodies in one-third of the 32 DMD boys, improving their locomotive ability and quality of life. The behavioral improvement subsided for insufficient dosage of myoblasts and continual upper body weakening. 4) the dosage used is more effective in the AF than in the KF, and is least effective in the KE; 5) more than 5 billion myoblasts are necessary to strengthen both lower limbs of a DMD boy between 6 to 14 yr of age; 6) the more degenerated proximal muscles will need more myoblasts per unit muscle volume than the distal muscles for HMGT to be effective; 7) HMGT is more effective in the younger, ambulatory subjects than in the older, nonambulatory subjects; 8) Cy is not responsible for the functional improvement, since muscle function continues to improve 9 mo after HMGT despite Cy withdrawal at 6 mo after HMGT; 9) Cy immunosuppression permits donor cell survival and development, without overt rejection symptoms, when properly managed; 10) myoblasts from frozen reserves are as effective as those from fresh muscle biopsies; 11) fifteen billion myoblasts can be cultured from a 1-g muscle biopsy; 12) billions of cultured myoblasts can be injected into subjects without tumor formation.

Better Repair Than Replenish

Although donor myoblasts are known to fuse among themselves, forming normal myofibers to replenish degenerated ones, this mechanism is much less efficient than repairing degenerating myofibers through incorporation of donor myonuclei as a result of the fusion of host and donor cells [95,140,145,147,148,169,171]. Myonuclear counts of myotubes and teased myofibers suggest that between 200 to 500 myoblasts fuse to form a single myofiber. The myotube so formed has to be attached, vascularized, and innervated, with subsequent differentiation and development of cytoarchitecture before being functional. On the contrary, the host degenerating myofibers already have functioning blood, nerve and tendon connections and require about 20% normal nuclei within them to change the dystrophic characteristics to normal [147]. Accordingly, HMGT should be more effective in younger patients in whom dystrophy has not destroyed as many myofibers.

Frozen or Fresh

Donor myoblasts derived from frozen reserves are as effective as those from fresh muscle biopsies. The majority of the subjects (22 out of 32) received myoblasts from frozen reserves, and improvement was observed in this group as in the remaining 10 subjects. This finding substantiates the idea that cell lines of superior myoblasts can be established and stored in cell banks [146].23 Furthermore, myoblasts can be harvested from culture, frozen, and transported for use in HMGT clinics away from the cell banks.

Cyclosporine

The use of Cy is necessary to provide initial suppression against rejection because myoblasts exhibit surface antigenic determinants. The latter are either masked or absent in myotubes and myofibers.29 The continuous increases in muscle function after Cy withdrawal in this study, and in the dy2Jdy2J mouse studies [146-148], indicate that it is not necessary to immunosuppress the HMGT recipient beyond 6 months, using half of Sandoz recommended Cy dosage (15 mg/kg/ body weight/day).

Cy is not responsible for the functional improvement because muscle function continued to improve in the AF and the KF despite Cy withdrawal. The progressive weakening of the KE further argues against the claim that Cy improves DMD muscle function. Other evidence indicating why Cy is not responsible for DMD functional improvement has already been reported [141].

Dosing of Cells

Judging from the results of this study, it will require at least 25 billion myoblasts to populate the major muscle groups of the whole body of a DMD boy for HMGT to show dramatic behavioral improvement and possible life-prolongation. Fewer myoblasts will be needed by the younger patients who demonstrate less degeneration and weakening. Perhaps the ideal age of HMGT treatment for DMD boys is between three to five years when muscle growth is rapid. The proximal muscles will require more myoblasts per unit volume muscle than the distal muscles for HMGT to be effective. Whereas many authors have questioned whether enough myoblasts can be cultured [94,96,127], it is obvious from this study that the demand can be met by taking two grams of donor muscle biopsy.

Indeed, HMGT is the first genetic treatment to have produced any functional improvement in humans, through incorporation of normal genes into genetically defective cells and through incorporation of genetically normal cells into the genetically abnormal organisms [101,141,191].

Controversies

Gene therapy encompasses interventions that involve deliberate alteration of the genetic material of living cells to prevent or to treat diseases [246]. According to this FDA definition, the first MTT/HMGT on a DMD boy on February 15, 1990 marked the first clinical trial on human gene therapy [197]. In addition to fulfilling their primary muscle-building mission, the myoblasts served as the source and the transfer vehicles of normal genes to correct the gene defects of DMD.

A pioneering work [207,217,243] is often considered as the "first human gene therapy"; correction of the ADA deficiency study began on September 14, 1990 [207], two months after the MTT/HMGT correction of the DMD gene defect was published [256]. In the ADA protocol, T cells from a patient with a severe combined immunodeficiency disorder (SCID) were transduced with functional ADA genes ex vivo and returned to the patient after expansion through culture. In the MTT/HMGT protocol, primary culture of myoblasts derived from a muscle biopsy of a normal donor was injected into a muscle of the DMD subject to produce in vivo nuclear complementation. Both gene therapies utilize cell transplantation to treat diseases. The ADA protocol involved genetic modification and correction of the patients T cells with the adenosine deaminase gene whereas in the DMD protocol normal donor cells were used which were not genetically modified ex vivo.

Six years after the foremost MTT/HMGT, dystrophin was found in the myoblast-injected muscle but not in the sham-injected muscle (Figure 8) [262]. Six years is the longest period through which any gene therapy has sustained positive results. Despite cyclosporine withdrawal at 3 months after MTT/HMGT, myofibers expressing foreign dystrophin were not rejected. This is because dystrophin is located in the inner surface of the plasma membrane, and because mature myofibers do not exhibit MHC-1 surface antigens. Not only has the result demonstrated MTT/HMGT overall safety and efficacy in this single case, it also shows stability in the integration, regulation and expression of the dystrophin gene. The presence of dystrophin in the myoblast-injected but not in the sham-injected muscle provided unequivocal evidence of the survival and development of donor myoblasts in the myoblast-injected muscle.

In a randomized double-blind study involving three subjects, myoblast-injected EDBs showed increases in tensions whereas sham-injected EDBs showed reductions [101,257]. Both immunocytochemical staining and immunoblot revealed dystrophin in the myoblast-injected EDBs. Dystrophic characteristics such as fiber splitting, central nucleation, phagocytic necrosis, variation in fiber shape and size, and infiltration of fat and connective tissues were less frequently observed in these muscles. Sham-injected EDBs exhibited significant structural and functional degeneration and no dystrophin. Throughout the study, there was no sign of erythema, swelling or tenderness at the injection sites. Serial laboratory evaluation including electrolytes, creatinine, and urea did not reveal any significant changes before or after MTT/HMGT.

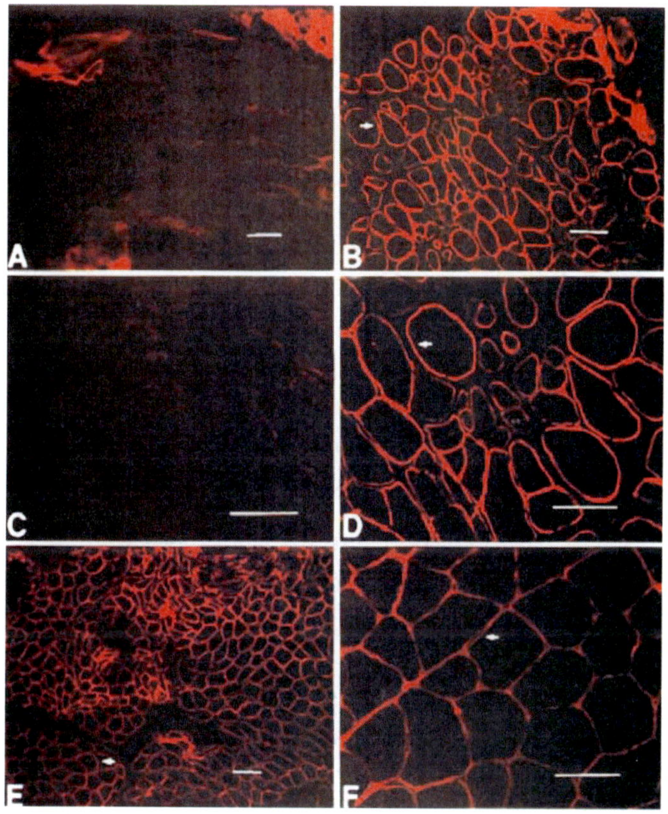

Figure 8. Immunocytochemical demonstration of dystrophin in DMD muscles 6 yr after MTT/HMGT. Dystrophin absent in sham- injected EDBmuscle (A, C), but present in the contralateral myoblaast-injected muscle (B, D). Dystrophin demonstrated at low (E) and high (F) magnification in normal control muscle. Arrows show dystrophin staining on sarcolemma. Cross-section; bar=100μm.

To reconcile these positive results with less convencing ones [102, 103, 105, 208, 210, 269, 271], several issues need to be addressed. To begin with, the use of large quantities of pure live myoblasts is a pre-requisite of successful MTT/HMGT. Except for Law's study [141,253], there is no published pictorial evidence to substantiate the purity, myogenicity and viability of the injected myoblasts as claimed.

Myoblast cultures are usually contaminated with fibroblast overgrowth. MTT/HMGT with such impure culture could lead to deposition of connective tissues rather than myofiber production. Culturing 50 billion pure human myoblasts for MTT/HMGT from two grams of muscle biopsy has only been reported by our team [261]. Other teams work at ranges of hundreds of millions of myoblasts.

In most studies [102, 103,105,208,210,269,271] myoblasts were transported frozen from the site of harvest, chilled for over two hours before being injected. Since myoblasts have a high metabolic rate, they could not have survived for two hours without significant nutrients, oxygen and proper pH, being closely packed in saline within a vial for transport. Determination of cell viability before HMGT were not conducted in these studies. Our myoblasts were injected into the subject within minutes of harvest, at the same location without transport.

HMGT studies that reported failure [102,103,105,208,210,269,271] subscribed to the fallacy of making 55 to 330 injections into a muscle the size of an egg, traumatizing indiscriminately the underlying nerves, muscle, and vasculature. These injection traumas boosted macrophage access and host immune responses [234]. They also induced fibrosis [221]. Surviving myoblasts fused within three weeks in small mouse muscles [95]. A nerve with multiple injuries could not regenerate soon enough through scar and connective tissues to innervate the newly-formed myotubes in a large human dystrophic muscle. Stabilization of muscle contractile properties in a similar situation is achieved by 60 days in the rat, and functional return is incomplete [219]. Non-innervated myotubes died within one week. Whatever few myotubes that developed in the unsuccessful MTT/HMGT studies could not compensate for the traumatized myofibers.

In the study yielding positive results, 5 to 8 x 10^8 pure myoblasts were delivered with eight injections into the biceps brachii without nerve injury [258,261]. Contrarily, in another study, 55 sites, each 5 mm apart, distributed in 11 rows and 5 columns, were injected throughout the depth of each biceps of 5- to 9- year old boys [369]. This was repeated monthly for six months. Axonal sprouts, myotubes and neuromuscular junctions that take six weeks to mature [233] were repeatedly traumatized by a total of 330 injections until the biceps, with or without myoblast/cyclosporine, were irreversibly damaged or destroyed. The result: no functional difference between myoblast- and sham-injected muscles [269].

Once injected, the myoblasts are subjected to scavenger hunt by macrophages for up to three weeks. This is because myoblasts exhibit MHC-1 surface antigens [190,232] that become absent after cell fusion. The latter occurs between one to three weeks after myoblast injection [95]. An allowance in the number of injected myoblasts has to be made to satisfy the unavoidable scavenger process. As reflected in the small numbers of myoblasts injected in unsuccessful studies, it appears that either such allowance was not considered or that the teams were not able to produce larger quantities of pure myoblasts. Although myoblast loss can be minimized by down-regulating macrophage activity [235], such additional compromisation of the host immune system may lead to higher risk of infection, since MTT/HMGT subjects are already taking immunosuppressants.

The less successful MTT/HMGT teams focused on immunosuppression to prevent T-lymphocyte proliferation and antibody production without overcoming the primary hurdle of providing enough pure and live myoblasts. A basic study indicates that cyclophosphamide did not permit myoblast engraftment in the mouse [290], and a MTT/HMGT clinical trial was conducted without success using cyclophosphamide immunosuppression [208]. Cyclosporine [256] and potentially FK506 [247] remain the immunosuppressants of choice for MTT/HMGT. Results could have been more positive if either was employed in the study of Tremblay et al. [103,105].

All of these single muscle MTT/HMGT studies had begun before the FDA established policies and regulations for cell/gene therapies. Our studies are the only ones that received permission for an investigational new drug application approval on MTT/HMGT for treatment of multiple muscles. As a cell/gene therapy, all American MTT/HMGT clinical trials must come under FDA purview.

Whole Body Treatment

Dose Escalation

Beginning with 8 million myoblasts into a small foot muscle, our team proceeded to test 5 billion cells into 22 leg muscles, 25 billion cells into 64 body muscles, and now 50 billion cells into 82 muscles. With over 280 procedures having been conducted to date, the complete safety of the MTT/HMGT procedure has been proven. There have been no adverse reactions or side effects.

Myoblasts	Muscles	Subjects
8 million	1	11
5 billion	22	32
25 billion	64	40
50 billion	82	197

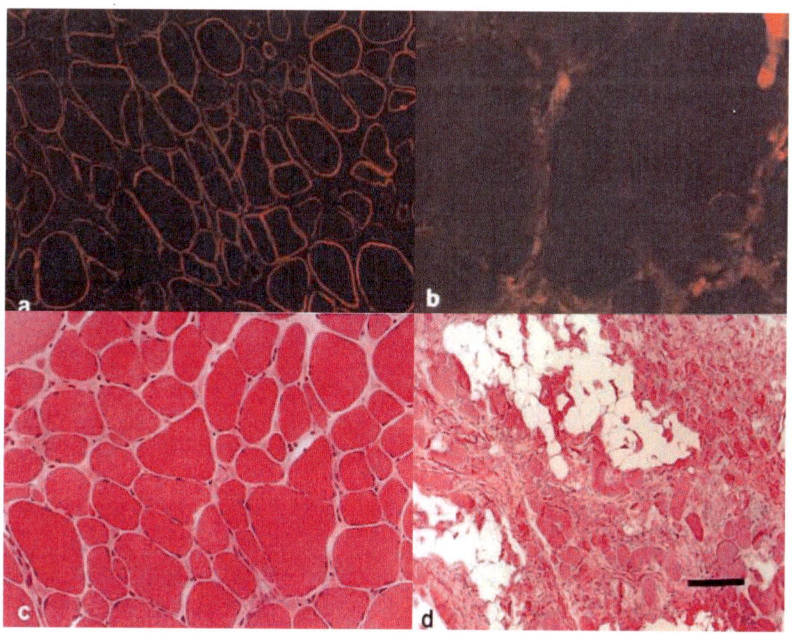

Figure 9. At 9 mo after HMGT, dystrophin was found in the myoblast- injected muscle (a) with histological improvement (c), but not in the placebo-injected muscle (b), which showed fat, connective tissues and minimal amount of myofibers (d).

The 25 Billion Myoblast Protocol

Under FDA purview, MTT/HMGT Phase II clinical trials began on DMD. The whole body trial (WBT) consisted of injecting 25 billion myoblasts in two MTT/HMGT procedures separated by 3 to 9 mo. Each procedure delivered up to 200 injections or 12.5 billion myoblasts to either 28 muscles in the upper body (UBT) or to 36 muscles in the lower body (LBT). A randomized double-blind portion of the study was conducted on the biceps brachii or quadriceps. Subjects took oral cyclosporine for 3 months after each MTT. One infantile facioscapulohumeral dystrophy and 40 DMD boys aged 6 to 16 received WBT in the past 36 months with no adverse reaction.

Nine months after MTT/HMGT immunocytochemical evidence of dystrophin were demonstrated in 18 of the 20 DMD subjects biopsied (Figure 9). Dystrophin positive sections showed less dystrophic characteristics than dystrophin-negative ones (Figure 9). Forced vital capacity increased by 33.3% and maximum voluntary ventilation increased by 28% at 12 months after UBT [261].

Plantar flexion showed an increase of 45% in maximum isometric contraction force in 12 months in the DMD subjects when compared to the natural deterioration. Behavioral improvements in running, balancing, climbing stairs and playing ball were noted [259-261,263,264]. Notable was a 16-yr-old DMD subject who continued to walk without assistance and capable of driving an automobile by himself.

The 50 Billion Myoblast Protocol

This study involves a one time injection of 50 billion myoblasts into 82 muscles with 179 skin punctures, approved by the FDA for subjects with DMD, Becker MD and Limb-girdle MD [339]. Over 190 subjects who underwent this protocol have experienced no adverse reaction. For the DMD subjects aged 5 to 16, there was a significant increase in the maximum isometric force generated by the plantar flexor muscles at 3, 6, and 9 months after HMGT.

This functional improvement is more pronounced with the 50-billion HMGT than with the 25-billion HMGT, indicating that it is dose-dependent. Thus, in the 25-billion HMGT, 800 million myoblasts were injected into the plantar flexors, producing a mean 61% increase in force at 15-months after MTT/HMGT. With the 50 billion HMGT, 50% more myoblasts were injected, projecting a 10% greater increase in force at 15 months after HMGT.

Elevated serum creatine kinase (CK) has traditionally been used to diagnose muscle degeneration, notable in DMD [300]. The initial 22 DMD subjects, mean ages 10.7-yr-old and, median age 9.9 yr-old, showed a 19.3% increase in serum CK within 3 months before HMGT. This trend was reversed after HMGT, and the serum CK declined at a steady rate of 48.7% over 12 months. This result provides strong evidence that HMGT repairs muscle cell membrane leakage of enzymes. This contention is further substantiated by similar findings with another muscle enzyme AST, aspartate aminotransferase.

The breakthrough came when a 29-yr-old Becker MD (BMD) subject began to walk, with his hands being held, beginning at 2.5 months after the 50-billion HMGT. He had previously been diagnosed repeatedly with BMD. He had been non-ambulatory and required the use of a wheelchair for over four years as documented in his medical record. He began walking with assistance a total of eight steps at 3 months after HMGT. This ability increased with time, reaching 60 steps at eight months after HMGT. He began to stand and walk with his crutches at four months after HMGT (Figure 10). Eventually he married, had a daughter and has led a quality life.

Gene Therapy and Molecular Biology Vol 1, page 357

Fig. 9. First muscular dystrophy subject ever to walk after wheelchair bound for years.

(A). The 29-yr-old BMD subject had been wheelchair-bound for over 4 years.

(B,C,D,E) He began to walk with his hands held at 2.5 months after the 50-billion MTT.

(F) At 4 months after MTT, he was able to walk on crutches for about 20 steps.

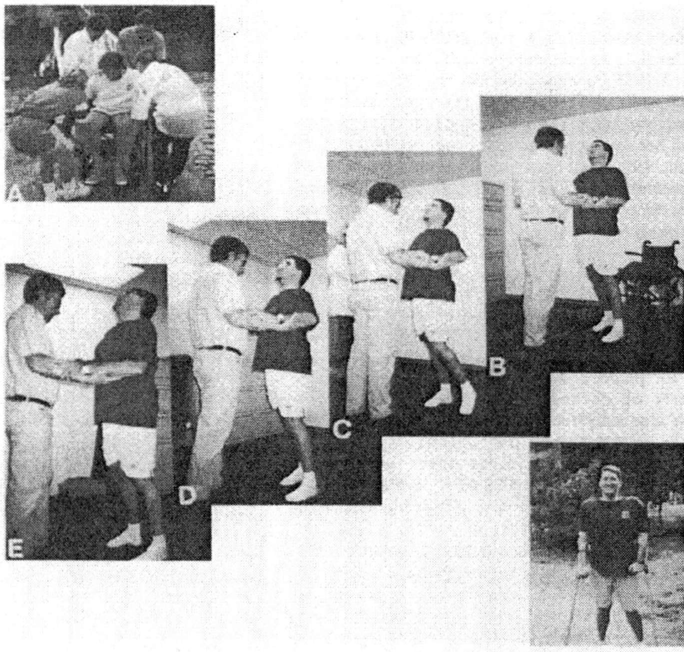

Figure 10. First muscular dystrophy subject ever to walk on crutches after wheelchair bound for 4 years.

HUMAN MYOBLAST GENOME THERAPY AND THE REGENERATIVE HEART

Heart muscle degeneration is the leading cause of debilitation and death in humans. It is the common pathway underlying congenital and infectious cardiomyopathies, myocardial infarction, congestive heart failure, angina, coronary artery disease and peripheral vascular disease, all of which constitute the cardiovascular diseases. Global healthcare spending on the latter topped $280 billion in 2001. In the United States alone, approximately $186 billion is spent every year in treating some 60 million cardiovascular disease patients. About 50% of the 5 million plus patients suffering congestive heart failure will die within 5 years of diagnosis.

Bioengineering the regenerative heart may provide a novel treatment for cardiovascular diseases. Through endomyocardial injections of cultured skeletal myoblasts, the latter spontaneously transfer their nuclei into cardiomyocytes to impart myogenic regeneration. Injected myoblasts trans-differentiate to become cardiomyocytes. Donor myoblasts also fuse among themselves to form new myofibers, depositing contractile filaments to improve heart contractility. These myofibers contain satellite cells with regenerative vigor to combat heart muscle degeneration. The proof of concept is described below.

Human myoblasts transduced with $VEGF_{165}$ and Angiopoietin-1 genes produced six times more capillaries in porcine myocardium than placebo. Xenograft rejection was not observed for up to 30 weeks despite cyclosporine discontinuation at 6 weeks. Cyclosporine usage may be changed to 2 weeks in view of new findings. First in Man (FIM) studies using cGMP-produced pure myoblasts of autogenic and allogenic origins are documented. Pros and cons of autografts vs. allografts are compared to guide future development of Heart Cell Therapy using HMGT.

Heart Muscle Regeneration

Heart muscle degeneration cascades with cardiomyocyte membrane leakage, uncontrolled Ca 2+ influx, mitochondrial ATP shutdown, inability to exude Ca 2+ through the cell surface and to reabsorb Ca 2+ into the sarcoplasmic reticulum, myofibrillar hypercontracture and disarrangement. Apoptosis ensues. Fibroblasts proliferate and infiltrate. The heart muscle, which was once populated by live cardiomyocytes with proteinaceous contractile filaments such as myosin, actin, troponin, tropomyosin, is now partially occupied by fibrous scars that are incapable of electric conduction, mechanical contraction and revascularization. These scars continue to exert a negative compliance on the heart, and the circulation, despite remodeling occurs after a myocardial infarction.

Natural Heart Muscle Repair

Ultimately heart muscle degeneration results in loss of live cardiomyocytes, contractile filaments, contractility, heart function and healthy circulation. The damaged heart responds by cell division of cardiomyocytes. However such regenerative capacity is hardly significant. Cardiomyocytes in culturo will undergo no more than three to five divisions, yielding an insufficient number of cells to repopulate any myocardial infarct.

Cardiomyocytes do not multiply significantly because the human telomeric DNA repeats [294] in these terminally differentiated cells are minimal. Telomerasing cardiomyocytes in vivo still remains a technical challenge. Without significant mitotic activity, surviving cardiomyocytes cannot provide enough new cells to deposit the contractile filaments necessary to maintain normal heart function.

The degenerative heart also transmits biochemical signals to recruit stem cells from the stroma and the bone marrow in an attempt to repair the muscle damage. This process can be augmented with stem cell transplants. Being pluripotent, embryonic or adult stem cells exhibit uncontrolled differentiation into various lineages to produce bone, cartilage, fat, connective tissue, skeletal and heart muscles (Figure 1). Because fibroblast growth factor level is elevated in the degenerative heart, much of the recruited stem cells differentiate to become fibroblasts instead of cardiomyocytes, thus depositing fibrous scars and not contractile filaments of muscles.

Despite the claimed success of transmyocardial revascularization using laser, angiogenic factors and genes, the damaged myocardium really needs additional live myogenic cells to deposit contractile filaments to regain heart function, preferably before fibroblast infiltration, which leads to scar formation.

Myocardial Cell Transplantation

Myocardial cell transplantation is intended to compensate for the loss of cardiomyocyte number and aims to limit and/or reverse the consequences of contractile dysfunction of a damaged left ventricle [295-298]. These effects could be related directly to the injected cells or mediated indirectly by angiogenic or growth factors secreted by transplant cells [299-301]. More recently, transplantation of cells with inherent ability to secrete growth factors or genetically modulated cells carrying angiogenic growth factors is being assessed with the added consequence of angiogenesis concomitant with myogenesis [302-307]. The techniques for cardiac regeneration appear promising and offer alternative methods for the treatment of ischemic heart disease [308,353].

Choise of Cell

Studies in cellular cardiomyoplasty have used embryonic stem cells, adult and fetal cardiomyocytes, fibroblasts, smooth muscle cells, mesenchymal stem cells, and myoblasts [309-316]. Adult and fetal cardiomyocyte implants produced viable and stable grafts in small mammals [317]. They form intercalated discs and gap junctions after grafting, suggesting electromechanical coupling of these cells with the host myocardium [318]. However, the use of embryonic stem cells and fetal cardiomyocytes in humans raises ethical and availability issues.

Myoblast Suitability

The advantages of using autologous skeletal myoblasts are availability, the lack of immunologic barriers to the transplantation process, which precludes the need for immunosuppression to allow donor cell acceptance by the host. Myoblasts can be genetically modified in vitro to deliver angiogenic cytokines and growth factors to encourage angiomyogenesis.

Animal studies have shown that grafted myoblasts form myotubes in the myocardium and eventually mature to become myofibers with contractile apparatus. These results have been confirmed in humans [319,320]. The regenerates acquire a fatigue-resistant slow-twitch muscle phenotype which is more suited to perform the cardiac work load. Transplantation of myoblasts produces significant functional improvement in damaged hearts [295,298,321-323]. There have been contradicting reports on the differentiation of myoblasts into cardiomyocyte-like cells with intercalated discs [324]. It is doubtful if the injected myoblasts make any meaningful electromechanical connections with host cardiomyocytes through gap junctions [325].

World's *First* Human Myoblast Transfer into the Heart

Being terminally differentiated, cardiomyocytes do not divide significantly to regenerate the myocardium. Stem cells transplant posts ethical issues of abortion and technical uncertainties as to whether these pluripotent cells will absolutely differentiate into cardiomyocytes and not osteoblasts, chondrocytes or others. We present the first successful endovascular transfer of human myoblasts into the porcine myocardium. Porcine hearts highly resemble those of humans and are more prompt to fibrillate upon minor injury.

Following International Animal Care guidelines, a 100-lb juvenile female pig was anesthetized. Access was obtained via the right femoral artery using cutdown technique. Catheter advance into the left ventricle through the aorta was guided with fluoroscopy followed by endomyocardial mapping with the electromagnetic NOGA system (Biosense Webster Inc, Johnson & Johnson).

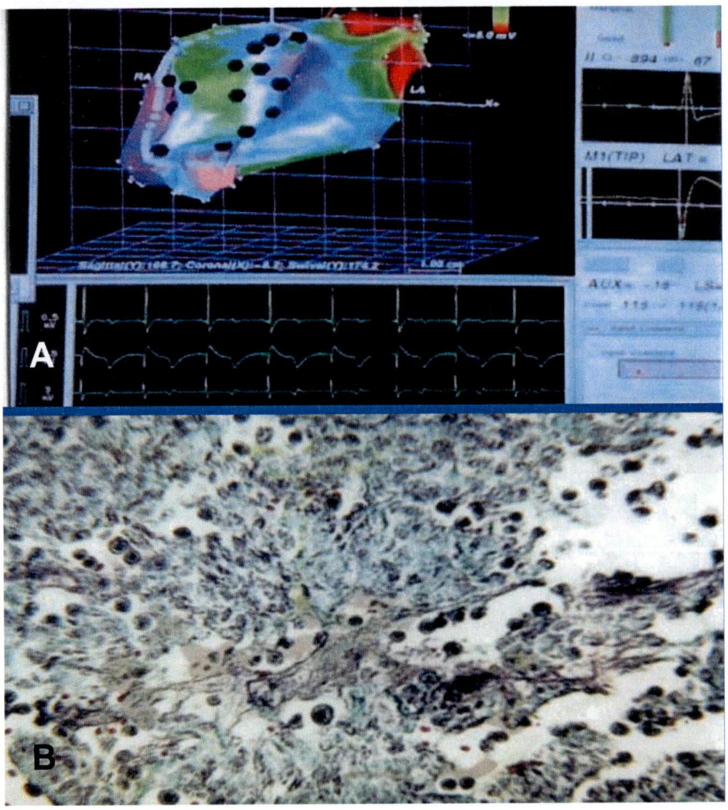

Figure 11. World's first human myoblast injections into the porcine heart to determine feasibility/safety and injection parameters for the MyoStar catheter and the process of heart cell therapy.

Approximately 10^9 human myoblasts were injected through a needle timed to protrude 6 mm from the tip of the catheter into the myocardium. Twenty injections were made at different locations within 40 minutes, having volumes of 0, 1, 0.2, 0.3, 0.5, 1.0 ml, and cell concentration of $100X10^6$ /mL (Figure 11A). Heart rate, electrocardiogram and temperature were continuously monitored. Other than transient short runs of ventricular ectopy, the pig remained in stable condition throughout the injection period. There was no significant change in the parameters monitored. Vital dye staining of the myoblasts before versus after the procedure showed no significant difference in cell viability. Furthermore, cell passage through the injection catheter showed less than 5% of cell death.

At the completion of the procedure, the pig was sacrificed and the heart processed for histological examination. Transmyocardial perforation was not observed. Numerous prominent round and mononucleated human myoblasts were found widely and evenly distributed throughout the apex and the lateral wall of the pig left ventricle where the myoblasts were injected (Figure 11B). This study provides the first direct evidence demonstrating the feasibility and safety of endovascular delivery of human myoblasts into the porcine heart using the NOGA catheter injection system.

This *first* human myoblast transfer into the heart revealed that it was safe to administer one billion myoblasts at $100x10^6$/ml through the Myostar catheter of the NOGA system (Biosense Webster Inc) using 20 injections at different locations inside the left ventricle of a swine [438]. It was determined that 0.3 ml to 0.5 ml would be the optimal volume per injection. EKG was normal throughout the study without arrhythmia.

The Regenerative Heart

The goal to bioengineer the regenerative heart seems to be within reach with HMGT. Five grams of muscle biopsies would be taken from both quadriceps of a heart patient with age ranging from 40 to 90, culturing approximately one billion myoblasts in four weeks and then injecting or surgically implanting these cells just inside the non-vascularized infracted myocardium.

Heart Cell Therapy (HCT) [328], as this is called, is administered with the vision that the myoblasts will survive, develop and function as aliens in the heart, and their nuclei as aliens within cardiomyocytes and myofibers. The myocardial aliens are newly formed skeletal myofibers that contribute to cardiac output through production of contractile filaments. They are donor in origin and as skeletal myofibers will have satellite cells and regenerative capability. The cardiomyocyte aliens are donor myoblast nuclei carrying chromosomes with long telomeric DNA subunits that are essential for mitosis. Upon injury of the trans-differentiated or heterokaryotic cardiomyocyte, the myoblast regenerative genome will be activated to produce foreign contractile filaments such as myosin and actin.

Proof of Concept

Human myoblasts were manufactured according to our inhouse SOP's and trade secrets with a license of the U.S. Patent No. 5,130,14110 and with a license of the Singapore Patent No. 3449011 (WO 96/18303). This method of culture yielded purity of 99% by human desmin immunostaining (Figure 12). From 2g of human muscle biopsy, 50 billion pure myoblasts are produced routinely in our GMP facilities in about 45 days (Figure 13).

In the myogenesis study, cultured myoblasts derived from satellite cells of human rectus femoris biopsies were transduced with retroviral vector carrying Lac-Z reporter gene, and about 75% of the myonuclei were successfully transduced. Trypan blue stain revealed >95% cell viability immediately before injection.

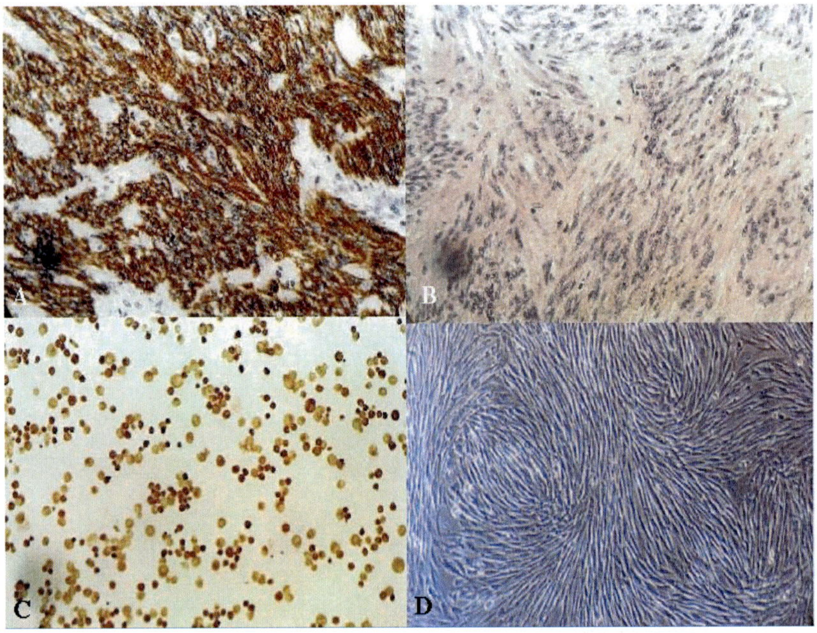

Figure 12. Human desmin immunostain for myoblast purity. (a) Positive control of leiomyosarcoma, staining brown. (b) Negative control. (c) Pure human myoblasts immunostained with desmin. (d) Pure human myoblasts in culture.

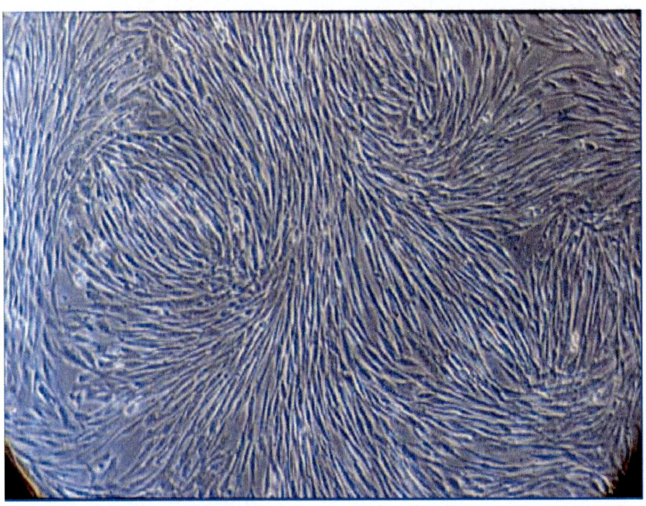

Figure 13. Pure myoblasts for MTT/HMGT.

A porcine heart model of chronic ischemia was produced by clamping an ameroid ring around the left circumflex artery. Four weeks later, the heart was exposed by left thoracotomy. Twenty injections (0.25ml each) containing 300 million myoblasts, or 5ml total volume of basal DMEM as control, were injected into the left ventricle intramyocardially. Left ventricular function was assessed using MIBI-Tc99m SPECT scanning one week before injection to confirm myocardial infarction, and at 6 weeks after injection.

Animals were maintained on cyclosporine at 5 mg/kg body weight from 5 days before, until 6 weeks after cell transplantation. The animals were euthanized at 6 weeks to 7 months post-operatively, and the hearts were processed for histological, immunocytochemical and ultrastructural studies.

Histological examination of myoblast-injected myocardium showed cardiomyocytes containing Lac-Z positive nuclei (of donor origin) after 12 weeks (Figure 14B). More than 80% of the Lac-Z positive cardiomyocytes immunostained positively for human myosin heavy chain (Figure 14A). The human genome was integrated (Lac-Z labelled nuclei) and expressed (myosin immunostain) in the porcine myocardium.

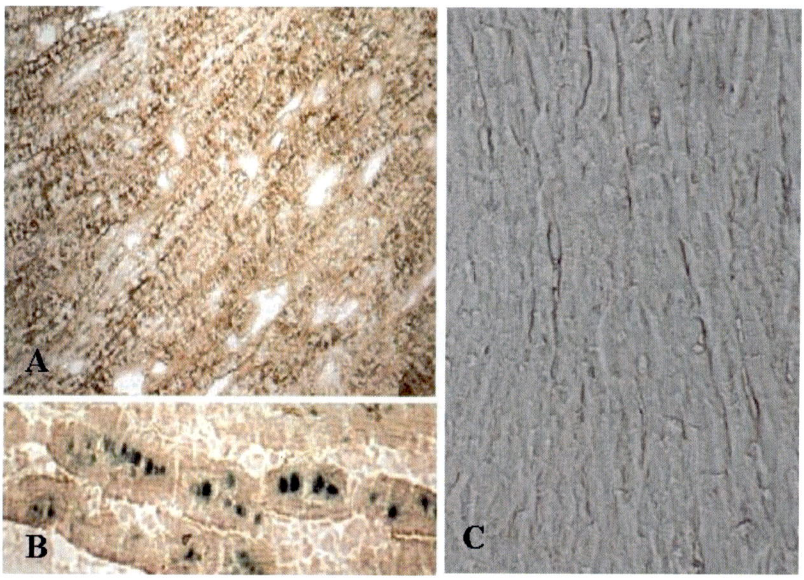

Figure 14. (A) Brownish immunostain of human myosin in porcine myocardium 12 weeks after human myoblast injection. (b) Cardiomyocytes each with 4 to 6 LacZ-positive nuclei and human myosin stain, indicative of their being donor or myoblastic in origin. (c) Negative immunostain (gray) of human myosin in porcine myocardium sham-injected without myoblasts.

Donor myoblasts trans-differentiated to become cardiomyocytes that are characterized by having four to five nuclei each. The control heart without myoblast injection did not show Lac-Z positive myonuclei nor human myosin (Figure 14C). Triple stain of myoblast-injected myocardia demonstrated multinucleated heterokaryons containing human and porcine nuclei with expression of human myosin (Figure 15). Electron microscopy demonstrated human myotubes and skeletal myofibers with satellite cells in the porcine myocardium (Figure 16).

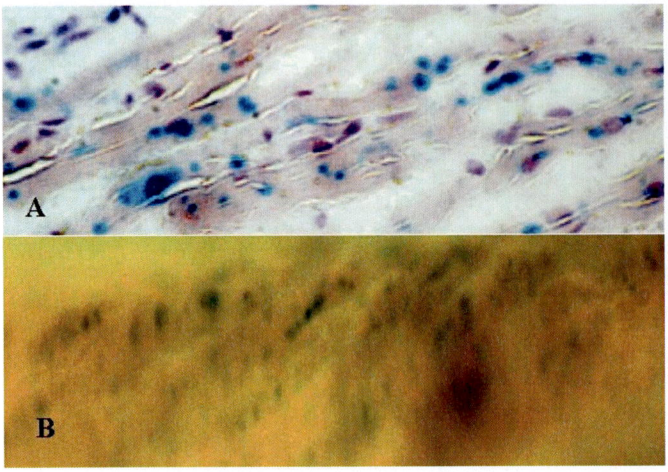

Figure 15. (a) Heterokaryons derived from fusion of porcine cardiomyocytes and human myoblasts showing LacZ-positive human myoblast nuclei (bluish green) and porcine cardiomyocyte nuclei (purple) in the heterokaryotic synytium. (b) These heterokaryons expressed human myosin heavy chain.

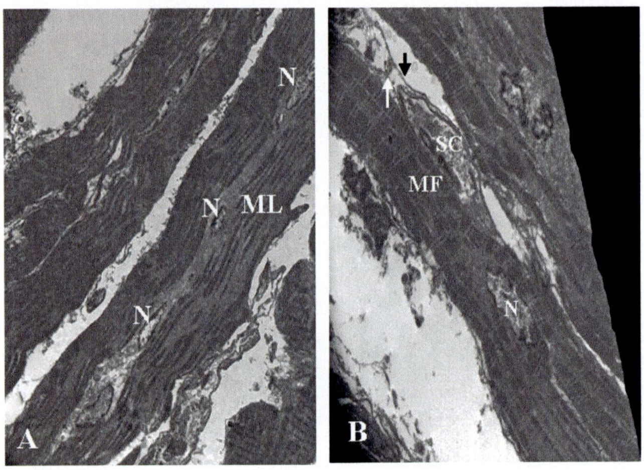

Figure 16. Electron microscopy of the myoblast-injected porcine myocardium showing (a) myotubes with central nuclei and myfibril (ML) deposits, and (b) skeletal myofiber with satellite cell (SC) and nucleus (N). The satellite cell is located between the basement membrane (black arrow) and the plasma membrane (white arrow). Sarcomeres show proper alignment of newly formed contractile filaments.

Three myogenesis mechanisms were elucidated as proof of concept with 50 human/porcine xenografts using cyclosporine as immunosuppressant. Some myoblasts trans-differentiated to become cardiomyocytes. Others transferred their nuclei into host cardiomyocytes through natural cell fusion. As yet others formed skeletal myofibers with satellite cells. De novo production of contractile filaments augmented heart contractility (Figure 17) [351,352].

Laser nuclear capture together with single nucleus RT-PCR was performed to delineate host and donor nuclei. In situ hybridization using fluorescent DNA probes specific for human Y-chromosome and chromosomes 1&10 for pig were used to demonstrate nuclear mosaicism (Figure 18). The implantation of human myoblasts derived from young males into female pigs provides additional demonstration of donor cell survival and development in the host muscles (Figure 19).

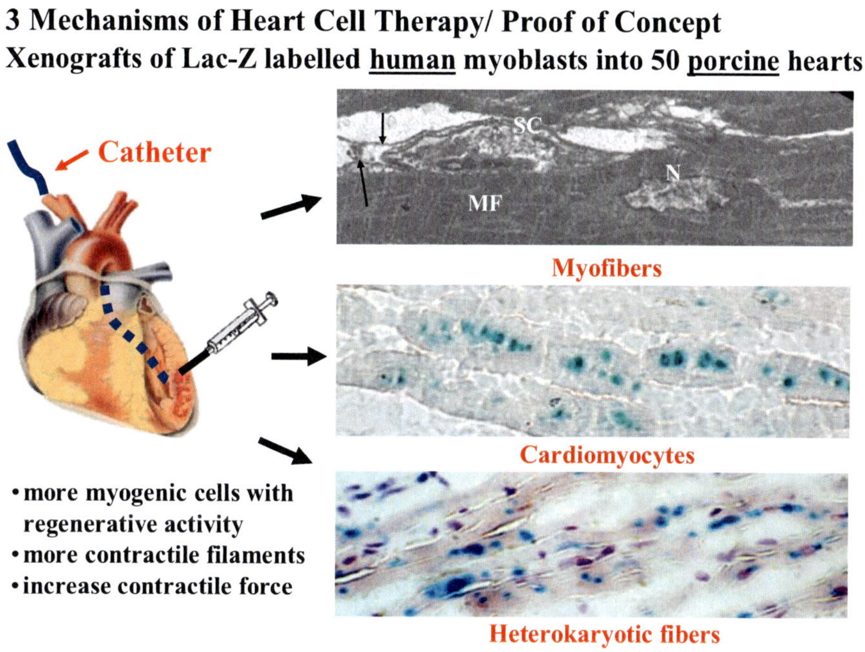

Figure 17. Proof of concept for HMGT to treat heart muscle degeneration, showing the fate of implanted human myoblasts into porcine myocardia. Three mechanisms were identified.

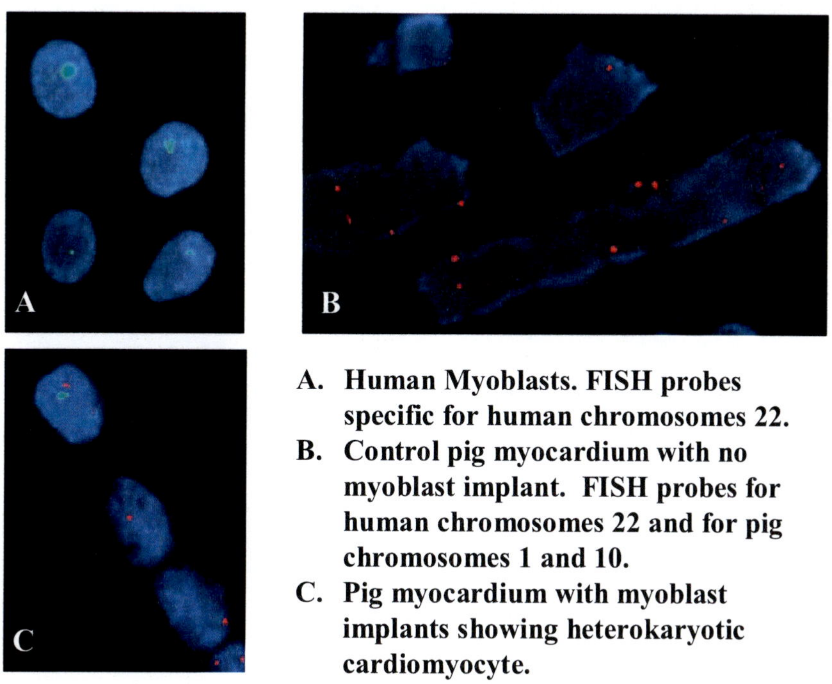

Figure 18. FISH probes to demonstrate heterokaryotic cardiomyocyte.

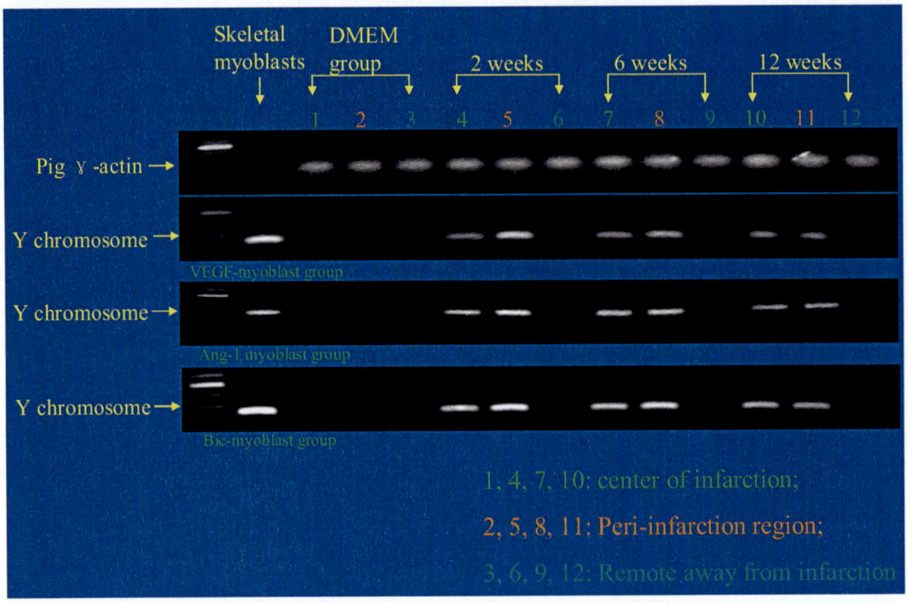

Figure 19. PCR analysis for detection of human Y chromosome in female pig heart. 1, 4, 7, 10: center of infarction; 2, 5, 8, 11: Peri-infarction region; 3, 6, 9, 12: Remote away from infarction.

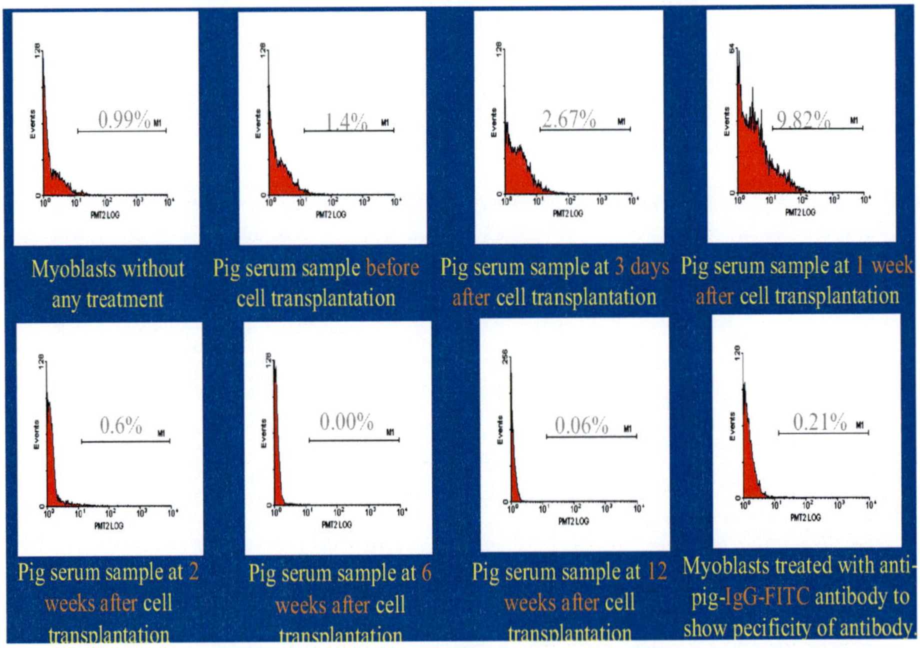

Figure 20. Pig immuno-tolerance with human myoblasts -transient presence of pig anti-human skeletal myoblast antibody. A. Myoblasts without any treatment. B. Pig serum sample before cell transplantation. C. Pig serum sample at 3 days after cell transplantation. D. Pig serum sample at 1 week after cell transplantation. E. Pig serum sample at 2 weeks after cell transplantation. F. Pig serum sample at 6 weeks after cell transplantation. G. Pig serum sample at 12 weeks after cell transplantation. H. Myoblasts treated with anti-pig-IgG-FITC antibody to show specificity of antibody.

Discontinuation of cyclosporine after 6 weeks prompted no xenograft rejection for up to 30 weeks. There was a transient elevation of the porcine anti-human- myoblast antibodies at one week after the xenograft (Figure 20). The antibody level subsided at the second week after HMGT, indicating that no more than two weeks of cyclosporine mmunosuppression would be necessary for human/pig xenografts. Modification of the cyclosporine administration protocol will be implemented accordingly in the future.

Therapeutic Angiogenesis

Therapeutic angiogenesis improves tissue ischemia by supplementing and supporting the intrinsic process of angiogenesis [354]. Besides direct administration of angiogenic growth factor proteins, injection of naked DNA, non-viral vectors and viral vector constructs carrying angiogenic genes have been used. A novel approach is to achieve therapeutic angiogenesis through cell mediated gene transfer. Genetically modulated cells carrying exogenous genes encoding for angiogenic factors and the cells with inherent ability to secrete angiogenic cytokines, such as bone marrow stem cells, embryonic stem cells and endothelial progenitor cells, have been used to achieve revascularization.

This review discusses proof of the concept pre-clinical studies and phase-I /II human trials using $VEGF_{165}$, and cellular angiogenesis at length in the light of the literature and analyzes the problems and considerations of these approaches as a treatment strategy in the clinical perspective for the treatment of ischemic heart disease.

Angiogenesis and vasculogenesis are primarily responsible for the development of the vascular system in the embryo. Whereas vasculogenesis involves angioblasts for the in situ new vessel development, angiogenesis occurs by sprouting of the already present primitive vasculature using previously differentiated cells [331]. A third mechanism that contributes to the development of collateral vessels is arteriogenesis through remodeling of the existing arterioles [329]. Angiogenesis is an integral part of many of the physiological and pathological processes, and angiogenesis is controlled by various factors that induce or inhibit blood vessel formation. Disruption of the natural balance between the pro- and anti-angiogenic factors results in pathological angiogenesis with abnormal blood vessel formation.

Therapeutic angiogenesis exploits the natural process for enhanced neovascularization [333]. Apart from direct administration of angiogenic proteins, naked plasmid and viral vector constructs carrying angiogenic genes have been used. Recent studies have suggested therapeutic angiogenesis through cell-mediated gene transfer and by stem cell transplantation. The transplantation of genetically modified cells serves as a reservoir to produce biologically active angiogenic factors in a localized and sustained pattern. Similarly, the application of stem cells not only can derive new vessel formation, but also differentiate into cardiomyocytes to restore injured myocardium.

This review summarizes the recent progress in therapeutic angiogenesis in conjunction with myogenesis, and critically analyzes the problems and considerations of this approach as a treatment strategy in the clinical setup for the treatment of myocardial infarction.

$VEGF_{165}$Protein Delivery

The imbalance between supply and demand for oxygen in the hypoperfused myocardium up-regulates the expression of pro-angiogenic factors and their receptors [334,335]. $VEGF_{165}$ is the major angiogenic factor involved in physiological as well as pathological angiogenesis [330]. The neovascularization is mainly achieved by endothelial cell proliferation triggered through $VEGF_{165}$ receptors, especially the $VEGF_{165}$ receptor-2 [337]. $VEGF_{165}$ induces nitric oxide production and cGMP accumulation in cultured endothelial cells through the activation of endothelial nitric oxide synthetase (eNOS) [338]. The important functions of these two mediators are vasodilatation, inhibition of smooth muscle proliferation, anti-platelet accumulation and inhibition of leukocyte adhesion, leading to vascular protection [336]. Some of the factors stimulate angiogenesis through the induction of $VEGF_{165}$.

We have studied the potential of human myoblasts carrying human $VEGF_{165}$ for myogenesis and angiogenesis in a porcine heart model of myocardial infarction [332].

Angiomyogenesis

In the angiomyogenesis study, the human myoblasts were transduced with retroviral and adenoviral vectors carrying Lac-Z and human $VEGF_{165}$ genes, respectively. The cells were characterized for $VEGF_{165}$ transduction and expression efficiency by immunostaining, ELISA, immunoblotting and RT-PCR. A porcine heart model of infarction was created in eight female swines by left circumflex artery ligation. The animals were grouped as control and myoblast-implanted. Angiography was performed to ensure complete occlusion of the blood vessel. Infarction was confirmed with MIBI-$Tc^{99}m$ SPECT scanning. Four weeks later, 5ml basal DMEM without or with $3x10^8$ human myoblasts carrying $VEGF_{165}$ and Lac-Z genes. The animals were maintained on cyclosporine (5 mg/kg body weight) for six weeks post-operatively. Hearts were then explanted and processed for immunocytochemical studies.

The transduction efficiency for Lac-Z and $VEGF_{165}$ was 75-80% and >95% respectively. The transduced myoblasts continued to secrete VEGF165 for longer than 18 days, at significantly higher level (37 ± 3ng/ml) than non-transduced ones (200 ± 30pg/ml). Dye exclusion test revealed >95% cell viability at the time of injection. Histological examination showed extensive survival of the grafted myoblasts expressing Lac-Z gene in and around the infarct. The SPECT scans showed improved perfusion in the infarcted region. Immunostaining for vWF-VIII expression showed significantly higher number (28.3 ± 1.8) of microvessels in the infarction region after transplantation of $hVEGF_{165}$ transduced myoblasts as compared to the DMEM injected (4.2 ± 0.4) pig hearts at low power (200x/field) (Figure 21C).

Most previous studies to achieve therapeutic angiogenesis have relied on a single growth factor protein or its gene administration. Angiogenesis involves interplay between numerous growth factors, their receptors, and intracellular signals. Recent studies have focused on the use of a combination of factors or their genes. The synergistic interaction in vivo between $VEGF_{165}$ and Ang-1 to produce functional and leak-resistant neovascularization support the suitability of their combined administration. Combining the synergistic effect between

$VEGF_{165}$ and Ang-1, together with myoblast therapy, has the additional advantage of inducing myogenesis together with angiogenesis for treatment of ischemic limbs [355]. The present study involved transduction of $VEGF_{165}$ and Ang-1 genes into autologous skeletal myoblasts using a new bicistronic adenoviral vector for the treatment of hind limb ischemia in a rabbit model.

Unlike the use of monocistronic vectors, bicistronic vector provides a less cumbersome delivery option by simplifying the construct design and enabling delivery of the two transgenes in a single administration [356]. Although the donor myoblasts survived adequately in vivo for 6 weeks or longer, as assessed with X-gal staining, mortality was still inadvertent, despite use of autologous cells. Similar to previous findings, the fate of our implanted myoblasts was either that they fused to mature fibers of the host, to regenerating muscle cells of the host, or to other implanted myoblasts, they remained quiescent as mononucleated cells, or they died [357,358].

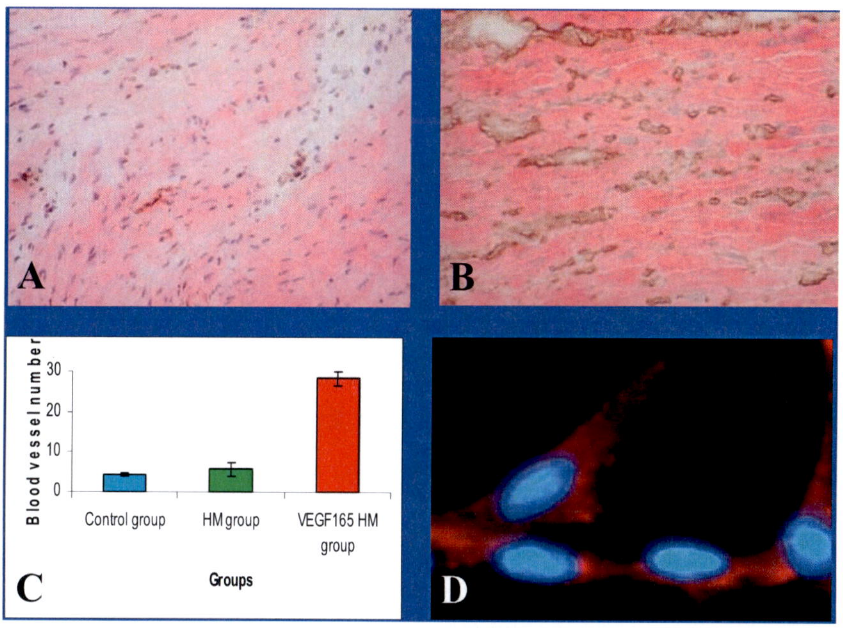

Figure 21. (a) Control infarcted myocardium immunostained for von Willebrand factor(vWF) VIII and counterstained with eosin to show capillaries and scar tissues. (b) VEGF $_{165}$transduced myoblasts produced increased vascular density, (c) seven times that of the control in the number of capillaries. (d) Immunofluorecent $VEGF_{165}$ myoblasts.

The donor myoblasts transduced with Ad-Bic is simultaneously expressed both VEGF and Ang-1, and significantly increased neovascularization in the ischemic porcine myocardium (Figure 22). Double fluorescent immunostaining of tissue sections for vWF and SMA revealed that a large proportion of the vessels observed were coated with smooth muscle layer. This observation is in agreement with previous studies that show the role of Ang-1 in development of mature, stable, non-leaky vessels [359]. This suggests that increased blood vessel density was genuinely the result of Ad-Bicis transduced myoblasts treatment.

First in Man (FIM) Studies

Myoblast Autograft

On May 14, 2002, a 55-year-old man suffering acute ischemic myocardial infarction received cGMP-produced pure myoblasts into his beating myocardium as an adjunct to off-pump coronary artery bypass grafting (CABG). Professor Eugene Sim of the National University Hospital of Singapore led his team in the operation.

Coronary angiography previously demonstrated a right dominant system, 80% occlusion of the mid-left anterior descending artery, 70% occlusion of the proximal first diagonal branch and 50% stenosis of the circumflex artery. The mid-right coronary artery was occluded with good collateral supply from the left coronary artery. Echocardiography 5 days later revealed akinetic apex, anterior wall and septum of the left ventricle, with a left ventricular ejection fraction of 31%.

Tomographic Tc99m tetrofosmin scan 25 days after the acute myocardial infarction revealed a large partially reversible defect on the anterior wall and a moderate-sized reversible defect on the inferior wall. After qualifying for the inclusion/exclusion criteria, signing patients' informed consent, and obtaining approval from the hospital ethics committee, the patient was included as a volunteer for myocardial cell transplantation adjunct to CABG at 2 months after the initial presentation.

Approximately 2.5g of rectus femoris muscle biopsy was taken under local anesthesia from the patient and cultured under cGMP for 32 days to produce one billion cells. Myoblast purity of >99% was ascertained by positive immunostain for human desmin.

Under direct vision and stabilization with the Octopus III tissue stabiliser, 4.65×10^8 autologous myoblasts in 3 mL serum were injected into the myocardium within 15 minutes using a 27-gauge needle. Twenty-five injections were made on the anterior wall, near the apex, on the posterior wall, in and around the infarcted areas. These injections of 0.1 mL to 0.2 mL each were performed at half- to one-minute intervals to observe for dysrhythmias.

The patient recovered well from the operation. Serum creatine kinase was 222 μg/L and the MB fraction was 6 μg/L at 3 hours post-operatively, and 385 μg/L and 6 μg/L respectively at 12 hours. Postoperative 24-hour Holter monitoring revealed no arrhythmia. The patient was discharged on the eighth post-operative day.

His EKG has shown no arrhythmia. He has shown good effort tolerance, no dyspnea or angina, and a significant increase in ejection fraction since three weeks post-operatively. This case study suggests that beating-heart cellular cardiomyoplasty can be a safe and viable option.

Myoblast Allograft

Myoblast allograft is being developed to prevent and to alleviate heart muscle degeneration [348]. We report its *first application in man.*

The feasibility and safety of myoblast allograft was assessed by injecting the infarcted myocardia of two men, aged 63 and 49, with 1.1 and 1.2 billion myoblasts respectively, using 2-month cyclosporine immunosuppression.

Donor myoblasts were manufactured in compliance with current Good Manufacture Practice (cGMP) and ISO 9001 conditions. About 2.18g of muscle biopsy was taken under local anesthesia from the rectus femoris of a 20 year-old pathogen-free male volunteer, after he had met muscle donor criteria and had signed informed consent. At harvest, the culture yielded 3.64×10^9 myoblasts that were 98.3% pure by positive desmin immunostain. It was 91.5% viable according to vital dye exclusion tests. The cells were potent in myogenecity in that myotubes comprised more than 99% of the culture in fusion medium. Throughout the culture and for the final injectates, the myoblasts were free of endotoxin, mycoplasma, and negative for sterility (14 day test) and gram stain (absence of gram positive or negative bacteria) according to certified laboratory analyses.

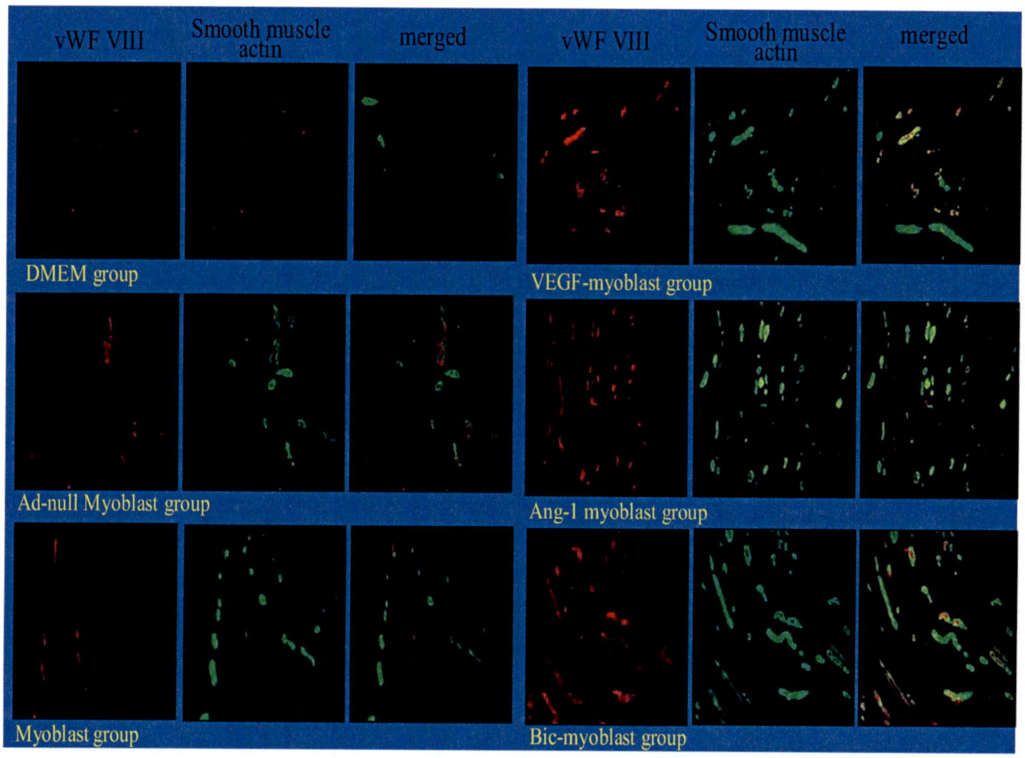

Figure 22. Neovascularization of infracted myocardia at 12 weeks after treatment. When compared to the DMEM group, both the ad-null myoblast group and the non-transduced myoblast group showed formation of new capillaries. These latters are less numerous and less mature than the transduced myoblast groups.

Both patients enrolled as clinical trial subjects after qualifying for inclusion/exclusion criteria, signing patients' informed consents, and obtaining institutional and academy approval. They had atherosclerosis, coronary artery disease, history of acute myocardial infarction, stable angina Functional Class IV (CCS), and arterial hypertension II (risk 4). Positron emission tomography (PET) with 18FDG revealed scarred myocardium in LV septal, apical, and anterior/posterior wall regions. Echocardiography showed regional akinesis/hypokinesis, LVEF being 41% and 38% for the subjects respectively. Single-photon emission computed tomography (SPECT) with 30 mCi ^{99}mTc-tetrofosmin was performed during exercise with bicycle and during rest. Ischaemic changes and scars were confirmed, with significant perfusion defective areas (Figures 23, 24). Total LVEF were 39% and 38%

for the two subjects respectively. Coronarography revealed calcination, stenosis, and occlusion of the left and right coronary arteries. Left ventriculography demonstrated hypokinesis in segments II, III, IV, and V in the older subject.

Our 15 years of experience of injecting allogenic myoblasts into 280 muscular dystrophy patients and 31 years of animal allograft/xenograft studies suggested that eight weeks of cyclosporine immunosuppression are sufficient to allow foreign myoblast engraftment in heart patients. The subjects took two oral doses of cyclosporine totaling 5 to 7 mg/kg body weight per day, beginning at five days before grafting, weaning at half-doses in the last two weeks, and off cyclosporine at eight weeks after grafting. Whole blood trough level of cyclosporine was monitored every three days at 23 hours after the morning dose. Doses were adjusted to maintain the level at about 250 ng/mL.

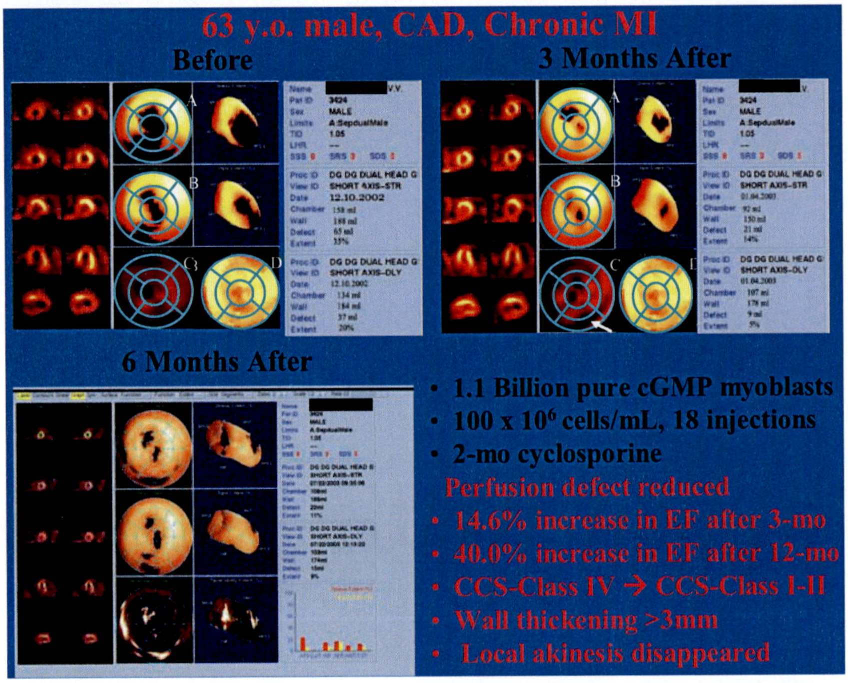

Figure 23. Polar views on PET and SPECT tomograms of the left ventricular myocardium before and at 3 and 6 months after myoblast allograft for the 63 y.o. patient. Before surgery (a) 35% perfusion defect during exercise and (b) 20% at rest, (c) significant decrease of systolic thickening and (d) preserved metabolism are indicative of scarred and hibernated myocardium. After surgery (a) 14% perfusion defect during exercise and (b) 5% at rest, (c) decrease in systolic thickening is less, and perfusion in the posterior wall (arrow) has been retrieved. (d) Metabolism is preserved and shows no change.

Academician Leo A. Bockeria of the Russian Academy of Medical Sciences is the world's first to have injected the hearts of the patients with allogenic myoblasts. On January 17, 2003, the subjects underwent bypass grafting immediately prior to myoblast implantation. They received 18/19 injections totaling 1.1/1.2 billion myoblasts respectively. The 10-min procedure was performed open-chest on an ice-chilled, non-beating heart with the subject under general anesthesia and on respirator. Cells were injected at $100x10^6$ /mL. The 0.5 mL injections were made under direct vision between the infarcted and viable myocardium, and into scarred tissue.

The subjects recovered from the general anesthesia with no rash or fever. Holter EKG monitor registered sporadic ventricular arrhythmias and ventricular extrasystoles that were eventually eliminated with Amiodarone treatment. Both patients were discharged after careful observation for 3 weeks. There was no evidence of overt blood pressure changes or renal failure.

Despite cyclosporine discontinuation at 2 months postoperatively, no sign of rejection was observed. At 3, 6 and 9-month follow-ups, both subjects were in stable condition at Class I-II (CCS). They no longer suffered angina or shortness of breath. Echocardiography showed significant increases in LVEF respectively with no local akinetic/hypokinetic regions. SPECT with [30] mCi 99mTc-tetrofosmin confirmed similar LVEF increases and demonstrated reduction on perfusion defective areas during exercise and rest (Figures 23, 24). ^{18}FDG accumulation was equable throughout LV myocardium, and glucose metabolism changes were not revealed.

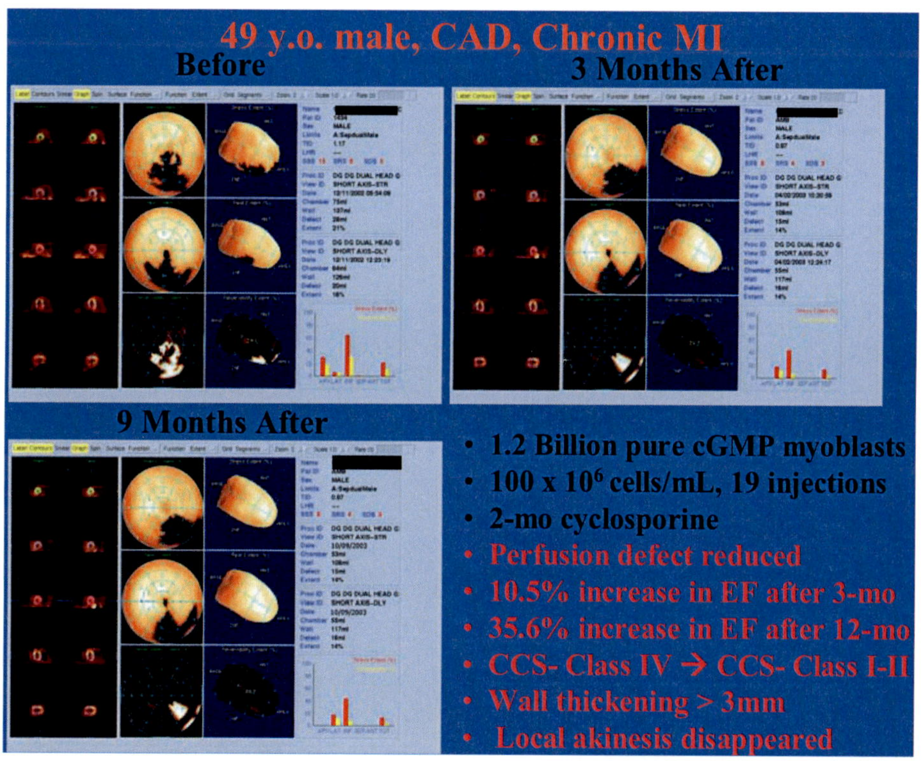

Figure 24. Polar views on PET and SPECT tomograms of the left ventricular myocardium before and at 3 and 9 months after myoblast allograft for the 49 y.o. patient showing similar improvement with time.

Human myoblast allografts may provide an alternative therapy for heart muscle degeneration and prevention of heart attack, with virtually unlimited cell availability and only 2-month immunosuppression. Cell potency from young donors is much higher than that of autografts from older heart patients. The regenerative heart is also a rejuvenated heart. Whereas good cells are readily available from cell banks in allografts, autograft patients have to be biopsied and wait about one month for their own myoblasts to grow to reach one billion in number. In acute myocardial infarction, the immediate allografts may prevent fibroblast infiltration and eventual scar formation.

Patients with infectious diseases such as AIDS can be treated with allografts without fear of contamination of the cell culture system [348,351,352]. Furthermore, heart patients with genetic diseases such as the muscular dystrophies can be treated with genetically normal allografts.

The survival of the two subjects without any sign of rejection after cyclosporine withdrawal confirmed our report on safety in the human/porcine cardiac xenografts [348,351,352]. Cyclosporine immunosuppression for 3 months allowed allogeneic myoblasts to survive, develop, and function in a muscle of a Duchenne muscular dystrophy patient, expressing the therapeutic protein dystrophin at 3 months [256] unto 6 years [262] after myoblast transfer. Myoblast autograft has been demonstrated to have survived and developed up to 17.5 months after implantation [319,369] in an ischaemic heart patient. The myoblast purity of this study was determined using CD56+, an antibody which reacts with stem cells, neurons, and fibroblasts rather than with myoblasts. CD56+ antibody yields false data on myoblast purity.

Clinical Trials to Date (Nov. 2003)

NTR = no test report on cGMP

START DATE	TEAM	TYPE of TRIAL	TOTAL NO. OF SUBJECTS	TOTAL NO. OF SURVIVALS	CELLS / TESTS	RESULTS	REFERENCE
6/15/00	Menasche / Paris	Open heart with bypass, autograft	11 patients	10	(0.5-1.2B) x 10^9, 65% pure, 95% viable, NTR	EF from 25% to 35%, donor cells survived, function improved	Lancet, 2001 Lancet, 2002
5/23/01	Smits/Bioheart (Rotterdam. Milan, Mt. Sinai)	Endovascular, autograft	24 patients	20	25-300 x 10^6, 35% - 90% pure, NTR	4 died; >50% patients developed arrhythmis	Am J Cardiol 92 supp (6), 2003
5/30/01	Dib / Diacrin, (Ariz HI/ Ohio State U., Cleveland Clinic UCLA)	Open heart with bypass, autograft	12 patients	12	10, 30, 100, 300 millions, 35-95% pure, NTR	EF from 22% to 34%, wall thickening, gap junction , no connexin 34, 4 Patients had arrhythmia before & after procedure	Am J Cardiol 92 supp (6), 2003
5/14/02	Law/Sim/NUH, Singapore	Open heart with bypass, autograft	1 patient	1	0.46x10^9, 98% pure, 99% viable, 95% potent, all cGMP tests completed	EF 31% to 34%; NYHA Level III to II	Handbook of Cardiovascular Cell Transplantation, Ch 17, 2003 Am J Cardiol 92 supp (6), 2003 J Mol Med, 2003
1/17/03	Law/ Bockeria/ Russian Academy of Medical Sciences	Open heart with bypass, allograft	2 patients	2	1x10^9, 98% pure, 80% viable, 95% potent, all cGMP tests completed	EF 39% to 47%; CCS III/IV to CS I; LV Defective area reduced, wall thickening.	Handbook of Cardiovascular Cell Transplantation, Ch 17, 2003 Am J Cardiol 92 supp (6), 2003 J Mol Med, 2003
2001	Hosp. Avellaneda	Surgery, autograft	1 patient	1	NTR	Unknown	BAM 13:17, 2003
2002	Siminiak / Poland	Open heart with bypass, autograft	10 patients	9	20 millions; NTR	EF 28% to 35%, wall thickening	Am J Cardiol 92 supp (6), 2003
2002	Canalejo/Univ. Navarra	Surgery, autograft	10 patients	Unknown	NTR	No adverse reaction	BAM 13:17, 2003
2003	Kao/ Nanjing U.	Surgery, autograft	3 patients	Unknown	NTR	Unknown	BAM 13:17, 2003

Figure 25. Phase I clinical trials of myoblasts for heart muscle degeneration.

Including those studied recently in the Magic trial, more than 150 patients have received myoblasts worldwide since the initial procedure in June 2000 (Figure 24). Mortality has been less than 10%. If future myoblast allografts prove to be safe and efficacious in treating heart dysfunction, the 2-month immunosuppression, when compared to life-long immunosuppression in heart transplant patients, will greatly improve life expectancy and quality of life of heart patients. Moreover, myoblast allografts can significantly reduce medical cost considering quality control and analysis tests on one lot of 50 billion myoblasts are much less costly than the same tests on 50 lots of one billion myoblasts. Future allograft studies are warranted. The cessation of akinesis/hypokinesis, reduction in perfusion defects, and increase in LVEF suggest that future myoblast allograft studies are warranted.

The future of HCT lies with myoblast allograft, 90% delivered endovascularly and 10% epicardially in adjunct to CABG. Myoblasts will likely be transduced with $VEGF_{165}$, Ang-1,TGF-β, or similar factors to allow concomitant angiomyogenesis.

Our ongoing clinical trial is based on unequivocal evidence of cGMP-produced pure human myoblasts and proof of concept for Heart Cell Therapy. Human myoblasts survived and integrated into porcine ischemic myocardium, allowing concomitant cell therapy and gene therapy to produce angiomyogenesis. Whereas the newly formed myofibers harbor satellite cells and impart regenerative capacity to the heart muscle, the genetic transformation of cardiomyocytes in vivo to become regenerative heterokaryons through myoblast genome transfer [328] constitutes the ultimate heart repair. The regenerative heart [349] also contains trans-differentiated cardiomyocytes of myoblastic origin. In all three scenarios, new contractile filaments are deposited to improve heart contractility. This latter can be translated into the improvement in the quality of life of heart patients and in the prevention of heart attacks.

It can be concluded that pure myoblasts transduced with Ang-1 and $VEGF_{165}$, when injected intramyocardially, are potential therapeutic transgene vehicles for concurrent angiogenesis and myogenesis to treat heart failure [348,350]. Immunosuppression using cyclosporine for six weeks is effective for long-term survival of xenografts or allografts. The feasibility and preliminary safety/efficacy observed in the world's first human myoblast allografts lead the way in developing a low-cost, easy-to-use treatment for heart failure and prevention of heart attacks, with virtually unlimited cell availability and short-term immunosuppression.

The World's First Myoblast Study of Type II Diabetic Patients

Type II diabetes, also called non-insulin-dependent diabetes mellitus (NIDDM), is characterized by high blood glucose resulting from the genetic defect of the GLUT4 genome. The latter is manifested in the diminished glucose uptake into skeletal muscles. In normal human beings, insulin combines with insulin receptors to change the membrane conformation of skeletal muscle fibers, allowing blood glucose to move down its concentration gradient into the fibers for metabolism.

A disorder called 'insulin resistance' exists in Type II diabetic patients in which normal or even elevated levels of plasma insulin would not elicit normal glucose uptake into the muscle fibers. This study hypothesizes that these diabetic fibers exhibit less insulin receptors, or that these receptors exhibit abnormal molecular conformation, or both. Considering that highly metabolic muscle fibers constitute more than 50% of the human body by volume and weight, failure of blood glucose to gain entry would undoubtedly lead to high blood glucose and result in various organ failures sequentially.This is not the only defect, but it is likely the primary and significant one.

A potential genetic treatment of the disease involves myoblast transfer therapy (MTT) which is a platform technology of cell transplantation, genome therapy and tissue engineering [326,328]. It consists of culturing immature muscle cells called myoblasts, derived originally from a 2g skeletal muscle biopsy from a healthy, young, male donor, and implanting them into the major muscle groups of the upper and lower extremities of the diabetic patients.

The myoblasts exhibit natural cell fusion, and transfer their nuclei carrying the normal human genome into the host skeletal muscle fibers to achieve genetic repair. Others fuse among themselves to form new myofibers that exhibit normal insulin receptors of donor origin. Through both mechanisms, new insulin receptors of donor origin that are genetically normal, will be produced in the skeletal myofibers of the host.

The survival, development, and functioning of the implanted allogeneic myoblasts have previously been demonstrated in studies involving about 260 muscular dystrophy subjects and two chronically myocardial infracted heart subjects with 100% safety and substantial efficacy results [349,351]. Immunorejection was minimized using two months of cyclosporine following MTT. In addition, over 120 ischemic heart patients have received autologous myoblasts in their hearts in 10 countries. Mortality rate has been less than 10% traversing the last four years, with efficacy data being collected in Phase II clinical trials in Europe. Reported here are the world's first genetic transplants of two Type II diabetic patients using allogeneic myoblasts.

Human myoblasts were manufactured according to the standard operating procedures (SOPs) of Cell Transplants Singapore Pte Ltd (CTS) with US patent no. 5,130,141 and Singapore patent no. 34490 (WO 96/18303) licenses. Cell production was in compliance with current good manufacturing practice (cGMP) and international organization for standardization (ISO) standard 9001 conditions. Around 2g of muscle biopsy was taken under local anesthesia from a 20-year-old, pathogen-free, male volunteer after he had met muscle donor criteria. Initial dissociation isolated approximately 10,000 satellite cells, which were then cultured accordingly to CTS's SOPs.

The culture yielded 47.4×10^9 myoblasts that were 100% pure by positive desmin immunostain, and 92.8% viable according to vital dye exclusion tests. The cells were potent in myogenecity in that numerous myotubes were observed within four days in a fusion medium. Throughout the culture and for the final injectates, the myoblasts were free of endotoxin (<1.0EU/ml) and mycoplasma, and negative for sterility (14-day test) and gram stain (absence of gram positive or negative bacteria) according to certified laboratory analyses outside CTS.

Patient 1 is 42 years old, 157cm tall, and weighs 68kg. Patient 2 is 36 years old, 158cm tall, and also weighs 68kg. Diagnosis showed about two years history of Type II diabetes and hypertension but otherwise normal in heart, lung, kidney, and liver function without obesity. The laboratory report revealed tests results for syphilis, hepatitis B surface antigens, antibodies to HIV and hepatitis C virus to be negative.

Both patients had previously been enrolled as clinical trial subjects after qualifying for inclusion/exclusion criteria, and signing patients' informed consents with institutional approval. The subjects took two oral doses of cyclosporine totalling 5–7mg/kg body weight per day, beginning two days before grafting, weaning at half-dosage in the last two weeks, and off cyclosporine at eight weeks after grafting. The whole blood trough level of cyclosporine was monitored every two weeks. Doses were adjusted in an attempt to maintain the level at about 250ng/ml.

Myoblasts were harvested and processed under biological safety cabinets (Class 100) inside a cleanroom. Having been washed thoroughly, they were suspended in the injection medium. Quality assurance/quality control processes ensued, and the final quality control release testforms were issued for each of the two subjects. The myoblasts were then carried in syringes within sterile enclosures into two surgical suites for simultaneous implantation into both subjects. These were the world's first cases of allogeneic myoblasts being injected into Type II diabetic patients.

The patients received 132 injections each and 24/23.4 billion myoblasts, respectively.The two hour procedure was performed with the patients under general anesthesia. Cells were injected at $50x10^6$/ml and the injections were made under direct vision into 54 major muscle groups of each subject. The patients were transferred to the intensive care unit, where they recovered from the anesthesia and routine monitors were administered.

The subjects recovered from the general anesthesia without rash or fever, and both patients were discharged at 48 hours post-operatively.

Most pertinent to the specific goal of this study was that, despite cyclosporine discontinuation at two months postoperatively, no sign of rejection was observed. The patients appear to have good general health before and after MTT. Plasma glucose and insulin levels data were collected.

This pioneering feasibility/safety study of myoblast allografts into the skeletal muscles of Type II diabetic patients leads the way in developing a genetic treatment for the disease [360]. The immunosuppressant cyclosporine used in this study is known to increase plasma glucose. The subjects were weaned off cyclosporine. They will be further monitored to determine the preliminary efficacy, if any, of the 24-billion PMGT. Meanwhile, the procedure appears to be safe for both subjects.

FUTURE PERSPECTIVES

Essence of Cell Therapy

The cell is the basic unit of all lives. It is that infinitely small entity which life is made of. With the immense wisdom and knowledge of the human race, we have not been able to produce a living cell from nonliving ingredients such as DNA, ions, and biochemicals.

Cell Culture is the only method known to man for the replication of cells in vitro. With proper techniques and precautions, normal cells can be cultured in sufficient quantity to repair, and to replenish degenerates and wounds [361].

Cell transplantation bridges the gap between in vitro and in vivo systems, and allows propagation of "new life" in degenerative tissues or organs of the living yet genetically defective or injured body.

Cell fusion transfers all the normal genes within the nucleus like delivering a repair kit to the abnormal cell. It is important to recognize that, for proper installation and future operation, the software packaged in the chromosomes needs other cell organelles as hardware to operate.

Correction of a gene defect occurs spontaneously at the cellular level after cell fusion. The natural integration, regulation and expression of the full complement of over 30,000 normal genes impart the normal phenotypes onto the heterokaryon. Protein(s) or factor(s) that were not produced by the host genome because of the genetic defect are now produced by the donor genome that is normal. Various cofactors derived from expression of the other genes corroborate to restore the normal phenotype.

Heart Muscle Genes

The development of CardioChip [347] allows early diagnosis of cardiovascular diseases using 10,368 expressed sequence tags (ESTs). In collaboration with Prof. C C Liew of Harvard University, heart muscle genes can be identified to provide the template for early diagnosis (Figures 26, 27). Subjects so identified can have muscle biopsy taken before any symptom occurs. Myoblasts can be processed and deposited in a cell bank for future transduced HMGT to prevent sudden heart attack.

The Ultimate Gene Transfer

The use of myoblasts as gene transfer vehicles dates as early as 1978.1 In mammals, myoblasts is the only cell type which divides extensively, migrates, fuses naturally to form syncytia, loses MHC-1 antigens soon after fusion, and develops to occupy 50% of the body weight in humans. These combined properties render myoblasts ideal for gene transfer.

Natural transduction of normal nuclei ensures orderly replacement of dystrophin and related proteins at the cellular level in DMD. This ideal gene transfer procedure is unique to muscle. After all, only myoblasts can fuse and only muscle fibers are multinucleated in the human body. By harnessing these intrinsic properties, MTT transfers all normal genes to achieve genetic repair. Since donor myoblasts also fuse among themselves to form normal fibers in MTT, the muscles benefit from the addition of genetically normal cells as well.

This differs significantly from the gene therapy format in which single copies of the down-sized dystrophin gene are transduced as viral conjugates into the mature dystrophic myofibers in which many proteins, both structural and regulatory, are lost. Multiple gene insertion is necessary to produce the lost proteins for the development of a cure or a treatment. More gene insertion is needed to produce the cofactors to regulate and to express these lost proteins in order to repair the degenerating cell.

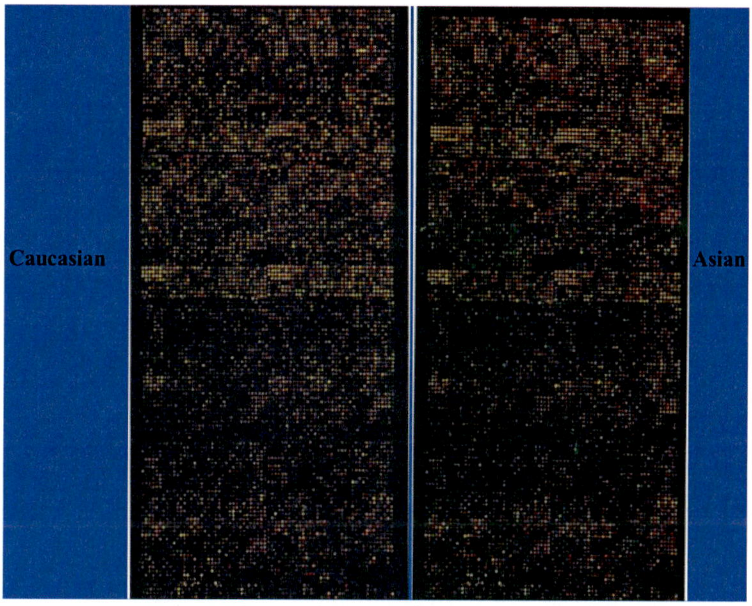

Figure 26. CardioChip analyses of Caucacian and Asian myoblast ESTs to determine 1) if there are any contaminants, 2) the consistency of myoblast production by the American team vs the Singapore team, 3) the myogenic genome for the heart.

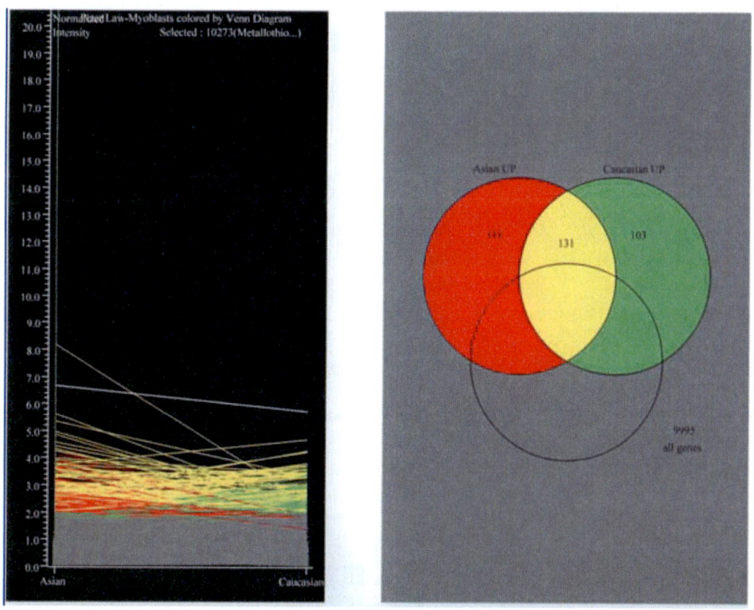

Figure 27. CardioChip analyses of Caucacian and Asian myoblasts identified 78 ESTs related to heart muscle in its early development. These ESTs can be used as a template for early diagnosis of heart muscle diseases.

Controlled Cell Fusion

It will be useful to be able to control, initiate or facilitate cell fusion once myoblasts are injected. Myoblasts fuse readily at low serum concentration in culture [362]. The process is more complex in the in vivo situation.

As the myoblasts are injected intramuscularly into the extracellular matrix (ECM), injection trauma causes the release of basic fibroblast growth factor (bFGF) and large chondroitin-6-sulfate proteoglycan (LC6SP) [363]. These latter growth factors stimulate myoblast proliferation. Unfortunately, they also stimulate the proliferation of fibroblasts that are already present in increased amount in the dystrophic or the infarcted muscles. That is why it is necessary to inject as pure as possible fractions of myoblasts in MTT without contaminating fibroblasts.

Controlled cell fusion can be achieved by artificially increasing the concentration of LC6SP over the endogenous level. In addition, insulin or insulin-like growth factor I (IGF-1) [364] may facilitate the developmental process, resulting in the formation of myotubes soon after myoblast injection. The use of bFGF, LC6SP and IGF-1 at optimal concentrations in the cell culture medium and in the injection medium will be examined.

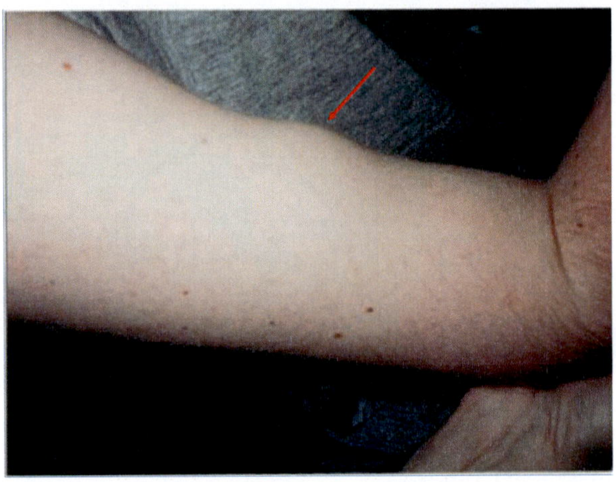

Figure 28. HMGT enhances the size, shape, consistency and strength of a biceps brachii muscle in the upper arm of a patient suffering Infantile Facioscapulohumeral Dystrophy.

Anti-Aging Cosmetic Enhancement

HMGT can significantly contribute to the field of gerontology and plastic surgery. With cell therapy, implantation of silicone could be avoided. The use of fat cells could be used in a much more natural way to replace silicone injections for facial, breast and hip augmentation. Myotubes are known to survive in adipose tissues [365]. Modified adipose tissue involving mixing and/or hybridization of myoblasts and fat cells can be used to control size, shape and consistency of body parts. Since muscle cells do not break down as easily as fat cells, good results may be long-lasting. Today, body builders are in search of increasing muscle mass at the right places. The use of myoblast transfer to boost muscle mass is a natural solution.

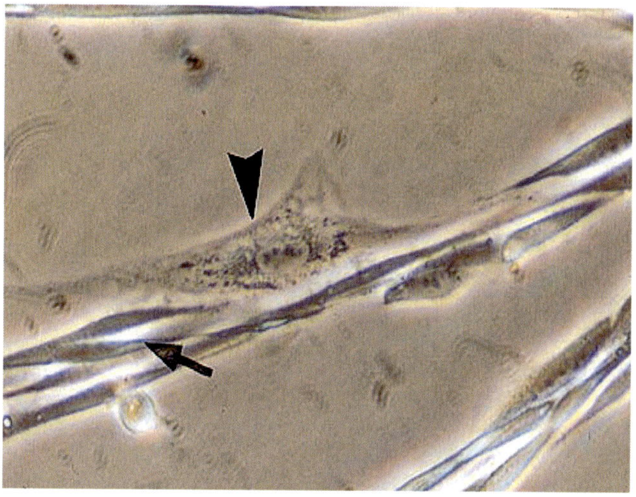

Figure 29. Myoblasts are small, spindle-shaped cells that can survive and develop in intercellular fluid without blood supply. Fibroblasts are coarse, polygonal cells that have shorter life.

By now, the survival, development and functioning of donor myoblasts in host muscles, either normal or scarred, have been repeatedly demonstrated by numerous laboratories. HMGT can thus modulate the size, shape, consistency and strength of all external body parts (Figure 28). Since myoblasts are small (8 to 15 μm)/ spindle-shaped / durable cells when compared to the relatively large/ polygonal/ rough skin fibroblasts (Figure 29), myoblasts can be "plated" over diseased /degenerative/old skins to cover wrinkles/blemishes/spot (Figure 30). The angiomyogenic HMGT will not only treat heart infarcts, but should also have application to sexual impotency, baldness, red lips and pink faces by providing more muscles, blood supplies and nutrients. This on-going development promises anti-aging aesthetica for health and beauty, and shall highly increase the quality of life of mankind.

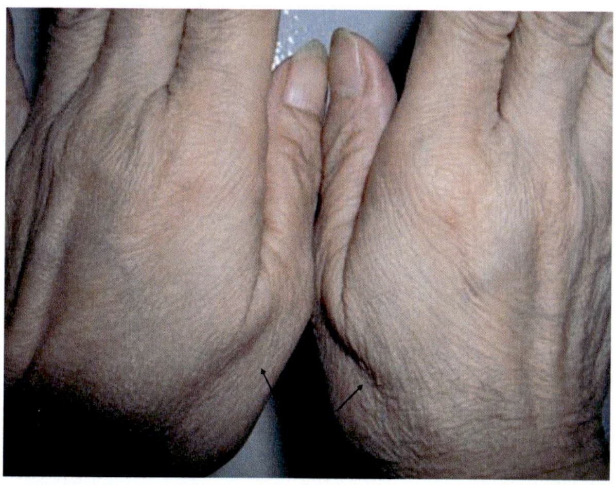

Figure 30. The left hand of a 60-yr-old female worker appeared "younger" with lighter tone and less wrinkles 2 weeks after external application of one billion allogeneic myoblasts without immunosuppression. The right hand received equal volume and number of application of the carrier solution.

Bone Degeneration

During embryonic development, mesenchymal progenitor cells differentiate into myoblasts, osteoblasts, chondrocytes, adipocytes, fibroblasts and cardiomyocytes under controls of various regulatory factors. Ectopic bone formation in muscle has been achieved through implantation of bone morphogenetic protein (BMP). BMP-2 was shown to convert the differentiation pathway of clonal myoblasts into the osteoblast lineage [245]. Our combined use of BMP-2, myoblasts and hydroxyapetite produced bone formation to fill a small hole drilled on a rabbit skull (Figure 31). This opens new ways to treat conditions of bone degeneration such as the degeneration of tooth pulp, hip, bone/joint, and long bone fractures. Given the ability to mass-produce myoblasts that can be transformed into osteoblasts, and potentially chondrocytes, the difficulty of proliferating osteoblasts and chondrocytes can be overcome. Cultured autologous chondrocytes can be used to repair deep cartilage defects in the femorotibial articular surface of the human knee joint [218].

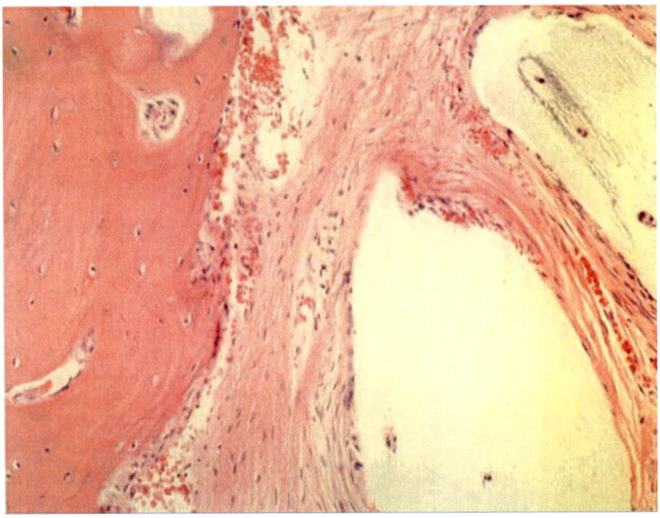

Figure 31. A combination of human myoblasts, BMP-2 and hydoxyappetite produced bone-like filling for a minute hole drilled on the skull of a rabbit (left). Without myoblasts, only fibroblasts infiltrated a control hole filled with BMP-2 and hydoxyappetite.

Myoblast: The Universal Gene Transfer Vehicles

Whereas MTT results in the formation of genetic mosaicism with gene transfer occurring in vivo, the production of heterokaryons in vitro has immense medical application. This can be achieved by controlled cell fusion with myoblasts.

This original program of research relates to the in culturo transfer of the normal nuclei with all of their normal genes from donor myoblasts into the genetically normal and/or abnormal cells, e.g. the cardiomyocytes. This development is especially important considering that cardiomyopathic symptoms develop in mid adolescence in about 10% of the DMD population [20]. By age 18, all DMD individuals develop caridomyopathy [21].

Undoubtedly, the ability to replenish degenerated and degenerating cardiomyocytes will have immense impact on heart diseases even in the normal population [366] where there is a great shortage of hearts for transplantation.

Normal cardiomyocytes exhibit very limited ability to proliferate in vivo or in vitro. The heart muscles damaged in heart attacks or in hereditary cardiomyopathy cannot repair themselves through regeneration. The integration of the skeletal muscle cell characteristic, mitosis, will enable the heterkaryotic cardiomyocytes to proliferate in vitro.

Controlled cell fusion between normal myoblasts and normal cardiomyocytes may result in heterokaryons exhibiting the characteristics of both parental myogenic cell types. Clones can be selected based on their abilities to undergo mitosis in vitro, to develop desmosomes, gap junctions, and to contract strongly in synchrony after cell transplantation.

These genetically superior cells can then be delivered through catheter pathways after mapping of the injured sites. With the ability to grow large quantity of these cardiomyocytes, the correction of structural, electrical and contractile abnormalities in cardiomyopathy can be tested first in dystrophic, cardiomyopathic hamsters [367,368] and, if safe and effective, in humans.

The genetic transfer of the mitotic property of myoblasts onto cardiomyocytes with in vitro controlled cell fusion enables the resulting heterokaryotic cardiomyocytes to multiply, yielding enough number of cells for the cell transplant to be effective.

The use of myoblasts as gene transfer vehicles does not stop with myogenic cells. Myoblasts have been genetically modified to provide systemic delivery of human growth hormone [120,121] and may in time provide long-term secretion of insulin for diabetes and various hormones and/or factors. Myoblasts tranduced to secrete opiode derivatives may be used to treat pain, depression and addictions.

Automated Cell Processors

The great demand for normal myoblasts, myotubes and young muscles, the labor intensiveness and high cost of cell culturing, harvesting and packaging, and the fallibility of human imprecision will soon necessitate the invention and development of automated cell processors capable of producing huge quantities of viable, sterile, genetically well-defined and functionally demonstrated biologics, an example of which are the myogenic cells.

This invention will be one of the most important offspring of modern day computer science, mechanical engineering and cytogenetics (Figure 32). The intakes will be for biopsies of various human tissues. The computer will be programmed to process tissue(s), with precision controls in time, space, proportions of culture ingredients and apparatus maneuvers. Cell conditions can be monitored at any time during the process and flexibility is built-in to allow changes. Different protocols can be programmed into the software for culture, controlled cell fusion, harvest and package. The outputs supply injectable cells ready for cell therapy or shipment. The cell processor will be self-contained in a sterile enclosure large enough to house the hardware in which cells are cultured and manipulated. Various cell types can be manufactured depending on the software used.

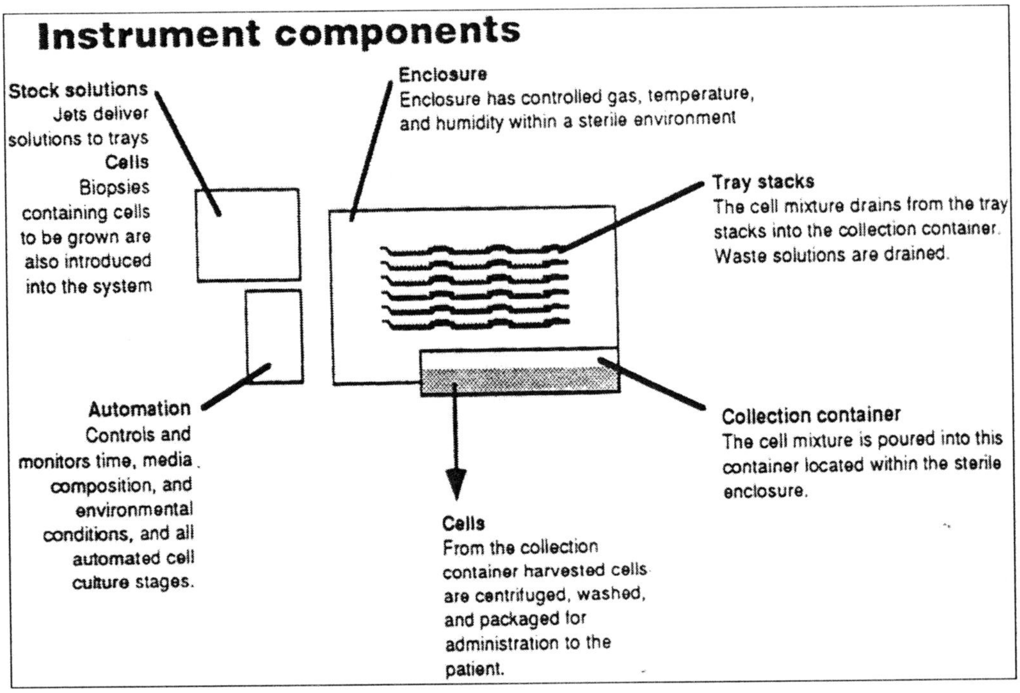

Figure 32. A basic design for the automated cell processor.

The automated cell processor will replace the current bulky inefficient culture equipment, elaborate manpower, and their mistakes. Its installation into clinics, hospitals and medical centers will allow customized medicine for individual patients by providing autologous HMGT. Fresh cells will be nearby. Cell costs will be reduced. Its contribution to human health is highly significant.

Cell Therapy is Now

Health is the well-being of all body cells. In hereditary degenerative diseases, sick cells need repairing and dead cells need replacing for health maintenance.

Cell culture is the way to generate new, live cells that are capable of surviving, developing and functioning in the body after transplantation, replacing degenerated cells that are lost.

Myoblasts are the only cells in the body capable of natural cell fusion. The latter allows the transfer of all of the normal genes into genetically defective cells to achieve phenotypic repair through complementation. HMGT/MTT on DMD is the first human gene therapy demonstrated to be safe and effective. The use of HMGT/MTT to transfer any other genes and their promoters/enhancers to treat other forms of diseases is underway. Endogenous to the human body, myoblasts will prove to be efficient, safe and universal gene transfer vehicles. Since a foreign gene always exerts its effect on a cell, cell therapy will always be the common pathway to health. After all, cells are what life is made of.

REFERENCES

[1] O'Rahilly R, Muller F. *Developmental stages in human embryos*. Meriden: Meriden-Stinehour Press, 1987:1-306.

[2] Jacobson M. *Developmental neurobiology*. New York: Holt, Rinehart & Winston Inc, 1970:1-465.

[3] Banker BQ, Przybylski RJ, Van der Meulen JP, Victor M, eds. *Research in muscle development and the muscle spindle*. Princeton: Excerpta Medica, 1971:1-226.

[4] Pearson ML, Epstein HF, eds. *Muscle development. Molecular and cellular control*. New York: Cold Spring Harbor Laboratory, 1982:339-96, 509-67.

[5] Davidson EH. *Gene activity in early development*. 2nd ed. New York: Academic Press, 1976:175-9.

[6] McComas AJ. Neuromuscular function and disorders. Boston: Butterworths, 1977:1-364.

[7] Hoffman, EP, Brown RH, Kunkel LM. Dystrophin: the protein product of the Duchenne muscular dystrophy locus. *Cell* 1987; 51:919-28.

[8] Monaco AP, Neve RL, Colletti-Feener C et al. Isolation of candidate cDNAs for portions of the Duchenne muscular dystrophy gene. *Nature* 1986; 323:646-50.

[9] Koenig M. Hoffman EP, Bertelson CJ et al. Complete cloning of the Duchenne muscular dystrophy (DMD) cDNA and preliminary genomic organization of the DMD gene in normal and affected individuals. *Cell* 1987; 50:509-17.

[10] Brooke MH, Fenichel GM, Griggs RC, et al. Clinical investigation of Duchenne muscular dystrophy: interesting results in a trial of prednisone. *Arch Neurol* 1987; 44:812-7.

[11] Brooke MH, Fenichel GM, Griggs RC, et al. Clinical investigation in Duchenne dystrophy: 2: Determination of the "power" of therapeutic trials based on the natural history. *Muscle Nerve* 1983; 6:91-103.

[12] Fenichel GM, Florence JM, Pestronk A, et al. Long-term benefit from prednisone therapy in Duchenne muscular dystrophy. *Neurology* 1991; 41:1874-7.

[13] Fenichel GM, Mendell JR, Moxley RT, et al. A comparison of daily and alternate-day prednisone therapy in the treatment of Duchenne muscular dystrophy. *Arch Neur* 1991; 48: 575-9.

[14] Mendell JR, Moxley RT, Griggs RC, et al. Randomized, double-blind six-month trial of prednisone in Duchenne's muscular dystrophy. *N Eng J Med* 1989; 320:1592-7.

[15] Stern LM, Fewings JD, Bretag AH, et al. The progression of Duchenne muscular dystrophy: clinical trial of allopurinol therapy. *Neurology* 1981; 31:422-6.

[16] Ziter FA, Allsop KG, Tyler FH. Assessment of muscle strength in Duchenne muscular dystrophy. *Neurology* 1977; 27:981-4.

[17] Walton JN, Gardner-Medwin D. Progressive muscular dystrophy and the myotonic disorders. In: Walton JN, Gardner-Medwin D, eds. *Disorders of voluntary muscle*. New York: Churchill Livingstone, 1981:487.

[18] Emery AE. Population frequencies of inherited neuromuscular diseases – a world survey. *Neuromuscular Disorders* 1991; 1:19.

[19] Iannaccone ST. Current status of Duchenne muscular dystrophy. *Ped Neurol* 1992; 39:879-94.

[20] Engel AG. Duchenne dystrophy. In: Banker BQ, Engel AG, eds. *Myology*. New York: McGraw Hill, 1986:1185.

[21] Nigro G, Comi LI, Politano L, et al. Treatment of cardiac involvement in late stage of Duchenne muscular dystrophy. *Acta Cardiomiologica* 1990; 2:13-24.

[22] Michelson AM, Russell ES, Harman PJ. Dystrophia muscularis: a hereditary primary myopathy in the house mouse. *Proc Nat Acad Sci* 1955; 41:1079-84.

[23] Meier H, West WT, Hoag WG. Preclinical histopathology of mouse muscular dystrophy. *Arch Path* 1965; 80:165-70.

[24] Law PK, Atwood HL. Nonequivalence of surgical and natural denervation in dystrophic mouse muscles. *Exp Neurol* 1972; 34:200-9.

[25] Law PK, Atwood HL, McComas AJ. Functional denervation in the soleus muscle of dystrophic mice. *Exp Neurol* 1976; 51:434-43.

[26] Mokri B, Engel AG. Duchenne dystrophy: electron microscopic findings pointing to a basic or early abnormality in the plasma membrane of the muscle fiber. *Neurology* 1975; 25:1111.

[27] Law PK, Saito A, Fleischer S. Ultrastructural changes in muscle and motor end-plate of the dystrophic mouse. *Exp Neurol* 1983; 80:361-82.

[28] Law PK, Cosmos E, Butler J, McComas AJ. The absence of dystrophic characteristics in normal muscles successfully cross-reinnervated by nerves of dystrophic genotype: physiological and cytochemical study of crosssed solei of normal and dystrophic parabiotic mice. *Exp Neurol* 1976; 51: 1-21.

[29] Law PK. "Myotrophic" influences on notoneurones of normal and dystrophic mice in parabiosis. *Exp Neurol* 1977; 54:444-52.

[30] Saito A, Law PK, Fleischer S. Study of neurotrophism with ultrastructure of normal/dystrophic parabiotic mice. *Muscle Nerve* 1983; 6:14-28.

[31] Carlson BM. A quantitative study of muscle fiber survival and regeneration in normal, predenervated, and Marcaine-treated free muscle grafts in the rat. *Exp Neurol* 1976; 52:421-32.

[32] Hakelius L, Nystrom B, Stalberg E. Histochemical and neurophysiological studies of autotransplanted cat muscle. *Scand J Plast Reconstr Surg* 1975;9:15- 24.

[33] Hall-Craggs ECB, Brand P. Effect of previous nerve injury on the regeneration of free autogenous muscle grafts. *Exp Neurol* 1977; 57:275-81.

[34] Mufti SA, Carlson BM, Maxwell LC, Faulkner JA. The free autografting of entire limb muscles in the cat: morphology. *Anat Rec* 1977; 188:417-30.

[35] Thompson N. A review of autogenous skeletal muscle grafts and their clinical applications. *Clin Plast Surg* 1974; 1:349-403.

[36] Studitsky AN. The restoration of muscle by means of transplantation of minced muscle tissue. *Dokl Akad Nauk SSSR* 1952; 84:389-92.

[37] Cosmos E. Muscle transplants: role in the etiology of herditary muscular dystrophy. In: Milhoart AP, ed. *Exploratory concepts in muscular dystrophy II*. Amsterdam: Excerpta Medica 1974:368-73.

[38] Carlson BM. The regeneration of skeletal muscle – a review. *Am J Anat* 1973; 137:119-50.

[39] Carlson BM. Regeneration of the completely excised gastrocnemius muscle in the frog and rat from minced muscle fragments. *J Morphol* 1968; 125:447-72.

[40] Allbrook D, Baker W de C, Kirkaldy-Willis WH. Muscle regeneration in experimental animals and in man. The cycle of tissue change that follows trauma in the injured limb syndrome. *J Bone Joint Surg* 1966; 48:153-69.

[41] Snow MH. Myogenic cell formation in regenerating rat skeletal muscle injured by mincing. I. A fine structural study. *Anat Rec* 1977; 188:181-200.

[42] Snow MH. Myogenic cell formation in regenerating rat skeletal muscle injured by mincing. II. An autoradiographic study. *Anat Rec* 1977; 188:201-18.

[43] Susheela AK, Hudgson P, Walton JN. Histological and histochemical studies of experimentally-induced degeneration and regeneration in normal and dystrophic mouse muscle. *J Neurol* 1969; 9:423-42.

[44] Mauro A. Satellite cells of skeletal muscle fibers. *J Biophys Biochem Cytol* 1961; 9:493-95.

[45] Carlson M, Hansen-Smith FM, Magon DK. The life history of a free muscle graft. In: Mauro A, ed. *Muscle Regeneration*. New York: Raven Press 1979:493-507.

[46] Cardasis CA, Cooper GW. A method for the chemical isolation of individual fibers and its application to a study of the effect of denervation on the number of nuclei per fiber. *J Exp Zool* 1975; 191:333-58.

[47] Burch T, Law PK. Normal development of muscle fibers and motor endplates in dystrophic mice. *Exp. Neurol.* *58*:570–574, 1978

[48] Rowe RWD, Goldspink G. Muscle fiber growth in five different muscles in both sexes of mice. I. Normal mice. *J Anat* 1969; 104:519-30.

[49] Rowe RWD, Goldspink G. Muscle fiber growth in five different muscles in both sexes of mice. II. Dystrophic mice. *J Anat* 1969; 104:531-8.

[50] Law PK. Reduced regenerative capability of minced dystrophic mouse muscles. *Exp Neurol* 1978; 60:231-43.

[51] Mastaglia FL, Kakulas BA. Regeneration in Duchenne muscular dystrophy: a histological and histochemical study. *Brain* 1969; 92:809-18.

[52] Carlson BM, Gutmann E. Development of contractile properities of minced muscle regenerates in the rat. *Exp Neurol* 1972; 36:239-49.

[53] Church JCT. Cell populations in skeletal muscles after regeneration. *J Embryol Exp Morphol* 1970; 23:531-7.

[54] Vracko R, Benditt EP. Basal lamina: the scaffold for orderly cell replacement. *J Cell Biol* 1972; 55:406-19.

[55] Mufti SA. Regeneration following denervation of minced gastrocnemius muscle in mice. *J Neurol Sci* 1977; 33:251-66.

[56] Wakayama Y. Electron microscopic study on the satellite cells in the muscle of Duchenne muscular dystrophy. *J Neuropathol Exp Neurol* 1976; 35:532-40.

[57] Cosmos E. Muscle-nerve transplants. Experimental models to study influences on differentiation. *Physiologist* 1973; 16:167-77.

[58] Gallup B, Dubowitz V. Failure of "dystrophic" neurones to support functional regeneration of normal or dystrophic muscle in culture. *Nature* 1973; 243:287-9.

[59] Hironaka T, Miyata Y. Transplantation of skeletal muscle in normal and dystrophic mice. *Exp Neurol* 1975; 47:1-15.

[60] Laird JL, Timmer RF. Homotransplantation of dystrophic and normal muscle. *Arch Pathol* 1965; 80:442-6.

[61] Salafsky B. Functional studies of regenerated muscles from normal and dystrophic mice. *Nature* 1971; 229:270-2.

[62] Guth L, Yellin H. The dynamic nature of the so-called "fiber types" of mammalian skeletal muscle. *Exp Neurol* 1971; 31:277-300.

[63] Gutmann E, Schiaffino S, Hanzlikova V. Mechanism of compensatory hypertrophy in skeletal muscle of the rat. *Exp Neurol* 1971: 31:451-64.

[64] Ianuzzo CD, Gollnick PD, Armstrong RB. Compensatory adaptations of skeletal muscle fiber types to a long-term functional overload. *Life Sci* 1976; 19:1517-24.

[65] Vrbova G. The effect of motoneurone activity on the speed of contraction of striated muscle. *J Physiol (Lond)* 1963; 169:513-26.

[66] Bergel DH, Brown MC, Butler RG, Zacks RM. The effect of stretching a contracting muscle on its subsequent performance during shortening. *J Physiol (Lond)* 225:21-2P.

[67] Goldspink G, Tabary C, Tabary JC, et al. Effect of denervation on the adaptation of sarcomere number and muscle extensibility to the functional length of the muscle. *J Physiol (Lond)* 1974; 236:733-42.

[68] Tomanck RJ. A histochemical study of postnatal differentiation of skeletal muscle with reference to functional overload. *Dev Biol* 1974; 42:305-14.

[69] Yellin H. Changes in fiber types of the hypertrophying denervated hemidiaphragm. *Exp Neurol* 1974; 42:412-28.

[70] Eisen A, Karpati G, Carpenter S, Danon J. The motor unit profile of the rat soleus in experimental myopathy and reinnervation. *Neurology* 1974; 24:878-84.

[71] Liberman AR. The axon reaction: a review of the principal features of perikaryal responses to axon injury. *Int Rev Neurobiol* 1971; 14:49-124.

[72] Mark RF. Matching muscles and motoneurones. A review of some experiments on motor nerve regeneration. *Brain Res* 1969; 14:245-54.

[73] Law PK, Caccia MR. Physiological estimates of the sizes and numbers of motor units in soleus muscles of dystrophic mice. *J Neurol Sci* 1975; 24:251-6.

[74] Anderson, CW (ed). *Genetic Interaction and Gene Transfer*. New York: Brookhaven National Lab, 1978.

[75] Constantini F, Jacnisch R (eds). *Genetic Manipulation of the Early Mammalian Embryo Banbury Report 20*. New York: Cold Spring Harbor Lab, 1985.

[76] Friedmann T. *Gene Therapy – Fact and Fiction in Biology's New Approaches to Disease*. New York: Cold Spring Harbor Lab, 1983.

[77] Hogan, BLM, Constantini F, Lacy E (eds). *Manipulating the Mouse Embryo*. New York: Cold Spring Harbor Lab, 1986.

[78] Orkin SH. Reverse genetics and human diseases. *Cell* 1986; 47:845-50.

[79] Worton RG, Duff C, Sylvester TE, et al. Duchenne muscular dystrophy involving translocation of the DMD gene next to ribosomal RNA genes. *Science* 1984; 224:1447-1449.

[80] Kunkel LM, Hoffman EP. Duchenne/Becker muscular dystrophy: an overview of the gene, the protein and current diagnostics. *Med Bull* 1989; 45:630-43.

[81] Harley HG, Brook JD, Rundle SA, et al. Expansion of an unstable DNA region and phenotypic variation in myotonic dystrophy. *Nature* 1992; 355:545-7.

[82] Buxton J, Shelbourne P, Davies J, et al. Detection of an unstable fragment of DNA specific to individuals with myotonic dystrophy. *Nature* 1992; 355:547-8.

[83] Aslandis C, Jansen G, Amemiya C, et al. Cloning of the essential myotonic dystrophy region and mapping of the putative defect. *Nature* 1992; 355:548-51.

[84] Wijmenga C, Frantz R, Brouwer OF, et al. Location of facioscapulohumeral muscular dystrophy gene on chromosome 4. *Lancet* 1990; 336:651-3.

[85] Beckmann J, Richard I, Hillaire D, et al. *A gene for limb-girdle muscular dystrophy maps to chromosome 15 by linkage*. Comptes Rendus de l'Academie des Sciences (Paris) 1991; 312 series III:141-8.

[86] Speer MC, Yamaoka LH, Gilchrist JM, et al. *Localisation of an autosomal dominant form of limb-girdle muscular dystrophy to chromosome 5q*. Human genome mapping 1991; 11:abstract 26929.

[87] Romeo G, Roncuzzi L, Sangiorgi S, et al. Mapping of the Emery-Dreifuss gene through reconstruction of crossover points in two Italian pedigrees. *Hum Genet* 1988; 80:59-62.

[88] Specht LA, Beggs AH, Korf B, et al. Prediction of dystrophin phenotype by DNA analysis in Duchenne/Becker muscular dystrophy. *Ped Neurol* 1992; 8:432-6.

[89] Ahn AH, Kunkel LM. The structural and functional diversity of dystrophin. *Nature* Genetics 1993; 3:283-91.

[90] Koenig M, Monaco AP, Kunkel LM. The complete sequence of dystophin predicts a rod-shaped cytoskeletal protein. *Cell* 1988; 53:219-26.

[91] Bushby KMD. Recent advances in understanding muscular dystrophy. *Arch Dis Childhood* 1992; 67:1310-2.

[92] Mongini T, Palmucci L, Doriguzzi C, et al. Absence of dystrophin in two patients with Becker type Xp21 muscular dystrophy. *Neurosci Letters* 1992; 147:37-40.

[93] Hoffman EP, Fischbeck KH, Brown RH, et al. Characterization of dystrophin in muscle biopsy specimens from patients with Duchenne's or Becker's muscular dystrophy. *New Engl J Med* 1988; 318:1363-8.

[94] Hoffman EP. Myoblast transplantation: What's going on? *Cell Transplantation* 1993; 2:49-57.

[95] Chen M, Li HJ, Fang Q, et al. Dystrophin cytochemistry in mdx mouse muscles injected with labeled normal myoblasts. *Cell Transplantation* 1992; 1:17–22.

[96] Partridge TA, Morgan JE, Coulton GR, et al. Conversion of mdx myofibers from dystrophin-negative to -positive by injection of normal myoblasts. *Nature* 1989; 337:176–9.

[97] Karpati G, Pouliot Y, Zubrzycka-Gaarn, et al. Dystrophin is expressed in mdx skeletal muscle fibers after normal myoblast implantation. *Am J Pathol* 1989; 135:27-32.

[98] Law PK, Bertorini TE, Goodwin TG, et al. Dystrophin production induced by myoblast transfer therapy in Duchenne muscular dystrophy. *Lancet* 1990; 336:114–5.

[99] Law PK, Goodwin TG, Fang Q, et al. Myoblast transfer therapy for Duchenne muscular dystrophy. *Acta Paediatr Jpn* 1991; 33:206–15.

[100] Law PK, Goodwin TG, Fang Q, et al. Pioneering development of myoblast transfer therapy. In: Angelini C et al. (eds). *Muscular Dystrophy Research*. New York: Elsevier Science Publishers, 1991:109–16.

[101] Law PK, Goodwin TG, Fang Q, et al. Long-term improvement in muscle function, structure and biochemistry following myoblast transfer in DMD. *Acta Cardiomiologica* 1991; 1:281–301.

[102] Gussoni E, Pavlath GK, Lanctot AM, et al. Normal dystrophin transcripts detected in Duchenne muscular dystrophy patients after myoblast transplantation. *Nature* 1992; 356:435–8.

[103] Tremblay JP, Malouin F, Roy R, et al. Results of a triple blind clinical study of myoblast transplantations without immunosuppressive treatment in young boys with Duchenne muscular dystrophy. *Cell Transplantation* 1993; 2:99–112.

[104] Huard J, Bourchard JP, Roy R, et al. Myoblast transplantation produced dystrophin-positive muscle fibers in a 16-year-old patient with Duchenne muscular dystrophy. *Clin Sci* 1991; 81: 287–8.

[105] Huard J, Bouchard JP, Roy R, et al. Human myoblast transplantation: preliminary results of 4 cases. *Muscle Nerve*. 1992; 15:550-60.

[106] Acsadi G, Dickson G, Love DR, et al. Human dystrophin expression in mdx mice after intramuscular injection of DNA constructs. *Nature* 1991; 352:815–8.

[107] Wolff JA, Malone RW, Williams P, et al. Direct gene transfer into mouse muscle in vivo. *Science* 1991; 247:1465–8.

[108] Lee CC, Pons F, Jones PG, et al. Mdx transgenic mouse – restoration of recombinant dystrophin to the dystrophic muscle. *Human Gene Therapy* 1993; 4:273-81.

[109] Dunckley MG, Wells DJ, Walsh FS, et al. Direct retroviral-mediated transfer of a dystrophin minigene into mdx mouse muscle in vivo. *Human Molecular Genetics* 1993; 2:717-23.

[110] Ragot T, Vincent N, Chafey P. Efficient adenovirus-mediated transfer of a human minidystrophin gene to skeletal muscle of mdx mice. *Nature* 1993; 361:647-50.

[111] Cooper BJ. The xmd dog: molecular and phenotypic characteristics. In: Kakulas BA, Howell JM, Roses AD (eds). *Duchenne Muscular Dystrophy. Animal Models and Genetic manipulations*. New York: Raven Press, 1992:109-11.

[112] Hyde SC, Gill DR, Higgins CF, et al. Correction of the ion transport defect in cystic fibrosis transgenic mice by gene therapy. *Nature* 1993; 362:250-5.

[113] Cournoyer D, Caskey CT. Gene therapy of the immune system. *Annu Rev. Immunol.* 1993; 11:297-329.

[114] Jiao S, Williams P, Berg RK, et al. Direct gene transfer into nonhuman primate myofibers in vivo. *Human Gene Therapy* 1992; 3:21-33.

[115] Wolff JA, Dority ME, Jiao S, et al. Expression of naked plasmids by cultured myotubes and entry of plasmids into t tubules and caveolae of mammalian skeletal muscle. *J Cell* Sci 1992; 103:1249-59.

[116] Xu H, Miller J, Liang BT. High-efficiency gene transfer into cardiac myocytes in vivo. *Trends Cardiovasc Med* 1991; 1:271-6.

[117] Leinwand LA, Leiden JM. Gene transfer into cardiac myocytes in vivo. *Trends Cardiovasc Med* 1991; 1:271-6.

[118] Caskey CT, Rossiter BJF. Molecular medicine. 9th Ernst Klenk Lecture. *Biol Chem Hoppe-Seyler* 1992; 373:159-70.

[119] Smith BF, Hoffman RK, Giger URS, Wolfe JH. Genes transferred by retroviral vectors into normal and mutant myobalasts in primary cultures are expressed in myotubes. *Mol Cell Bio* 1990; 10:3268-71.

[120] Dhawan J, Pan LC, Pavlath GK, et al. Systemic delivery of human growth hormone by injection of genetically engineered myoblasts. *Science* 1991; 254:1509-12.

[121] Barr E, Leiden JM. Systemic delivery of recombinant proteins by genetically modified myoblasts. *Science* 1991; 254:1507-9.

[122] Culver KW, Osborne WR, Miller AD, et al. Correction of ADA deficiency in human T lymphocytes using retroviral-mediated gene transfer. *Transplant Proc* 1991; 23: 170-1.

[123] Dickson G, Love DR, Davies KE, et al. Human dystrophin gene transfer: production and expression of a functional recombinant DNA-based gene. *Hum Genet* 1991; 88:53-8.

[124] Law PK. A consideration of the theoretical strategies which may minimize, ameliorate or correct the defect in DMD. In: Kakulas BA, Masaglia FL (eds). *Pathogenesis and Therapy of Duchenne and Becker Muscular Dystrophy*. New York: Raven Press, 1991:190.

[125] Acsadi G, Jani A, Massie B, et al. A differential efficiency of adenovirus-mediated in vivo gene transfer into skeletal muscle cells of different maturity. *Hum Mol Gen* 1994; 3:579-84.

[126] Neumeyer AM, DiGregorio DM, Brown RH Jr. Arterial delivery of myoblasts to skeletal muscle. *Neurology* 1992; 42:2258-62.

[127] Watson JD, Gilman M, Witkowski J, et al (eds). *Recombinant DNA*. New York: WH Freeman and Co., 1992:576.

[128] Lee CC, Pearlman JA, Chamberlain JS, et al. Expression of recombinant dystrophin and its localization to the cell membrane. *Nature* 1991; 349:334-6.

[129] Campbell KP, Kahl SD. Association of dystrophin and an integral membrane glycoprotein. *Nature* 1989; 338:259-62.

[130] Ervasti JM, Ohlendieck K, Kahl SD, et al. Deficiency of a glycoprotein component of the dystrophin complex in dystrophic muscle. *Nature* 1990; 345:315-9.

[131] Ervasti JM, Campbell KP. Membrane organization of the dystrophin-glycoprotein complex. *Cell* 1991; 66:1121-31.

[132] Ibraghimov-Beskrovnaya O, Ervasti JM, Leveille CJ, et al. Primary structure of dystrophin-associated glycoproteins linking dystropin to the extracellular matrix. *Nature* 1992; 355:696-702.

[133] Ohlendieck K, Matsumura K, Ionasescu VV, et al. Duchenne muscular dystrophy: deficiency of dystrophin-associated proteins in the sarcolemma. *Neurology* 1993; 43:795-800.

[134] Matsumura K, Nonaka I, Campbell K. Abnormal expression of dystrophin-associated proteins in Fukuyama-type congenital muscular dystrophy. *Lancet* 1993; 341:521-2.

[135] Matsumura K, Ervasti JM, Ohlendieck K, et al. Association of dystrophin-related protein with dystrophin-associated proteins in mdx mouse muscle. *Nature* 1992; 360:588-91.

[136] Roberds SL, Ervasti JM, Anderson RD, et al. Disruption of the dystrophin-glycoprotein complex in the cardiomyopathic hamster. *J Biol Chem* 1993; 268:11496-9.

[137] Hasty P, Bradley A, Morris JH, et al. Muscle deficiency and neonatal death in mice with a targeted mutation in the myogenin gene. *Nature* 1993; 364:501-6.

[138] Nobeshima Y, Hanaoka K, Hayacaka M, et al. Myogenin gene disruption results in perinatal lethality because of severe muscle defect. *Nature* 1993; 364:532-5.

[139] Anderson WF. Human gene therapy. *Science* 1992; 256:808-13.

[140] Matsumura K, Tome FMS, Ionasescu V, et al. Deficiency of dystrophin-associated proteins in Duchenne muscular dystrophy patients lacking COOH-terminal domains of dystrophin. *J Clin Invest* 1993; 92:866-71.

[141] Law PK, TG Goodwin, Q Fang, et al. Feasibility, safety, and efficacy of myoblast transfer therapy on Duchenne muscular dystrophy boys. *Cell Transplant* 1992; 1:235–44.

[142] Gage F, Christen Y (eds). *Gene transfer and therapy in the nervous system*. New York: Springer-Verlag; 1992.

[143] Kakulas BA, Mastaglia FL, eds. *Pathogenesis and Therapy of Duchenne and Becker Muscular Dystrophy*. New York: Raven Press, 1990: 1-273.

[144] Griggs RC, Karpati G, eds. *Myoblast Transfer Therapy. Adv Exp Med Biol*. New York: Plenum Press, 1990; 280:1-316.

[145] Law PK, Goodwin TG, Wang MG. Normal myoblast injections provide genetic treatment for murine dystrophy. *Muscle Nerve* 1988; 11:525-33.

[146] Law PK, Goodwin TG, Li HJ. Histoincompatible myoblast injection improves muscle structure and function of dystrophic mice. *Transplant Proc* 1988; 20:1114-9.

[147] Law PK, Li JH, Goodwin TG, et al. Pathogenesis and treatment of hereditary muscular dystrophy. In: Kakulas, BA, Mastaglia FL, eds. *Pathogenesis and therapy of Duchenne and Becker muscular dystrophy*. New York: Raven Press, 1990: 101-18.

[148] Law PK, Goodwin TG, Li HJ, et al. Myoblast transfer improves muscle genetics/structure/function and normalizes the behavior and life-span of dystrophic mice. In: Griggs RC, Karpati G, eds. *Myoblast Transfer Therapy*. New York: Plenum Press, 1990: 75-87.

[149] Konigsberg IR. Clonal analysis of myogenesis. *Science* 1963; 140:1273-84.

[150] Bischoff R. Regeneration of single muscle fibers in vitro. *Anat Rec* 1975; 182:215-26.

[151] Hauschka SD, Linkhart TA, Clegg C, et al. Clonal studies of human and mouse muscle. In: Mauro A, ed. *Muscle Regeneration*. New York: Raven Press, 1979: 311-22.

[152] Lyon MF. Gene action in X-chromosomes of the mouse (M. musculus L). *Nature* 1961; 190:372-3.

[153] Peterson AC. Chimaera mouse study shows absence of disease in genetically dystrophic muscle. *Nature* 1974; 248:561-4.

[154] Peterson AC, Pena S. Relationship of genotype and in vitro contractility in mdg/mdg – +/+ "mosaic" myotubes. *Muscle Nerve* 1984; 7:194-203.

[155] McComas AJ, Sica REP, Campbell MJ. 'Sick' motoneurones. A unifying concept of muscle disease. *Lancet* 1971; 1:321-5.

[156] Sica REP, McComas AJ. The neural hypothesis of muscular dystrophy. A review of recent experimental evidence with particular reference to the Duchenne form. *Can J Neurol Sci* 1978; 5:189-97.

[157] Bradley WG, Jaros E. Involvement of peripheral and central nerves in murine dystrophy. *Ann NY Acad Sci* 1979; 317:132-42.

[158] Okada E, Mizukira V, Nakamura H. Dysmyelination in the sciatic nerves of dystrophic mice. *J Neurol Sci* 1976; 28:505-20.

[159] Meier H, Southard JL. Muscular dystrophy in the mouse caused by an allele at the dy locus. *Life Sci* 1970; 9:137-44.

[160] Bulfield G, Siller WG, Wight PAL, et al. X chromosome-linked muscular dystrophy (mdx) in the mouse. *Proc Natl Acad Sci USA* 1984; 81:1189-92.

[161] Dangain J, Vrbova G. Muscle development in mdx mutant mice. *Muscle Nerve* 1984; 7:700-4.

[162] Coulton GR, Morgan JE, Partridge TA, et al. The mdx mouse skeletal muscle myopathy: I, a histological, morphometric and biochemical investigation. *Neuropath Appl Neurobiol* 1988; 14:53-70.

[163] Coulton GR, Curtin NA, Morgan JE, et al. The mdx mouse skeletal muscle myopathy: II, contractile properties. *Neuropath Appl Neurobiol* 1988; 14:299-314.

[164] Cooper BJ, Winand NJ, Stedman H, et al. The homologue of the Duchenne locus is defective in X-linked muscular dystrophy of dogs. *Nature* 1988; 334:154-6.

[165] Kornegay JN, Tuler SM, Miller DM, et al. Muscular dystrophy in a litter of golden retriever dogs. *Muscle Nerve* 1988; 11:1056-64.

[166] Sugita H, Arahata K, Tsukahara T, et al. Immunohistochemistry of dystrophin in various neuromuscular diseases. In: Kakulas BA, Mastaglia FL, ed. *Pathogenesis and Therapy of Duchenne and Becker Muscular Dystrophy*. New York: Raven Press, 1990:33-44.

[167] Partridge TA, Sloper JC. A host contributions to the regeneration of muscle grafts. *J Neurol Sci* 1977; 33:425-35.

[168] Partridge TA, Grounds M, Sloper JC. Evidence of fusion between host and donor myoblasts in skeletal muscle grafts. *Nature* 1978; 273:306-8.

[169] Law PK, Yap JL. New muscle transplant method produces normal twitch tension in dystrophic muscle. *Muscle Nerve* 1979; 2:356-63.

[170] Jones PH. Implantation of cultured regenerate muscle cells into adult rat muscle. *Exp Neurol* 1979; 66:602-10.

[171] Law PK. Beneficial effects of transplanting normal limb-bud mesenchyme into dystrophic mouse muscle. *Muscle Nerve* 1982; 5:619-27.

[172] Ontell M. Muscular dystrophy and muscle regeneration. *Hum Path* 1986; 17:673-82.

[173] Watt DJ. Factors which affect the fusion of allogeneic muscle precursor cells in vivo. *Neuropath Appl Neurol* 1982; 8:135-47.

[174] Watt DJ, Lambert K, Morgan JE, et al. Incorporation of donor muscle precursor cells into an area of muscle regeneration in the host mouse. *J Neurol Sci.* 1982; 57:319-31.

[175] Watt DJ, Morgan JE, Partridge TA. Long-term survival of allografted muscle precursor cells following a limited period of treatment with cyclosporin A. *Clin Exp Immunol* 1984; 55:419-26.

[176] Morgan JE, Watt DJ, Sloper JC, et al. Partial correction of an inherited biochemical defect of skeletal muscle by grafts of normal muscle precursor cells. *J Neurol Sci.* 1988; 86:137-47.

[177] Morgan JE. Practical aspects of myoblast implantation: investigations on two inherited myopathies in animals. In: Griggs RC, Karpati G, eds. *Myoblast Transfer Therapy.* New York: Plenum Press, 1990:89-96.

[178] Kornegay JN, Prattis SM, Bogan DJ, et al. Results of myoblast transplantation in a canine model of muscle injury. In: Kakulas BA, Howell JMC, Roses AD, eds. Duchenne muscular dystrophy. *Animal models and genetic anipulation.* New York:Raven Press, 1992:203-12.

[179] Carter ND, Parr CW. Isoenzymes of phosphoglucose isomerase in mice. *Nature* 1967;216:511.

[180] Chapman VM, Whitten WK, Ruddle FH. Expression of aternal glucose phosphate isomerase-1 (Gpi-1) in preimplantation stages of mouse embryos. *Dev. Biol* 1971;26:153-8.

[181] Padua RA, Bulfield G, Peters J: Biochemical genetics of a new glucosephosphate isomerase allele (Gpi-1c) from wide mice. *Biochem Genet* 1978;16:127-43.

[182] Gearhart JD, Mintz B. Clonal origins of somites and their muscle derivations: evidence from allophenic mice. Dev Biol 1972; 29:27-37.

[183] Watt DJ, Morgan JE, Partridge TA. Use of mononuclear precursor cells to insert allogeneic genes into growing mouse muscles. *Muscle Nerve* 1984; 7:741-50.

[184] Frair PM, Peterson AC. The nuclear-cytoplasmic relationship in "mosaic" skeletal muscle fibers from mouse chimaeras. *Exp Cell Res* 1983; 145:167-78.

[185] Blau HM, Chiu CP, Pavlath G, et al. Muscle gene expression in heterokaryons. *Adv. Exp. Med. Biol.* 1985;182:231-47.

[186] Wang MG, Law PK. Activity levels and preparative electrophoresis of GPI-1BB and GPI-1CC in host and donor mouse solei. *Proc Intl. Union. Physiol. Sci.* 1986;16:14.

[187] Li HJ, Wang MG, Goodwin TG, et al. Purification of mouse muscle GPI-1BB and GPI-1CC for antibody production. *Soc Neurosci Abst* 1986; 12:1289.

[188] Li HJ, Chen M, Goodwin TG, et al. Immunocytochemical differentiation of glucosephosphate isomerase allotypes. *Soc Neurosci Abst* 1990;16:347.

[189] Fang Q, Chen M, Li HJ, et al. Vital marker for muscle nuclei in myoblast transfer *Can J Physiol Pharmacol* 1991;69:49-52.

[190] Friedlander M, Fischman DA. Immunological studies of the embryonic muscle cell surface. Antiserum to the pre-fusion myoblast. *J Cell Biol.*1979;81:193-214.

[191] Law PK, Goodwin TG, Fang Q, et al. Cell transplantation as an experimental treatment for Duchenne muscular dystrophy. *Cell Transplantation* 1993;2:485-505.

[192] Kahan BD, Bach JF, eds. Proceedings of the second international congress on cyclosporine. Therapeutic use in transplantation. *Transplant Proc* 1988; 20:1-1137.

[193] Starzl TE, Thomson AW, Todo SN, et al., eds. Proceedings of the first international congress on FK 506. *Transplant Proc* 1991; 23:2709-3380.

[194] Weller B, Massa R, Karpati G, et al. Glucocorticoids and imunosuppressants do not change the prevalence of necrosis and regeneration in mdx skeletal muscles. *Muscle Nerve* 1991; 14:771-4.

[195] Law, P.K., Li, H., Chen, M. et at. (1994). Myoblast injection methods regulates cell distribution and fusion. *Tranplant. Proc.* 26, 3417-3418.

[196] Law PK, Goodwin TG, Li HJ, et al. Plausible structural/functional/behavioral/ biochemical transformations following myoblast transfer therapy. In: Griggs R, Karpati G, eds. *Myoblast Transfer Therapy*. New York: Plenum Press, 1990:241-50.

[197] Hooper C. Duchenne therapy trials starting in U.S., Canada. *J NIH Res* 1990; 2:30.

[198] Law PK, Goodwin TG, Fang Q, et al. Myoblast transfer therapy for Duchenne muscular dystrophy. *Adv Clin Neurosci* 1992; 2:463-70.

[199] Roelofs RK, deArengo GS, Law PK, et al. Treatment of Duchenne muscular dystrophy and D,L-penicillamine: results of a double-blind trial. *Arch Neurol* 1979; 36:266-8.

[200] McComas AJ, Brandstater ME, Upton ARM, et al. Sick motor neurons and dystrophy: A reappraisal. In: Rowland LP, ed. *Pathogenesis of Human Muscular Dystrophies*. Amsterdam: Excerpta Medica, 1977:180-6.

[201] Thomson AW, ed. Sandimmune (Cyclosporin). *Mode of action and clinical application*. Boston: Kluwer, 1989:1-364.

[202] Kahan BD. Cyclosporine and transplantation. *Transplant Immunol Letter* 1988; 4:1-120.

[203] Hoffman EP, Watkins SC, Slayter HS, et al. Detection of a specific isoform of alpha-actinin with antisera directed against dystrophin. *J Cell Biol* 1989; 108:503-9.

[204] Bonilla E, Samitt CE, Miranda AF, et al. Duchenne muscular dystrophy: deficiency of dystrophin at the muscle cell surface. *Cell* 1988; 54: 447-52.

[205] Kolata G. Cell transplant found effective in muscle disease. Muscular dystrophy patient showed strength increase in first human test. *The New York Times, Sunday, June 3,* 1990; A1.

[206] Beardsley T. Profile:Gene doctor. W. French Anderson pioneers gene therapy. *Sci Am* 1990; 263:33-4.

[207] Anderson, WF: Editorial. September 14, 1990: The Beginning. *Hum Gene Ther* 1990; 1:371-2.

[208] Karpati G, Ajdukovic D, Arnold D, et al. Myoblast transfer in Duchenne muscular dystrophy. *Ann Neurol* 1993; 34:8-17.

[209] Karpati G. Myoblast transfer in Duchenne muscular dystrophy. A perspective. In: Angelini C, et al, eds. *Muscular Dystrophy Research*. New York: Elsevier Science Publishers, 1991:101-7.

[210] Miller RG, Pavlath G, Sharma K, et al. Myoblast implantation in Duchenne muscular dystrophy: the San Francisco study. *Neurology* 1992; 42:189-90.

[211] Caplan A, Carlson B, Faulkner J, et al. Skeletal muscle. In: Woo SLY, Buckwalter J, eds. *Injury and repair of the musculoskeletal soft tissues*. Park Ridge, IL: American Academy of Orthopaedic Surgeons, 1987:209-91.

[212] Aiameddine, H.S., Louboutin, J.P., Dehaupas, M. et al. (1994). Functional recovery induced by satellite cell grafts in irreversibly injured muscles. *Cell Transplantation 3*, 3-14.

[213] Albert, N., Tremblay, J.P. (1992). Evaluation of various gene transfection methods into human myoblast clones. *Transpl. Proc. 24*, 2784-2786.

[214] Alton, E.W.F.W. and Geddes D.M. (1994). Gene therapy for cystic fibrosis, a clinical perspective. *Gene Therapy 2*, 88-95.

[215] Anderson, W. F. (1995). Gene Therapy. *Scient. Amer. 273*, 96-98B.

[216] Appleyard, S. T., Dunn, M. 1., Dubowitz, V.et al. (1985). Increased expression of l-ILA ABC class I antigens by muscle fibres in Duchenne muscular dystrophy, inflammatory myopathy, and other neuromuscular disorders. *Lancet* 1, *36*1-363.

[217] Brenner, M.K. (1995). Human somatic gene therapy progress and problems. *Int. Med. 237*, 22~-239.

[218] Brittberg, M., Lindahl, A., Nilsson, A. et al. (1994). Treatment of deep cartilage defects in the knee with autologous chondrocyte transplantation. *N. Engl. J. Med. 331*, 889-895.

[219] Carlson, B.M. (1983). *The regeneration and transplantation of skeletal muscle*. In, Seil, F. ed. Nerve, organ, and tissue regeneration, Research perspectives. Academic Press, New York. pp. 431-454.

[220] Chang, P. L. (ed). (1994). *Somatic Gene Therapy*. CRC Press. New York.

[221] Chen, 5.5., Chien, C.H., Yu, H.S. (1988). Syndrome of deltoid and/or gluteal fibrotic contracture; an injection myopathy. *Acta Neurol. Scand. 78*, 167-176.

[222] Chiu, R.C.J., Zibaitis, A., and Kao, R.L. (1995). Cellular cardiomyoplasty, myocardial regeneration with satellite cell implantation. *Ann. Thorac. Surg. 60*, 12-18.

[223] Coolican, S.A., Samuel, D.S., Ewton, D.Z. et al., (1997). The MTT ogenic and myogenic actions of insulin-like growth factors utilize distinct signaling pathways. 1. *Biol. Chem. 272*, 6653-6662.

[224] Cornetta, K., Morgan, R.A., Anderson, W.F. (1991). Safety issues related to retrovirus-mediated gene transfer in humans. *Hum. Gene Ther. 2*, 5-14.

[225] Coutelle, C., Caplen, N., Hart, S. et al. (1994). *Towards gene therapy for cystic fibrosis*. In Dodge, J.A., Brock, J.H., and Widdicombe, J.H., eds. Cystic Fibrosis Current Topics. John Wiley and Sons. New York. 2, 33-54.

[226] Crystal, R.G., McElvaney, N.G., Chu, C.S. et al. (1994). Administration of an adenovirus containing the human CFTR c DNA to the respiratory tract of individuals with cystric fibrosis. *Nature Genet. 8*, 42-5 1.

[227] Culver, K. W. (1996). *Gene Therapy, A Primer for Physicians*. Mary Ann Liebert, Inc.. Larchmont.

[228] Curiel, D. T., Pilewski, J. M., Albelda, S.M. (1996). Gene therapy approaches for inherited and acquired lung diseases. *Am. J. Respir. Cell Mol.* Bol. *14*, 1-18.

[229] Daar, A. S., Fuggle, S. Y., Fabre, 1. W. et al. (1984). The detailed distribution of l-ILA-A, B, C, antigens in normal human organs. *Transplantation 38*, 287-298.

[230] Dai, Y., Roman, M., Naviaus, R.K. et al. (1992). Gene therapy via primary myoblasts, long term expression of factor IX protein following transplantation in vivo. *Proc. Natl. Acad. Sci. USA 89*, 10892-10895.

[231] Davis, H.L., Whalen, R.G., Demeneix, B.A. (1993). Direct gene transfer into skelet al muscles in vivo, factors affecting efficiency of transfer and stability of experssion. *Hum. Gene Ther. 4*, 151- 159.

[232] Fang, Q., Chen, M., Li, H.J. etat. (1994). MHC-l antigens on cultured human myoblasts. *Transpl. Proc. 26*, 3467.

[233] Fex, S. and Jirmanova, I. (1969). Innervation by nerve implants of fast' and "slow" skelet al muscles of the rat. *Acta Physiol. Scand. 76*, 257-269.

[234] Guerette, B., Asselin, I., Vilquin, J.T, et al. (1995). Lymphocyte infiltration follwing allo. and xenomyoblast transplantation in mdx mice. *Muscle Nerve 18*, 39-51.

[235] Guerette, B., Skuk, D., Celestin, F. (1997). Prevention by Anti-LFA-1 of acute myoblast death following transplantation. *J. Immunol. 159*, 2522-2531.

[236] Hamamori, Y., Samal, B., Tian, J. etat. (1994). Persistent erythropoiesis by myoblast transfer of erythropoietin cDNA. *Hum. Gene Ther. 5*, 1349-1356.

[237] Heyck. H., Laudahn, G., Carsten, P.M. (1996).Enzymaktivitatsbestimmungen bei Dystrophia musculorum progressiva. In MTT teilung. *Klinisce Wochenschrift 44*, 695.

[238] Hillman, A. (1996) Gene Therapy: Socioeconomic and Ethical Issues: A Roundtable Discussion. Hum. *Gene Ther. 7*, 1139-1144.

[239] Hoeben, R.C. (1995). Gene therapy for the haemophilias current status. *The International Association o f Biological Standardization. 23*, 27-29.

[240] Hoffman, E.P., Brown, R.H., Kunkel, L.M. (1987). Dystrophin, the protein product of the Duchenne muscular dystrophy locus. *Cell 51*, 919-928.

[241] Huber, B. E., Lazo, J. S. (eds.) (1994). *Gene Therapy for Neoplastic Diseases*. The New York Academy of Sciences. New York.

[242] Jiao, S., Cheng, L., Wolff, I.A. et al. (1993). Particle mbardment-mediated gene transfer and expression in rat brain tissue. *Bio/Technology 11*, 497-502.

[243] Karlsson, S. (1991). Treatment of genetic defects in hematopoietic cell function by gene transfer. *Blood 78*, 248 1-2492.

[244] *Gene Therapy and Molecular Biology Vol 1*, 17

[245] Katagiri, T. Yamaguchi, A., Komaki, M. et al. (1994). Bone morphogenetic protien-2 converts the differentiation pathway of C2C12 myoblasts into the osteoblast lineage. *J. Cell Biol. 127*, 1755-1766.

[246] Kessler, D.A., Siegel, J.P., Noguchi, P.D. et al. (1993). Regulation of somatic cell-therapy and gene therapy by the Food and Drug Administration. *N. Engl. J. Med. 329*, 1169-1173.

[247] Kinoshita, I., Vilquin, J.T., Guerette, B. et al. (1994). Very efficient myoblast allotransplantation in mice under FK506 immunosuppression. *Muscle Nerve 17*, 1407-1415.

[248] Knowles, M.R., Hohneker, K., Zhou, Z.Q. et al. (1995). A double blind vehicle-controlled study of adenoviral mediated gene transfer in the nasal epithelium of patients with cystic fibrosis. *N. Eng. J. Med. 333*, 823-831.

[249] Koh, G.Y., Soonpaa, M.H., Klug, M.G. et al. (1995). Stable fet al cardiomyocyte grafts in the hearts of dystrophic mice and dog. *J. Clin. Invest. 96*, 2034-2042.

[250] Langer, R, Vacanti, J.P. (1993). Tissue engineering. *Science 260*, 920-926.

[251] Lau, H.T., Yu, M., Fontana, A. et al. (1996). Prevention of islet allograft rejection with engineered myoblast expressing FasL in mice. *Science 273*, 109-112.

[252] Law, P.K. (1982). Beneficial effects of transplanting normal limb-bud mesenchyme into dystrophic mouse muscle. *Muscle Nerve 5*, 619-627.

[253] Law, P.K. (1994). *Myoblast Transfer: Gene Therapy for Muscular Dystrophy*. R.G. Landes Company, Austin, TX, p. 139-154.

[254] Law, P.K. and Atwood, H.L. (1972). Nonequivalence of surgical and natural denervation in dystrophic mouse muscle. *Exp. Neurol. 34*, 200-209.

[255] Law, P.K., Atwood, H.L., McComas, A.J. (1976). Functional denervation in the soleus muscle of dystrophic mice. *Exp. Neurol. 51*,434-443.

[256] Law, P.K., Bertorini, T.E., Goodwin, T.G. et al. (1990a). Dystrophin production induced by myoblast transfer therapy in Duchenne muscular dystrophy. *Lancet 336*, 114-115.

[257] Law, P.K., Goodwin, T.G., Fang, Q. et al. (1991b). Myoblast transfer therapy for Duchenne muscular dystrophy. *Acta Paediatr. Jpn. 33*, 206-215.

[258] Law, P.K., Goodwin, T.G. Fang, Q. et al. (1994a). Whole body myoblast transfer. *Transpl. Proc. 26*, 3381-3383.

[259] Law, P.K., Goodwin, T.G., Fang, Q. et al. (1995). Myoblast transfer, gene therapy for muscular dystrophy. *J. Cell Biochem*. p. 367.

[260] Law, P.K., Goodwin, T.G., Fang, Q. et al. (1996). Human gene therapy with myoblast transfer. *Mol. Biol. of the Cell. 7*, 3639.

[261] Law, P.K., Goodwin, T.G., Fang, Q. et al. (1997a). Human gene therapy with myoblast transfer. *Transpl. Proc. 29*, 2234-2237.

[262] Law, P.K., Goodwin, T.G., Fang, Q., et al. (199 7b). First human myoblast transfer therapy continues to show dystrophin after 6 years. *Cell Transplantation 6*, 95-100.

[263] Law, P.K., Goodwin, T.G., Fang, Q. et al. (1997c). Myoblast transfer therapy (MTT) phase II clinical trials. *J Physiol. Biochem. 53*, 80.

[264] Law, P.K., Goodwin, T.G., Fang, Q. (1997d). Advances in clinical trials of myoblast transfer therapy (MTT*). J. Neurol. Sci. 150*, 5253.

[265] Law, P.K., Goodwin, T.G., Wang, M.G. (1988 b). Normal myoblast injections provide genetic treatment for murine dystrophy. *Muscle Nerve 11*, 525-533.

[266] In Kakulas, B.A., Mastaglia, F.L. eds. *Gene Therapy and Molecular Biology Vol 1*, 18

[267] Li, R. K., Jia, Z.Q., Weisel, R.D. et al. (1996). transplantation improves heart function. *Cardiomyocyte Ann Thorac. Surg. 62*, 654-661.

[268] Massimino, M.L., Rapizzi, E., Cantini, M. et al. (1997). ED2+ macrophages increase selectively proliferation in muscle cultures. *Biochem. Res. Comm. 235*,

[269] Mendell, J.R., Kissel, J.T., Amato, A.A. et. al. (1995). Myoblast transfer in the treatment of Duchenne muscular dystrophy. *N. Engl. 1. Med. 333*, 832-838.

[270] Miyanohara, A., Johnson, P.A., Elam, R.L. et al. (1992). Direct gene transfer to the liver with herpes simplex virus Type 1 vectors, transient production of physiologically relevant levels of circulating factor IX. *New Biol. 4*, 23 8-246.

[271] Morandi, L., Bemasconi, P., Gebbia, M. et al. (1995). Lack of mRNA and dystrophin epxression in DMD patients three months after myoblast transfer. *Neuromusc. Disord. 5*, 291-295.

[272] Morishita, R., Gibbons, G.H. Horiuchi, M. et al. (1995). A gene therapy strategy using a transcription factor decoy of the E2F binding site inhibits smooth muscle proliferation in vivo. *Proc. Natl. Acad. Sci. USA 92*, 5855-5859.

[273] Morsy, M.A., Caskey, C.T. (1997). Safe gene vectors made simpler. *Nature Biotech. 15*, 17.

[274] Murry, C.E., Wiseman, R.W., Schwartz, S.M. et al. (1996). Skeletal myoblast transplantation for repair of myocardial necrosis. *J. Clin. Invest. 98*, 2512-2523.

[275] Nerem, R.M., Sambanis, A. (1995). Tissue engineering from biology to biological substitutes. *Tissue Engineering 1*, 3-12.

[276] Partridge, T.A., Grounds, M., Sloper, J.C. (1978). Evidence of fusion between host and donor myoblasts in skelet al muscle grafts. *Nature 273*, 306-308.

[277] Puchalski, R.B., Fahl, W.E. (1992). Gene transfer by electroporation, lipofection, DEAE.dextran transfection compatibility with cell-sorting by flow cytometry. *Cytometry 13*, 23-30.

[278] Ray, I., Gage, F.H. (1992). Gene transfer in established and primary fibroblast cell lines, comparision of transfection methods and promoters. *Biotechniques 13*, 598-603.

[279] Robinson, S.W., Cho, P.W., Levitsky, H.I. et al. (1996). *Cell Transplantation 5*, 77-91.

[280] Rosenfeld, M.A. and Collins, F.S. (1996). Gene therapy for cystic fibrosis. *Chest 109*, 241-252.

[281] Sautter, C., Waldner, H., Neuhaus-Url, G. et al. (1991). Micro-targeting, high efficency gene transfer using a novel approach for the acceleration of micro-projectiles. *Rio/Technology 9*, 1080-1085.

[282] Schwartz, E.R. (1997). Tissue engineering focused AIP program. *Tissue Engineering 3*, 5-17.

[283] Smith, T.A.G., Mehaffey, M.G. Kavda, D.B. et al. (1993). Adenovirus mediated expression of therapeutic plasma levels of human factor IX in mice. *Nat. Genet. 5*, 397-402.

[284] St. Louis, D., Verma, I.M. (1988). An alternative approach to somatic cell gene therapy. *Proc. Natl. Acad. Sc i. USA. 85*, 3150-3154.

[285] Stewart, M.J., Plautz, G.E., Del Buono, L. et al. (1992). Gene transfer in vivo with DNA-liposome complexes safety and acute toxicity in mice. *Hum. Gene Ther. 3*, 267-275.

[286] Sunada, Y., Bernier, S.M., Utani, A. et al. (1995). Identification of a novel mutant transcript of laminin a 2 chain gene responsible for muscular dystrophy and 2J dysmyelination in dy mice. *Hum. Mol. Genet. 4*, 1055-1061.

[287] Teboul, L., Gaillard, D., Staccini, (1995). Thiazolidinediones and fatty myogenic cells into adipose-like cells. *J. Biol. Chem. 270*, 28183-28187.

[288] Trubetskoy, V.S., Torchilin, V.P., Kennel, S.J. et al. (1992). Cationic liposomes enhance targeted delivery and expression of exogenous DNA mediated by N-terminal modified poly-L-lysine-antibody conjugate in mouse lung endothelial cells. Biochem. *Biophy. Acta. 1131*,311-313.

[289] Van Meter, C.H., Claycomb, W.C., Delcarpio, J.B. et al. (1995). Myoblast transplantation in the porcine model a potential technique for myocardial repair. 1. Thorac. *Cardiovasc. Surg. 110*, 1442-1448.

[290] Vilquin, J.T., Kinoshita, I, Roy, R. et al. (1995). Cyclosphosphamide immunosuppresion does not perMTT successful myoblast allotranplantation in mouse.Neuromus. *Disord. 5*, 511-517.

[291] Wolft~ I.A., Malone, R.W., Williams, P. et al. (1990). Direct gene transfer into mouse muscle in vivo. *Science 247*, 1465-1468.

[292] Wolff, I.A., Williams, P., Ascadi, G. et al. (1991). Conditions affecting direct gene transfer into rodent muscle in vivo. *Biotechniques 11*, 474-485.

[293] Yao, S.N., SMTTh, K.J., Kurachi, K. (1994). Primary myoblast-mediated gene transfer, persistent expression of human factor IX in mice. *Gene Ther. 1*, 99-107.

[294] Ishikawa F et al. "Nuclear proteins that bind the pre-mRNA 3' splice site sequence r(UUAG/G) and the human telomeric DNA sequence d(TTAGGG)In", *Molecular Cell Biology, 13*: 4301-4310,1993.

[295] Atkins B Z et al. Myogenic cell transplantation improves in vivo regional performance in infarcted rabbit myocardium. *J. Heart Lung Transplant 18*: 1173 – 1180,1999.

[296] Murry C E et al. Muscle cell grafting for the treatment and prevention of heart failure. *J. Card. Fail. 8*: S532 – S541,2002.

[297] Tang G H, et al. Cell transplantation to improve ventricular function in the failing heart. *Eur. J. Cardiothoracic Surg. 23*: 907 – 916,2003.

[298] Taylor D A, et al. Regenerating functional myocardium: Improved performance after skeletal myoblast transplantation. *Nat. Med. 4*: 929 – 933, 1998.

[299] Hamano K, et al. Therapeutic angiogenesis induced by local autologous bone marrow cell implantation. *Ann. Thoracic Surg. 73*: 1210 – 1215, 2002.

[300] Chedrawy E G et al. Incorporation and integration of implanted myogenic and stem cells into native myocardial fibers: Anatomic basis for functional improvements. *J. Thorac Cardiovasc Surg 124*: 584 – 590, 2002.

[301] Tomita S et al. Improved heart function with myogenesis and angiogenesis after autologous porcine bone marrow stromal cell transplantation. *J. Thorac Cardiovasc Surg 123*: 1132 – 1140, 2002.

[302] Chekanov V, and Kipshidze, N. Angiogenesis by means of endothelial cell transplantation. *J. Thorac Cardiovasc Surg 125*: 441 – 442,2003.

[303] Hamano K et al. Local implantation of autologous bone marrow cells for therapeutic angiogenesis in patients with ischemic heart disease. *Jpn. Circ. J. 65*:845 – 847, 2001.

[304] Sakai T et al. The use of ex vivo gene transfer based on muscle-derived stem cells for cardiovascular medicine. *Trends Cardiovasc Med 12*: 115 – 120, 2002.

[305] Suzuki K et al. Cell transplantation for the treatment of acute myocardial infarction using vascular endothelial growth factor-expressing skeletal myoblasts. *Circulation 104*: I207 – I212,2001.

[306] Tse H F et al. Angiogenesis in ischemic myocardium by intra-myocardial autologous bone marrow mononuclear cell implantation. *Lancet 361*: 47 – 49,2003.

[307] Yau T M et al. Enhanced myocardial angiogenesis by gene transfer with transplanted cells. *Circulation 104*: I218 – I222,2001.

[308] Atkins B Z. Results of cellular therapy for ischemic myocardial dysfunction. *Minerva Cardioangiol. 50*: 333 – 341,2002.

[309] Reinlib, L., and Field, L. Cell transplantation as future therapy for cardiovascular disease? A Workshop of the National Heart, Lung and Blood Institute. *Circulation 101*:e182 – e187,2002.

[310] Al Attar N et al. Long-term (1 year) functional and histological results of autologous skeletal muscle cells transplantation in rat. *Cardiovasc Res. 58*: 142 – 148,2003.

[311] Henningson C T et al. Embryonic and adult stem cell therapy. *Allergy Clin. Immunol. 111(2 Suppl.)*: S745 – S753,2003.

[312] Hutcheson K A et al. Comparison of benefits on myocardial performance of cellular cardiomyoplasty with skeletal myoblasts and fibroblasts. *Cell Transplant 9*: 359 – 368,2000.

[313] Kim W G et al. Autologous cardiomyocyte transplantation in an ovine myocardial infarction model. *Int. J. Artif. Org. 25*: 61 – 66,2002.

[314] Scorsin M et al. Comparison of the effects of fetal cardiomyocyte and skeletal myoblast transplantation on post infarction left ventricular function. *J. Thorac Cardiovasc.Surg 119*: 1169 – 1175,2000.

[315] Yoo K J et al. Autologous smooth muscle cell transplantation improved heart function in dilated cardiomyopathy. *Ann. Thorac Surg. 70*: 859 – 865, 2000.

[316] Yoo K J et al. Smooth muscle cells transplantation is better than heart cells transplantation for improvement of heart function in dilated cardiomyopathy. *Yonsei Med. J. 43*: 296 – 303,2002.

[317] Reffelmann T et al. Transplantation of neonatal cardiomyocytes after permanent coronary artery occlusion increases regional blood flow of infarcted myocardium. *J Mol Cell Cardiol 35*: 607 – 613,2003.

[318] Rubart M P et al. Physiological coupling of donor and host cardiomyocytes after cellular transplantation. *Circ. Res. 92*: 1217 – 1224,2003.

[319] Hagege A A et al. Viability and differentiation of autologous skeletal myoblast grafts in ischemic cardiomyopathy. *Lancet 361*: 491 – 492,2003.

[320] Pagani F D et al. Autologous skeletal myoblasts transplanted to ischemiadamaged myocardium in humans: Histological analysis of cell survival and differentiation. *J. Am. Coll. Cardiol. 41*: 879 – 888,2003.

[321] Ghostine S et al. Long-term efficacy of myoblast transplantation on regional structure and function after myocardial infarction. *Circulation 106*: I131 – I136,2002.

[322] Jain, M., et al. (2001). Cell therapy attenuates deleterious ventricular remodeling and improves cardiac performance after myocardial infarction. *Circulation 103*:1920 – 1927.

[323] Rajnoch, C., Chachques, J. C., Berrebi, A., Bruneval, P., Benoit, M. O., and Carpentier,A. (2001). Cellular therapy reverses myocardial dysfunction. *J. Thoracic Cardiovasc. Surg. 121*: 871 – 878.

[324] Reinecke, H., Poppa, V., and Murry, C. E. (2002). Skeletal muscle stem cells do not trans-differentiate into cardiomyocytes after cardiac grafting. *J. Mol. Cell. Cardiol. 34*:241 – 249.

[325] Leobon, B., Garcin, I., Menasche, P., Vilquin, J. -T., Audiant, E., and Charpak, S. (2002).Myoblasts transplanted into rat infarcted myocardium are functionally isolated from their host. *Proc. Natl. Acad. Sci. USA 100*: 7808 – 7811.

[326] P K Law, et al., "Myoblast transfer as a platform technology of gene therapy", Gene Therapy and Molecular Biology, 1: 345-363, 1998.

[327] P K Law, "Myoblast transfer as a platform technology of gene therapy", Regulatory Affairs Focus, (Technology), 4: 25-27, 1999.

[328] P K Law, "Nuclear transfer and human genome therapy", Business Briefing – Future Drug Discovery (Genomics), pp. 38-42, Dec. 2001.

[329] Chilian WM (2003) Editorial Comment on Arteriogenesis – is this terminology necessary? *Basic Res Cardiol 98*: 6–7

[330] Ferrara N (2001) Role of vascular endothelial growth factor in regulation of physiological angiogenesis. *Am J Physiol Cell Physiol 280*: C1358–C1366

[331] Freedman SB, Isner JM (2002) Therapeutic angiogenesis for coronary artery disease. *Ann Int Med 136*: 54–71

[332] Haider Kh H et al. (2002) Human myoblast carrying hVEGF165 as cellular assist device for myocardial therapeutics. *Circulation 106*: II-148

[333] Hockel M, Schlenger K, Doctrow S, Kissel T, Vaupel P (1993) Therapeutic angiogenesis. *Arch Surgery 128*: 423–429

[334] Hojo Y et al. (2000) Expression of vascular endothelial growth factor in patients with acute myocardial infarction. *J Am Coll Cardiol 35*: 968–973

[335] Iwakura A et al. (2000) Myocardial ischemia enhances the expression of acidic fibroblast growth factor in human pericardial fluid. *Heart and Vessels 15*:112–116

[336] Kalka C et al. (2000) VEGF gene transfer mobilizes endothelial progenitor cells in patients with inoperable coronary disease. *Ann Thorac Surg 70*: 829–834

[337] Shibuya M, Ito N, Claesson-Welsh L (1999) Structure and function of vascular endothelial growth factor receptor-1 and-2. Curr Top and functional restoration. *Eur Heart J 24*: 404–411

[338] Zachary I, Mathur A, Yla-Herttuala S, Martin J (2000) Vascular protection: a novel non-angiogenic cardiovascular role for vascular endothelial growth factor. *Arterioscler Thromb Vasc Biol 20*:1512–1520

[339] Venter, J C et al., "The sequence of the human genome", *Science* 2001, 291: 1,304-351.

[340] *"Nuclear Structure and Gene Expression"*, ed. by R C Bird, G S Stein, J B Lian, J L Stein, Academic Press 1997, pp. 1-296.

[341] Law, P K et al., "World's first human myoblast transfer into the heart", *F*rontiers in Physiology, p. A85, 2000.

[342] Law P K, "Compositions for and methods of treating muscle degeneration and weakness*", U.S. Patent No. 5,130,141, issued July 14*, 1992.

[343] Law P K, "Myoblast therapy for mammalian diseases", *Singapore Patent No. 34490 (WO 96/18303), issued August 22*, 2000.

[344] Menasche P, et al., "Early results of autologous skeletal myoblast transplantation in patients with severe ischemic heart failure", *Circulation, 104*: II-598, 2001.

[345] Law P K, "Automated Cell Processor*", U.S. Patent No. 6,261,832, issued July 17*, 2001.

[346] Law P K, "Instrument for Cell Culture", *Singapore Patent No. 74036, issued December 14*, 2001.

[347] Barnes J D,Stamatiou D, Liew C C, "Construction of a human cardiovascular cDNA microarray: portrait of the failing heart", *Biochem Biophy Res Comm, 280*: 964-969, 2001.

[348] Law P K, et al., *"Mechanisms of myoblast transfer in treating heart failure"*, in Kimchi A ed. Advances in Heart Failure. New York: Medimont. 2002: 43-48.

[349] Law P K, "The regenerative heart", Business Briefing: PharmaTech 2002, pp. 65-70, Apr. 2002.

[350] Law P K, "Concomitant angiogenesis/myogenesis in the regenerative heart", Business Briefing: Future Drug Discovery, pp. 64-67, October 2002.

[351] Law, P K, et al., " *Myoblast genome therapy and the regenerative heart"*, in : Kipshidze N N and Serruys P W ed. Handbook of Cardiovascular Cell Transplantation. Martin Dunitz,UK. (2004) pp.241-258.

[352] Law, P K, et al., Human VEGF165-myoblasts produce concomitant angiogenesis/ myogenesis in the regenerative heart. *Mol Cell Biochem 263*: 173-178. 2004.

[353] Husnain Kh. Haider,et al., Myoblast Transplantation for Cardiac Repair: A Clinical Perspective. Mol Ther 9:14, 2004.

[354] Ye, L.,et al.,Therapeutic angiogenesis *Basic Res Cardiol 99*: 121 – 132, 2004.

[355] Niagara, M I, et al. Autologous skeletal myoblasts transduced with a new adenoviral bicistronic vector for treatment of hind limb ischemia. *J Vas Surg. 40*:775-785,2004.

[356] Arsic N, Zentilin L, Zacchigna S, Santoro D, Stanta G, Salvi A, et al. Induction of functional neovascularization by combined VEGF and angiopoietin-1 gene transfer using AAV vectors. *Mol Ther* 2003;7:450-9.

[357] Rando TA, Blau HM. Primary mouse myoblast purification, characterization, and transplantation for cell-mediated gene therapy. *J Cell Biol* 1994;125:1275- 87.

[358] Smythe GM, Grounds MD. Exposure to tissue culture conditions can adversely affect myoblast behaviors in vivo in whole muscle grafts: implications for myoblast transfer therapy. *Cell Transplant* 2000;9:379-93.

[359] Thurston G, Rudge JS, Ioffe E, Zhou H, Ross L, Croll SD, et al. Angiopoietin-1 protects the adult vasculature against plasma leakage. *Nat Med* 2000;6:460-3.

[360] Law P K et al. The world's first myoblast study of Type II diabetic patients. Bussiness Briefing : North American Pharmacotherapy 2004 - Issue 2,pp.

[361] Law PK. Myoblast transfer : gene therapy for muscular dystrophy. Austin: R G Landes 1994, 1-164.

[362] Yaffe D, Shainberg A, Dym H. Studies on the prefusion stage during formation of multinucleated muscle fibers in vitro. In: Banker BQ, Przybylski RJ, Van Der Meulen JP, et al., eds. *Research in muscle development and the muscle spindle.* Amsterdam: Excerpta Medica 1972: 110-21.

[363] Young HE, Carrino DA, Caplan AI. Histochemical analysis of newly synthesized and accumulated sulfated glycosaminoglycans during musculogenesis in the embryonic chick leg. *J Morph* 1989; 201:85-103.

[364] Vandenburgh HH, Karlisch P, Shansky J, et al. Insulin and IGF-I induce pronounced hypertropy of skeletal myofibers in tissue culture. *Amer J Physiol* 1991; 260:C475-84.

[365] Satoh A, Labrecque C, Tremblay JP. Myotubes can be formed within implanted adipose tissue. Transpl Proc 1992; 24:3017-19.

[366] Myerburg RJ, Kessler KM, Interian A, et al. Clinical and experimental pathophysiology of sudden cardiac death. In: Zipes DP, Jalife J, eds. *Cardiac electrophysiology. From cell to bedside.* Philadelphia: WB Saunders Company, 1990: 666-78.

[367] Iwata Y, Nakamura H, Mizuno Y, et al. Defective association of dystrophin with sarolemmal glycoproteins in the cardiomyopathic hamster heart. *FEBS Lett* 1993; 329:227-31.

[368] Iwata Y, Nakamura H, Fujiwara K, et al. Altered membrane-dystrophin association in the cardiomyopathic hamster heart muscle. *Biochem Biophys Res Com* 1993; 190:589-95.

[369] P Menasche, et al., "Myoblast transplantation for heart failure", *Lancet, 357*: 279-280, 2001.

In: Genome Research Advances
Editor: Alberto P. Bellini, pp. 161-177
ISBN: 978-1-60021-805-7

Chapter III

"Delayed Division Delay" Revealed by Time-lapse Photographic Study: A New Endpoint of Radiation-induced Genomic Instability

Hiroshi Sasaki*
Kyushu University, Fukuoka 812-8582, Japan.

Abstract

LET-dependent features of cell abnormality leading to death, lethal sectoring, genomic instability and cell cycle change were compared between α- and X- irradiated HeLa cells by means of a time-lapse photography. Cell abnormality induced by α irradiation tended to be manifested in the form of death without division (mitotic or interphase death), rather than incomplete division accompanied by anaphase bridge derived from dicentric chromosome (two-hit type damage) that was predominant for X irradiation. Frequency of incomplete division at irradiated generation was decreased remarkably (25.8% → 6.8%) by splitting X ray dose (4 Gy → 2 × 2 Gy), correlating well with the increase in cell survival (6.8% → 14.0%). However, in the case of α irradiation especially at low dose, two split doses (0.6 Gy → 2 × 0.3 Gy) significantly decreased cell survival (12.3% → 8.8%) with the accompanying increase in the frequency of incomplete division (7.2% → 13.5%) (inverse split dose effect). This suggests that an interaction between the damage by the 1st dose and the one by the 2nd dose gives rise to 2-hit type damage owing to irreparability of the damage and enhances cell killing at low dose where interphase cell death is not predominant.

Lethal sectoring is the process for survival by which lethal damage remaining in irradiated cells is eliminated as lethal sector (offspring without reproductive integrity). This process occurs through the postirradiation 1st to 4th divisions, resulting in the appearance of a clonogenic progenitor harboring no lethal damage. However, nonlethal damage remaining in the clonogenic progenitors led to an elevated incidence of delayed

* Correspondence concerning this article should be addressed to Hiroshi Sasaki, 2-2-8 Iikura, Sawara-ku, Fukuoka 814-0161. Japan. Phone/Fax: +81-92-822-8354, E-mail: h.sasaki@hiz.bbiq.jp.

cell death in their progeny. From an analysis of the pedigrees of cells surviving α (0.45 Gy) or X irradiation (3 Gy) (20% survival dose) which were constructed by means of a long-term (~150 h) time-lapse observation, it became clear that the mean incidence was higher for α-particle (16.3%) than X-ray survivors (8.3%), indicating the greater potentiality for genomic instability by α particles. Evidence is available to suggest the intrinsic difference in the mechanisms for the induction of genomic instability by these two radiations; the association of misrepaired clustered DNA damage (CD) with α particles and the fixation of unrepaired potentially lethal damage (PLD) with X rays.

A novel phenomenon "delayed division delay (DDD)" was noticed, though occasionally (~10% per clonogenic progeny; ~30% per pedigree), during the postirradiation 1st – 3rd generations. DDD was much longer in α- (mean: ~11 h) than X-irradiated cells (~4 h). DDD might be triggered by delayed break and rejoining of DNA double-strand at the fragile misrepaired sites. This demonstrates the usefulness of DDD as a new endpoint of radiation-induced genomic instability. Contrary to DDD, a significant shortening (~5 h) of cell cycle time at the postirradiation 1st generation was found with the progeny of X-ray survivors harboring no lethal nor nonlethal damage.

ABBREVIATIONS

CD,	clustered DNA damage;
DDD,	delayed division delay;
GI,	genomic instability;
LET,	linear energy transfer;
LS,	lethal sectoring;
PLD,	potentially lethal damage

INTRODUCTION

Radiobiological studies of cultured mammalian cells are resulting in abundant knowledge of radiation damage at the cellular level. In most of these studies, random or synchronized cell population was used as a material. For example in the classical studies, the loss of colony-forming ability has been usually employed as a criterion of radiation-induced cell death. However, such studies are concerned only with the initial and final effects of radiation on proliferating cell populations; they do not clarify how irradiated cells divide and die, and how some progeny continue to divide and form a viable colony. The better method of studying such diverse phenomena is a time-lapse observation of individual cells.

Important findings on cell abnormality leading to death have been obtained with this method. Incomplete division followed by cytoplasmic cell fusion mostly between sister cells born after X irradiation, first reported by Marin & Bender [1], resulted in binucleate cells without colony-forming ability, therefore playing an important part in X-ray-induced reproductive cell death. Later, we detected a nuclear thread bridging the sister cells prior to cell fusion by means of autoradiography with ^{3}H-thymidine and scanning electron microscopy [2]. Hurwitz & Tolmach [3] found cell death at mitosis (mitotic death or mitotic catastrophe) in X-irradiated HeLa cells; cells were arrested at mitosis for 8 - 10 h before disintegration.

Based on long-term time-lapse observations of X-irradiated mouse L cells, Thompson & Suit [4] constructed cell pedigrees and examined the colony-forming process of surviving cells and recognized the phenomenon "lethal sectoring" , which had been reported by James et *al.* [5,6] who constructed the pedigrees of yeast surviving UV or X irradiation by isolating individual offspring, using a micromanipulator and found that the offspring without the colony-forming ability appeared during several postirradiation generations.

I have tried to elucidate the characteristic cellular responses after irradiation with α particles or accelerated high-LET charged particles compared to low-LET X rays. "Inverse split dose effect" was found with HeLa cells by splitting low dose of α particles (0.6 Gy → 2 × 0.3 Gy) ; cell survival decreased with the accompanying increase in the frequency of incomplete division (2 hit-type cell damage) [7]. "Dependence of induction of interphase cell death on LET" with maximum RBE (~3) around LET 200 keV/μm was recognized with CHO cells exposed to accelerated charged particles of various LETs [8]. I also constructed the pedigrees of HeLa cells surviving α or X irradiation but they remained unexplored for more than 20 years. Taking into consideration "genomic instability" and "clustered DNA damage" both of which are apt to be induced by α particles, these pedigrees were analyzed in detail. Fortunately, I found a novel phenomenon "delayed division delay" [9].

MATERIALS AND METHODS

Cell and Culture

Human cervical carcinoma (HeLa S3-9 IV: a clone established by T. T. Puck) cells, kindly provided by Dr. T. Terasima, were used in most of the experiments. The cells were grown in F10 medium supplemented with 0.05% heart infusion broth (Difco) and 20% fetal calf serum. Chinese hamster ovary (CHO-K1) cells, provided by the RIKEN Cell Bank, were used in the experiments shown in Figures. 6, 7 and 14. The cells were grown in Ham's F12 medium supplemented with 10% fetal calf serum.

A temperature-sensitive mutant (tsBN2) derived from wild-type BHK21 (baby hamster kidney) cell line, kindly provided by Dr. T. Nishimoto, was used in the experiment shown in Figure 13. The cells were cultured in Dulbecco's modified Eagle medium (DME) supplemented with 10% fetal calf serum at 33.5°C (permissive temperature). Premature chromosome condensation (PCC) is induced irrespective of the cell cycle by the incubation at nonpermissive temperature (40°C) in this ts mutant which has a single ts defect in the regulator for chromosome condensation (RCC1).

Irradiation

The radiation sources used were 200 kVp X rays (HVL, 0.92 mm Cu; dose rate, 0.7 Gy/ min), α particles from ^{241}Am point source (fluence rate, ~1.5×10^6/ cm^2/ min; dose rate, ~0.3

Gy/ min) and charged particles of various LETs accelerated by RIKEN Ring Cyclotron (RRC). At the time of α irradiation, a culture dish was put on an α source upside down after removal of the culture medium. The air path length of 12 mm from the α source to the cell surface and the thickness of the cells reduced the energy of α particles to 3.97 MeV (LET ~125 keV/μm in water), from 5.49 MeV.

Time-lapse Photography

For time-lapse observations of α or X irradiated cells, a plastic dish on which a few cells (~50) were grown in the center (diameter; ~1 cm) was placed in a glass container with ventilation ports and mounted on the stage of a phase-contrast microscope. In the case of irradiation with accelerated charged particles, cells were grown in a culture flask with ventilation ports at its screw cap. A small amount of 5% CO_2-95% air was supplied through these ports (Figure 1). Photographs were taken every 15 min (HeLa cells) or 30 min (CHO cells) with a 35-mm motor-driven camera. Since the exposures were of the same microscopic field during preirradiation and postirradiation period, the cell stage of asynchronously growing cells at irradiation could be estimated by elapsed time after cell division. The pedigrees of irradiated individual cells were constructed by reviewing the films. They were divided into survivor and nonsurvivior groups.

Scanning Electron Microscopy

In the experiment shown in Figure 7-a and 7-b, HeLa cells were grown on cover slips. After fixation in 3% glutaraldehyde (sodium phosphate buffer, pH 7.3), followed by 1% osmium tetroxide, the samples were dehydrated in graded concentrations of ethanol and dried using a critical point drying device (HCP-2) and carbon dioxide as the transitional liquid. The specimens were coated with carbon and gold under rotation *in vacuo* (lower than 5×10^{-5} Torr) and were examined with a "JSM-2" scanning electron microscope at 25 kV accelerating voltage.

Autoradiography

In the experiment shown in Figure 7-c, HeLa cells were grown on cover slips, labeled with ^{3}H-thymidine (0.02 μCi/ ml, 15 Ci/ m mole) for 24 h and rinsed with fresh medium before X irradiation. Cells were fixed with Carnoy solution, stained with acetoorcein and coated with NR-M_2 dipping emulsion (Konishiroku Photo Industries Co. Ltd.). The specimens were exposed for about 6 months to β rays from ^{3}H-thymidine incorporated into cellular DNA.

RESULTS

Lethal Hit Number

In the investigations using the loss of colony-forming ability as an indicator of cell death, the effects of high- but not low-LET radiation are characterized by the reduction or absence of an initial shoulder in the cell survival curve (Deering & Rice [10]), diminished or no recovery from sublethal damage (Ngo *et al.* [11]), a lowered value of OER (Oxygen enhancement ratio) (Barendsen *et al.* [12]), and a smaller variation of radiosensitivity during the cell cycle (Hall *et al.*[13]). These radiobiological differences between high- and low-LET radiations have been explained reasonably on the basis of a linear-quadratic model which entails one- and two-hit lethal events and the predominance of the former for high-LET radiation (Kellere & Rossi [14]).

Survival curves of HeLa cells irradiated with α particles and X rays are shown in Figure 2 [7]. The X-ray curve has an initial shoulder (D_q= 1.25 Gy) followed by an exponential part with a D_0 of 1.1 Gy, whereas the α-particle curve is completely exponential with a D_0 of 0.25 Gy. The fluence required to reduce the survival to e^{-1} (F_0) is $1.17 \times 10^{-2}/\ \mu m^2$; therefore the inactivation cross-section (σ) is calculated to be 85 μm^2 according to the formula $\sigma = 1/\ F_0$. Since an average geometrical cross-section of cell nucleus is about 220 μm^2 for the HeLa cells (Konishi *et al.* [15]), it is postulated that 2.6 hits (220/ 85) to the cell nucleus give rise to one lethal event. The reason why more than 1 hit of α particles are required in order to inactivate the reproductive integrity of the cell will be explained by the following facts; 1) some of the lethal damage is removed during several postirradiation cell divisions (lethal sectoring) and only irremovable lethal damage inactivates the cell, 2) some of the damage is nonlethal and responsible for genomic instability or completely repaired during division delay at irradiated generation.

Cell Abnormality and Death

Illustration of cell abnormality leading to death of irradiated cells (Figure 3) is based on the time-lapse observations mainly with HeLa cells including other cell lines such as CHO cells. According to the definition of interphase (premitotic) and reproductive (postmitotic) death, (A)-(B)-(E) and (C)-(D)-(F) correspond to the former and the latter, respectively. However, in HeLa but not CHO cells, death at mitosis (B: mitotic death or mitotic catastrophe) is characterized by a prolonged arrest at mitosis (8 - 10 hours) before disintegration and therefore is different from (A) and (E). Hopwood & Tolmach [16] considered deficient DNA synthesis as a cause of mitotic death. Death at interphase is divided into cell-cycle arrest-independent (A) and dependent (E) interphase death. The latter was predominantly induced by high-LET radiation [8]. Cells destined to undergo this type of interphase death were arrested before mitosis for more than 50 h for HeLa cells and 24 h for CHO cells and were transformed into giant cells (Figure 4-a and b). The morphological change in the cells that was associated with the initiation of interphase death was "rounding up" from the spread configuration (Figure 4-c and d). followed by fragmentation into multiple parts (Figure 4-e) and fusion, resulting in the formation of giant cells containing multiple micro nuclei, in regard to the CHO cells. These giant cells showed apoptotic features including DNA fragmentation. They then gradually disintegrated, but disintegration occurred from rounded configuration in the HeLa cells.

Dose response curve for the induction of interphase death in the CHO cells exposed to Ne ions (228 keV/μm) is different from the one exposed to X rays (Figure 5); the former increases linearly with dose, whereas the latter begins to increase when the dose exceeds ~10 Gy (threshold). This indicates the requirement, for the induction of interphase death, of multiple damage on the chromosome which may be accomplished by a single hit of heavy ion but demand higher dose in the case of X irradiation.

An incomplete division (C in Figure 3) accompanied by protoplasmic cell fusion resulted in binucleate cell formation. This type of abnormal division took place frequently during postirradiation several generations in the progeny of the cells that could not survive, and predominantly after exposure to low-LET ionizing radiation such as X or γ rays (Hurwitz & Tolmach [17]). An intercellular thread bridging the sisters of HeLa cells prior to fusion was always noticed with time-lapse photographic observations under a phase-contrast microscope (Figure 6). A chromatin thread-like structure (~100 μm; ~0.7 μm in diameter) was confirmed by scanning electron microscopy as well as by autoradiography with ^{3}H-thymidine (Figure 7). From the examination of 248 cases, fusion took place most frequently between sister cells (94%), occasionally between cousin or niece cells (5%) and infrequently between the cells derived from a common ancestor cell three or four generations past (1%). The repetition of incomplete division and fusion led to the formation of a multinucleate cell that died also apoptosis. Small giant cells resulting from a fairly long delay in division often underwent multipolar division (D) (Dodson *et al.* [18]), most of which was tripolar for HeLa cells, although its frequency was much lower than that of incomplete division. Fusion tended to occur between two of the three sisters.

Inverse Split Dose Effect

The effects of two split doses on cell survival were compared between α particles and X rays (Figure 8). When two split doses of X rays (2 × 2 Gy) were given with various time interval (0－4 h), cell survival increased about two fold within 2－4 h because of the recovery from sublethal damage (Elkind & Sutton [19]). In the case of α particles (2 × 0.3 Gy), the survival decreased significantly (12.3% → 8.8%) during time interval between doses. However, in a larger dose of α particles (1.2 Gy), there was no significant difference in survival between single and two split exposures (single, 1.05±0.08%; two split with 3-h interval, 0.9±0.1%).

The fate of cells at irradiated generation with respect to their ability or inability to divide after exposure to single or split doses of α particles or X rays was examined with time-lapse photography (Table 1). After a single exposure to X rays, an incomplete division is predominant but the frequency was reduced remarkably by splitting the dose (25.8% [4 Gy] → 6.8% [2×2 Gy]; 28.7% [6 Gy] → 10.9% [2×3 Gy]) with a concomitant increase in the frequency of normal (bipolar) division (68.0% [4 Gy] → 84.3% [2×2 Gy]; 61.7% [6 Gy] →

79.7% [2×3 Gy]). This correlated with an increase in cell survival by split dose (Figure 8). On the other hand, the frequency of incomplete division and mitotic death was significantly increased by two split doses of 0.6 Gy α particles with the accompanying decrease in bipolar division which seemed to correlate with a slight but significant decrease in cell survival (inverse split dose effect). In the case of 1.2 Gy α particles, interphase death was predominantly induced by single dose but dramatically reduced by split dose with a concomitant increase in the frequency of mitotic death as well as bipolar division. However, the survival was not increased nor decreased by two split doses of 1.2 Gy α particles. This suggested the change of type from interphase to reproductive death; about 20% cells corresponding to the fraction of bipolar division increased by split dose exposure (67.3 – 48.8 = 18.5%) should divide but might die afterwards.

A chromatin thread bridging the sister cells prior to cell fusion is considered to originate in dicentric chromosome that is a typical 2-hit type chromosome damage and formed by an illegitimate rejoining between two broken ends of centric chromosomes. Therefore, the decrease in incomplete division and the increase in bipolar division with a concomitant increase in survival after exposure to slit doses of X rays can be explained by the repair of damage induced by the 1st dose, which leads to chromosome break if not repaired, during the time interval between doses. On the other hand, the reverse split dose effect noticed with low dose (0.6 Gy) α particles will be interpreted reasonably by taking irreparability of the damage into account; if the damage leading to chromosome break is not repaired and a broken end of centric chromosome is produced during the time interval, its interaction with the damage by the second dose will result in the increase of 2-hit type damage.

Inverse phenomenon with high- relative to low-LET radiation has been identified also with respect to nonlethal radiation effect. Inverse low dose rate or split dose effect on the transformation of C3H 10T1/2 cells was reported by Hill *et al.* [20] who found that small doses (< 2 Gy) of fission neutrons greatly increased transformation efficiency when applied at low dose rates or fractionated over several hours. According to Lubin *et al.* [21], at high levels of cumulative exposure to α particles from radon as experienced by many underground miners of uranium, the risk of lung cancer followed an inverse dose rate pattern.

Another significant change of cell fate by split doses of α particles was the enhanced mitotic death that was noticed both in low and high doses (0.6 and 1.2 Gy). Though data were not shown here, I previously found the difference in the cell-cycle dependence of division delay between X rays and α particles; the largest delay at G1/ S boundary and the smallest delay at late S/ early G2 for X rays, whereas a smaller delay in G1 than S phase with maximum at late S/ G2 except for late G2 past G2 checkpoint beyond which no delay in division was recognized similarly as for X rays. Furthermore, division delay time was shortened by two split doses of X rays especially with regards to radioresistant late S cells, but such recovery was not found with α particles [7]. Using laser scanning cytometry, Goto *et al.* [22] reported that, relative to X rays, exposure to iron ions (LET 1,000 keV/μm) resulted in a substantially greater inhibition of S-phase progression as a result of a higher frequency of DNA replication-blocking clustered DNA damage. These findings suggested that an exposure of the 2nd dose to the cells arrested in S phase by the 1st dose of α particles might enhance inhibition of DNA replication, and the cells unable to complete DNA replication might die when entered mitosis.

Lethal Sectoring

When an irradiated cell forms a colony consisting of more than 50 cells within 1 – 2 weeks, the cell has been generally judged as a survivor. Optimum time for colony-formation will be 9 – 10 generations. During this period, an unirradiated cell forms a colony of about 500 – 1,000 cells. If a small colony of about 50 cells derives from a single clonogenic progenitor (clonogen), it should appear in the progeny until the 3rd or 4th postirradiation division, since an exponential proliferation for 6 generations (2^6= 64) is required for the number of cells per colony to exceed the critical value of 50. Lethal sectoring is the process for survival by which lethal damage remaining in irradiated cells is eliminated during the postirradiation divisions as offspring without reproductive integrity (lethal sector) with the accompanying appearance of clonogen that may harbor nonlethal but no lethal damage. This process tended to occur during the postirradiation 1st - 3rd divisions after exposure to α particles, and the 1st - 4th divisions for X irradiation.

The colony-forming process was explored in detail by analyzing the pedigrees of HeLa cells surviving α (0.45 Gy) and X irradiation (3 Gy) which were constructed by means of long-term (~150 h, corresponding to 7 - 8 postirradiation generations) time-lapse observations (Figure 9). Although the observation period was shorter than optimum time for colony-formation (9 - 10 generations), the judgment on 'an irradiated cell was survivor or nonsurvivor' and 'the offspring was clonogen or lethal sector' was done successfully by taking the cell fate and the change in cell-cycle time into consideration. In these pedigrees, lethal sector was indicated as asterisk, and clonogen was indicated as open symbol (stable type) or open triangle (unstable type). At first glance, pedigree (B) is simple and pedigree (A) is complicated. In pedigree (B), lethal sectoring was completed by the postirradiation 1st division; two sister cells appeared, the one retaining the reproductive integrity with a low incidence of delayed cell death (stable clonogen; two death and 60 normal divisions → 3.2% [2/62]) in the progeny, and the other whose descendants were dying after division (lethal sector). In the pedigree (A) lethal sectoring did not take place but the clonogen appeared at the 1st division. The process that was accompanied with the appearance of clonogen instead of lethal sector was defined as clonogenic sectoring. This clonogen divided further at the next division into stable and unstable ones (incidence of delayed cell death: >20%). As for the upper half of this pedigree, two successive lethal sectorings occurred at the 2nd and 3rd divisions.

The pattern of lethal sectoring is illustrated on the basis of total 54 pedigrees of survivors (Figure 10). Survivors are divided into LS positive and negative subgroups depending on the presence and absence of lethal sectoring in the pedigrees. No lethal damage is remaining in the latter because of the repair of the damage during division delay at irradiated generation. The LS negative survivor by itself is clonogen (A). Here the survivor or its offspring harboring lethal damage yet retaining the reproductive integrity is defined as a latent cell (closed circle). In this sense, the LS positive survivor is the latent cell. Four combinations of sister cells are expected after division of the latent cell; (I) lethal sector and clonogen, (II) lethal sector and latent cell, (III) clonogen and latent cell, (IV) two latent cells. Consequently, many patterns of lethal sectoring are possible when the postirradiation 1st - 3rd divisions are taken into account (B - G). The process of lethal sectoring persists until the latent cell generates the lethal sector and the clonogen coincidentally (e.g., until the 2nd division in C). (E) is when only one clonogen appeared at the 3rd division and formed a colony consisting of slightly more than 50 cells until the 9th generation. (F) shows that 3 clonogens and one lethal sector appeared, demonstrating the predominance of clonogenic sectoring.

The number of pedigrees available for survivor groups was 31 and 23 for α and X irradiation, respectively. Lethal sectoring was found in 25 pedigrees (25/31: 81%) (α-particle LS positive) but not in 6 pedigrees (6/31: 19%) (α-particle LS negative) for the former, and it was found in 15 pedigrees (15/23: 65%) (X-ray LS positive) but not in 8 pedigrees (8/23: 35%) (X-ray LS negative) for the latter. Although a definite conclusion could not be drawn with the limited number of pedigrees available, the following differences seemed to be noted in the process of lethal sectoring between α-particle and X-ray survivors; 1) the process completed earlier for α-particle (until the postirradiation 3rd divisions) than for X-ray survivors (until the 4th divisions), 2) clonogenic and lethal sectoring occurred similarly in α-particle survivors, but the former predominated over the latter in X-ray survivors, 3) mean number per pedigree was similar with respect to lethal sector (~1.5) both in α-particle and X-ray survivors, but the number of stable type clonogen was larger in X-ray (~2) than α-particle survivors (~1). These facts imply that the colony-size of the X-ray survivors is larger than that of α-particle survivors when compared at the same survival. This is consistent with the report of Walker *et al.* [23] that heavy ions caused a dose- and LET-dependent reduction in colony-size.

Delayed Cell Death

The incidence of delayed cell death in the clonogenic progeny was examined separately for LS positive and negative survivors (Table 2). A comparison between LS positive α-particle and X-ray survivors demonstrates that the mean incidence of dying cells is about two fold higher in the former (16.3% [194/1,190]) than in the latter (8.3% [90/1,079]). For α particles the value obtained with LS negative survivors (14.3% [45/324]) was similar to that obtained with LS positive survivors (16.3%). However, for X rays the value of LS negative survivors (4.7% [16/337]) was significantly lower than that of LS positive survivors (8.3%) and was close to that of unirradiated control cells (3.6% [8/222]). These results clearly show that genomic instability was induced in LS positive but not in negative survivors by X rays, whereas the instability was induced both in LS positive and negative survivors by α particles, suggesting an intrinsic difference in the mechanisms of genomic instability induction between these two radiations.

Thus the incidence of delayed cell death in LS positive survivors was about two fold higher for α than X irradiation when compared at 20% survival. The RBE of α particles for the induction of delayed cell death was estimated to be 9 or 14 when the RBE for cell killing (D_0 ratio = 4.4) or the ratio of exposed dose (X rays [3 Gy]/ α particles [0.45 Gy] = 6.7) was taken into consideration. The value will approach to the radiation weighting factor (w_R = 20) that has been recommended for α particles by ICRP [24], if LS negative survivors are included in RBE estimation.

Delayed Division Delay and A Shortening of Cell Cycle Time

A novel phenomenon "delayed division delay (DDD)" was recognized with the progeny during the postirradiation 1st to 3rd generations (Figure 11). Cells irradiated with α particles (0.45 Gy) (A, B) or X rays (3 Gy) (C, D) showed a remarkable division delay at irradiated generation. The increase in cell cycle time from preirradiation control (pre-α irradiation: ~17 h, pre-X irradiation: ~19 h) was 12 - 16 h and ~10 h (29 - 33 h and ~30 h as actual cell cycle time) after α and X irradiation, respectively. The cell cycle time at the 1st generation recovered to a normal level in (A), (B) and (D), but it did not recover because of DDD in one of the sister cells born at the 1st division as shown in (C). DDD also occurred at the 2nd generation (A, D) and the 3rd generation (B). Generally, DDD is similar to division delay at irradiated generation for α irradiation, but it is much shorter than the latter in the case of X irradiation. DDD was found with the latent cells just prior to lethal sectoring (A [30 with star marks], C and D [22 with star marks]) and with the clonogens immediately after clonogenic sectoring (A [32 with star marks]) or the completion of lethal sectoring (B [33 with star marks]).

A prolonged cell cycle time caused by DDD is shown for the individual cases (Table 3). The mean DDD was much longer in α (10.8 h) than X irradiated cells (4.0 h). The mean frequency of DDD was ~30% per pedigree and ~10% per clonogen or latent cell for both radiations when a detectable minor cell cycle change (1 - 3 h) was taken into account for its identification. It seemed that DDD was found with clonogen (~15% [6/40]) more predominantly than with latent cell (~7% [2/28]) for α irradiation and the converse was the case for X irradiation. In the present study, DDD could not identified in the progeny later than the 3rd generation after α irradiation because of limited observation period (~150 h), and so possibly α-particle-induced DDD can happen at later generations.

Contrary to DDD, a significant shortening (~5 h) of cell cycle time corresponding to the time difference between preirradiation (19.4±1.3 h) and postirradiation 1st generation (mean±SD: 14.7±1.1 h) was found with the progeny of X-ray LS negative survivors. The shortened time interval was roughly identical to division delay at irradiated generation (mean 5.4 h). Some cellular components synthesized in an extra amount during division delay might enhance subsequent cell cycle progression in the progeny of the cells harboring no lethal nor nonlethal damage which would have an inverse influence.

DISCUSSION

Features of the Damage Induced by α Particles Relative to X rays

Traversals of high-LET α particles (~125 keV/μm) through the higher-order structure of chromatin fibers may result in the clustering of damage on the chromosome (Goodhead [25]) and the formation of clustered DNA damage (CD) (Rydberg [26]) that is difficult to repair correctly (Rydberg *et al.* [27]), whereas low-LET X rays produce isolated sublethal (SLD) [19] or potentially lethal damage (PLD) (Phillips & Tolmach [28]) that can be repaired correctly by dose splitting or delayed plating. The fact that average 2.6 hit of α particles are required to give rise to one lethal event (Figure 2) indicates the probability of cell death by a single hit of α particles to nucleus of HeLa cells to be about 0.3 (~70% survival). Moreover, a high incidence of delayed cell death in the clonogenic progeny of α-particle survivors harboring no lethal damage (α-particle LS negative survivor) (Table 2) suggests that α particles are very efficient in inducing nonlethal damage, and a single hit of α particle to cell nucleus is sufficient to induce genomic instability in the progeny. However, no induced incidence of delayed cell death in the clonogenic progeny of X-ray LS negative survivors indicates complete repair of the damage (no production of lethal nor nonlethal damage) during division delay at irradiated generation.

Radiation-induced lethal damage is manifested in the form of cell abnormality leading to death. Incomplete division followed by fusion mostly between sister cells is caused by dicentric chromosome; a typical 2 hit-type chromosome damage that is predominantly induced by low-LET radiation such as X or γ rays. Cell abnormality characteristic for high-LET radiation such as α particles is interphase or premitotic death; cell death before reaching postirradiation 1st mitosis. This type of death seems to be induced by unrejoined chromosome breaks which arrest the cells at G2 checkpoint (Qin & Li [29]) for a long time and commit them to undergo apoptosis. Significant fraction (37%) of HeLa cells irradiated with 1.2 Gy of α particles (~1% survival dose) underwent interphase death (Table 1). More detailed study with CHO cells exposed to high-LET heavy-ions suggests the requirement of multiple (>10) traversals of heavy-ions (e.g., Ne-ions) through cell nucleus for the induction of interphase death in the majority (63%) of cells [8]. Remarkable decrease in the frequency of interphase death in HeLa cells by two split doses of α particles (1.2 Gy → 2 × 0.6 Gy with 3 h interval) supports this possibility. On the other hand, split dose experiment at low dose of α particles (0.6 Gy; ~10% survival) revealed the inverse split dose effect on cell killing which correlated with the increase in incomplete division (Figure 8). A similar experiment on genomic instability remains to be done.

Mechanisms of Radiation-induced Genomic Instability

Although the greater potentiality of delayed cell death as an endpoint of genomic instability (GI) was recognized after exposure to α particles compared to X rays, the intrinsic difference was also found between these two radiations. GI could be induced by α particles both in LS positive and negative survivors; however, it was inducible in LS positive but not in negative survivors in the case of X rays (Table 2). This fact suggests the different mechanisms involved in the induction of GI by these two radiations. I would like to propose the fixation of unrepaired PLD (including sublethal damage) -mediated mechanism for X rays and the misrepaired CD-mediated one for α particles. In Figure 12, the presumptive pathways from primary radiation damage to the induction of LS and GI are shown. Here it is postulated that the fixation of remaining unrepaired PLD by a higher-order conformational change of DNA/chromatin fibers during the cell cycle (e.g., chromosome condensation during the postirradiation 1st mitosis; see Figure 13 and the explanation in next paragraph) gives rise to both lethal and nonlethal damage and that the latter damage is responsible for GI induction after X irradiation (LS positive [GI+]). However, minor contributions of misrepaired PLD to the production of these damage can not be denied. When PLD is completely repaired, both lethal and nonlethal damage is not formed; there is no induction of genomic instability in X-ray LS negative survivors (LS negative [GI−]). On the other hand, both lethal and nonlethal latent damage will be produced at irradiated generation, mainly by misrepair and partly by the fixation of unrepaired CD after α-particle exposure (LS positive [GI+]). On the assumption that nonlethal damage is induced more predominantly than lethal damage by α particles and a single hit of α -particle to cell nucleus is sufficient to induce nonlethal damage, this damage can be produced in the cells harboring no lethal damage (LS negative [GI+]). Thus genomic instability is inducible both in LS positive and negative survivors for α particles, whereas it is inducible in LS positive but not in negative survivors for X rays. This explanation agrees well with the finding of Roy *et al.* [30] that delayed cell death was suppressed through PLD recovery by the delayed plating of X-irradiated human embryo cells. It is also supported by the finding [23] that the cross section for the effect of an LET-dependent reduction of colony-size was 100-300 μm^2. This was at least as large as the cell nuclear area and up to 3 times as large as the cross section for cell killing. A higher potentiality for GI induction by high-LET radiation will be one of the causes for a heavy-ion induced reduction of colony-size.

Figure 13 shows the PLD fixation experiments using tsBN2 mutant (Sasaki & Nishimoto [34]) in which premature chromosome condensation (PCC) occurs irrespective of the cell cycle at nonpermissive temperature (40°C). Although the incubation at 40°C for 3 h alone did not lower the survival of unirradiated tsBN2 cells, the cell killing effect of X rays was enhanced by this treatment immediately after irradiation for the cells synchronized at various stages of the cell cycle except those at around the late G2/M phase, where chromosome condensation occurred at a permissive temperature (33.5°C). This suggests that the enhanced interaction between PLDs (isolated DNA DSBs) during chromosome condensation may fix PLD as lethal damage.

Mechanisms of Delayed Division Delay

This is the first report demonstrating the occurrence of delayed division delay (DDD) in the progeny of cells irradiated with ionizing radiation, although Thompson & Humphrey [31] previously reported on the second-division delay that occurred mainly because of DNA replication block by pyrimidine dimers at the postirradiation 1st generation in the progeny of late S/G2 mouse L-P59 cells exposed to UV. Since cell cycle checkpoints arrest the cell cycle upon the occurrence of DNA double-strand breaks (DSB) in irradiated cells [29], DDD serves as an indicator of delayed DNA DSB and its rejoining. According to the suggestion of Suzuki *et al.* [32] that DNA DSB may create nonlethal potentially unstable chromosome regions (PUCR) at the break-rejoining sites through DNA DSB repair, α particles are assumed to generate clustered PUCR via CD, causing a persistent and pronounced DDD. Delayed DNA DSB at PUCR sites with the accompanying reactivation of DNA damage responding gene products (p53, NBS1, CHK2) [33] may trigger DDD.

At present, there is no evidence suggesting an intracellular presence of the enzyme system that specifically breaks the fragile misrepaired sites of DNA double-strand structure.

Therefore, I would like to propose a hypothesis that delayed DNA DSB is induced at the fragile sites by mechanical stress associated with a higher-order conformational change of DNA/chromatin fibers during late G2/M transition.

SUGGESTIONS FOR THE FUTURE STUDY

Time-lapse Photography

Present study indicates the usefulness of DDD as a new endpoint of radiation-induced genomic instability, but a long-term time-lapse observation of irradiated cells is required to detect DDD. Since the field size of photograph taken with 35mm camera under low magnification (×10 objective and ×5 ocular lens) is about 1.5 mm^2 and large enough to include about 500 HeLa cells, the colony-forming process for about 10 surviving cells can be followed by reviewing time-lapse photographic records on cell fates during at least 7 postirradiation generations. However, the cells must be cultured under the condition where the inhibition of cell proliferation by cell to cell contact is avoided.

More important information may be obtained by combining a time-lapse photography with immunostaining. For instance, immunostaining of the cells with anti-phosphorylated H2AX antibody immediately after the ending of time-lapse observations will provide some information concerning the relation of lethal sectoring and genomic instability with DNA damage remaining in the progeny of surviving cells, since Suzuki *et al.* [35] observed that phosphorylated histone H2AX foci remained in interphase cells even 96 h after exposure to X rays.

Radiation Source

A point α source of ^{241}Am is easy to handle and its α particles (LET~125 keV/μm at air path length of 12 mm) are very effective in inducing DDD and delayed cell death, therefore are recommended to be used in the study on genomic instability using cultured cells as materials.

Recently, the study on mutagenic or carcinogenic effect of accelerated high-LET Fe-ions on astronauts during space flight has been becoming increasingly important. Presence of non-hit cells, identified with a time-lapse observation (Mehnati *et al.* [36]), among hit cells is one of the features characteristic for exposure to very high-LET Fe-ions (Figure 14). Bystander effect may occur in non-hit cell (Little [37]). This effect will be ascertained if heavy-ion's hit or non-hit to individual cells can be judged accurately e.g. by using CR-39 plastics, on which cells are growing, before beginning a time-lapse observation.

Selection of the Cells Undergoing Delayed Division Delay

Delayed DNA DSB should arrest the cells at the checkpoint located in G2 until the rejoining of DNA DSB is completed. During delay time that is more than 10 h for α irradiation, the cells should become significantly larger than the G2 phase cells undergoing no DDD. The fraction of the cells undergoing DDD may be small, but can be increased by selecting larger cells using a pertinent method. Immunostaing with anti-phosphorylated H2AX antibody may be useful in the recognition of the cells undergoing DDD.

Conclusion

Both lethal and nonlethal damage originated in the same primary damage; CD with α irradiation and PLD with X irradiation. Lethal damage tended to be manifested in the form of the cell cycle arrest-dependent interphase death and an incomplete division followed by fusion after exposure to α particles and X rays, respectively. Lethal damage remaining in the progeny of surviving cells is eliminated as lethal sector during the postirradiation 1st - 4th divisions. Nonlethal damage remaining in the clonogenic progeny is responsible for the induction of genomic instability. This damage is supposed to correspond to the fragile sites on DNA double-strand structure created by misrejoining of DNA DSB through the misrepair of CD and the fixation of unrepaired PLD. Delayed DNA DSB at the fragile sites and its rejoining should trigger DDD in the clonogenic progeny of surviving cells. DDD is much longer in α- (mean: ~11 h) than X-irradiated cells (~4 h), correlating well with a higher incidence of delayed cell death in the progeny of α-particle survivors. Contrary to DDD, a significant shortening (~5 h) of cell cycle time at the postirradiation 1st generation was found with the progeny of X-ray survivors harboring no lethal nor nonlehtal damage, indicating the enhancement of subsequent cell cycle progression after the complete repair of radiation damage.

Acknowledgments

I thank the Radiation Research Society and Japan Radiation Research Society for permitting me to reproduce figures and tables of our papers published in Radiation Research and Journal of Radiation Research, respectively.

References

[1] Marin, G. and Bender, M. A. (1966) Radiation-induced mammalian cell death: Lapse-time cinemicrographic observations. *Exp. Cell Res. 43*: 413-423.

[2] Sasaki, H. and Hayashi, H. (1979) A nuclear thread bridging the sister cells prior to radiation-induced cell fusion. *Radiat. Res. 77*: 577-585.

[3] Hurwitz, C. and Tolmach, L. J. (1969) Time lapse cinemicrographic studies of X-irradiated HeLa S3 cells. I. Cell progression and cell disintegration. *Biophysical J. 9*: 607-633.

[4] Thompson, L. H. and Suit, H. D. (1969) Proliferation kinetics of X-irradiated mouse L cells studied with time-lapse photography. II. *Int. J. Radiat.Biol. 15*: 347-362.

[5] James, A. P. and Werner, M. M. (1966) Radiation-induced lethal sectoring in yeast. *Radiat. Res. 29*: 523-536.

[6] James, A. P., Werner, M. M., Saunders, A. S. and Harris, M. A. (1968) Persistence of X-ray induced lethal sectoring in yeast. *Radiat. Res. 34*: 475-487.

[7] Sasaki, H. (1984) Cell killing and division delay in asynchronous and synchronized HeLa cells irradiated with alpha particles or x rays. *Radiat. Res. 99*: 311-323.

[8] Sasaki, H., Yatagai, F., Kanai, T., Furusawa, Y., Hanaoka, F., Zhu, W.-G. and Mehnati, P. (1997) Dependence of induction of interphase death of Chinese hamster ovary cells exposed to accelerated heavy ions on linear energy transfer. *Radiat. Res. 148*: 449-454.

[9] Sasaki, H. (2004) Lethal sectoring, genomic instability, and delayed division delay in HeLa S3 cells surviving alpha- or X-irradiation. *J. Radiat. Res. 45*: 497-508.

[10] Deering, R. A. and Rice, R. (1962) Heavy ion irradiation of HeLa cells. *Radiat. Res. 17*: 774-786.

[11] Ngo, F. Q. H., Blakely, E. A. and Tobias, C. A. (1981) Sequential exposures of mammalian cells to low and high-LET radiations. *Radiat. Res. 87*: 59-78.

[12] Barendsen, G. W., Koot, C. J., van Kersen, G. R., Bewley, D. K., Field, S. B. and Parnell, C. J. (1966) The effect of oxygen on impairment of the proliferative capacity of human cells in culture by ionizing radiations of different LET. *Int. J. Radiat. Biol. 10*: 317-327.

[13] Hall, E. J., Gross, W., Dvorak, R. F., Keller, A. M. and Rossi, H. H. (1972) Survival curves and age response functions for Chinese hamster cells exposed to X-rays or high-LET alpha-particles. *Radiat. Res. 52*: 88-98.

[14] Keller, A. M. and Rossi, H. H. (1972) The theory of dual radiation action. *Curr. Top. Radiat. Res. Q. 8*: 85-158.

[15] Konishi, T., Takeyasu, A., Yasuda, N., Natsume, T., Nakajima, H., Matsumoto, K., Asuka, T., Sato, Y., Furusawa, Y. and Hieda, K. (2005) Number of Fe ion traversals through a cell nucleus for mammalian cell inactivation near the Bragg peak. *J. Radiat. Res. 46*: 415-424.

[16] Hopwood, L. E. and Tolmach, L. J. (1971) Deficient DNA synthesis and mitotic death in X-irradiated HeLa cells. *Radiat. Res. 46*: 70-84.

[17] Hurwitz, C. and Tolmach, L. J. (1969) Time-lapse cinemicrographic studies of X-irradiated HeLa S3 cell. II. Cell fusion. *Biophysical J. 9*: 1131-1143.

[18] Dodson, H., Bourke, E., Jeffers, L. J., Vangnarelli, P., Sonoda, E., Takeda, S., Earnshaw, W. C., Merdes, A. and Morrison, C. (2004) Centrosome amplification induced by DNA damage occurs during a prolonged G2 phase and involves ATM. *EMBO J. 23*: 3864-3873.

[19] Elkind, M. M. and Sutton, H. (1960) Radiation response of mammalian cells grown in culture. I. Repair of x-ray damage in surviving Chinese hamster cells. *Radiat. Res. 13*: 556-593.

[20] Hill, C. K., Carnes, B. A., Han, A. and Elkind, M. M. (1985) Neoplastic transformation is enhanced by multiple low doses of fission-spectrum neutrons. *Radiat. Res. 102*: 404-410.

[21] Lubin, J. H., Boice, J. D., Edling, C., Hornung, R. W., Howe, G., Kunz, E., Kusiak, R. A., Morrison, H. I., Radford, E. P. and Samet, J. M. (1995) Radon-exposed underground miners and inverse dose-rate (protraction enhancement) effects. *Health Phys. 69*: 494-500.

[22] Goto, S., Watanabe, M. and Yatagai, F. (2002) Delayed cell cycle progression in human lymphoblastoid cells after exposure to high-LET radiation correlates with extremely localized DNA damage. *Radiat. Res. 158*: 678-686.

[23] Walker, J. T., Todd, P. and Walker, O. A. (2002) Heritable non-lethal damage to cultured human cells irradiated with heavy ions. *J. Radiat. Res. 43 (Suppl.)*: S187-S191.

[24] ICRP (2003) Relative biological effectiveness (RBE), quality factor (Q), and radiation weighting factor (w_R). *ICRP Publication 92*, Ann. ICRP 33 (4).

[25] Goodhead, D. T. (1992) Track structure consideration in low and low dose rate effects of ionizing radiation. *Adv. in Radiat. Biol. 16*: 7-44.

[26] Rydberg, B. (1996) Cluster of DNA damage induced by ionizing radiation: Formation of short DNA fragments. II. Experimental detection. *Radiat. Res. 145*: 200-209.

[27] Rydberg, B., Cooper, B., Cooper, P. K., Holley, W. R. and Chatterjee, A. (2005) Dose-dependent misrejoining of radiation-induced DNA double-strand breaks in human fibroblasts: Experimental and theoretical study for high- and low-LET radiation. Radiat. *Res. 163*: 526-534.

[28] Phillips, R. A. and Tolmach, L. J. (1966) Repair of potentially lethal damage in x-irradiated HeLa cells. *Radiat. Res. 29*: 413-432.

[29] Qin, J. and Li, L. (2003) Molecular anatomy of the DNA damage and replication checkpoints. *Radiat. Res. 159*: 139-148.

[30] Roy, K., Kodama, S., Suzuki, K. and Watanabe, M. (1999) Delayed cell death, giant cell formation and chromosome instability induced by X-irradiation in human embryo cell. *J. Radiat. Res. 40*: 311-322.

[31] Thompson, L. H. and Humphrey, R. M. (1970) Proliferation kinetics of mouse L-P59 cells irradiated with ultraviolet light: A time-lapse photographic study. *Radiat. Res. 41*: 183-201.

[32] Suzuki, K., Ojima, M., Kodama, S. and Watanabe, M. (2003) Radiation-induced DNA damage and delayed induced genomic instability. *Oncogene 22*: 6988-6993.

[33] Suzuki, K., Yokoyama, S., Waseda, S., Kodama, S. and Watanabe, M. (2003) Delayed reactivation of p53 in the progeny of cells surviving ionizing radiation. *Cancer Res. 63*: 936-941.

[34] Sasaki, H. and Nishimoto, T. (1987) Chromosome condensation may enhance X-ray-related cell lethality in a temperature-sensitive mutant (tsBN2) of baby hamster kidney cells (BHK21). *Radiat. Res. 109*: 407-418.

[35] Suzuki, M., Suzuki, K., Kodama, S. and Watanabe, M. (2006) Phosphorylated histone H2AX foci persist on rejoined mitotic chromosomes in normal human diploid cells exposed to ionizing radiation. *Radiat. Res. 165*: 269-276.

[36] Mehnati, P., Morimoto, S., Yatagai, F., Furusawa, Y., Kobayashi, Y., Wada, S., Kanai, T., Hanaoka, F. and Sasaki, H. (2005) Exploration of 'over kill effect' of high-LET Ar- or Fe-ions by evaluating the fraction of non-hit cell and interphase death. *J. Radiat. Res. 46*: 343-350.

[37] Little, J. B. (2003) Genomic instability and bystander effects: a historical perspective. *Oncogene 22*: 6978-6987.

In: Genome Research Advances
Editor: Alberto P. Bellini, pp. 179-191

ISBN: 978-1-60021-805-7

Chapter IV

VIRAL INTEGRATION INTO HUMAN GENOME

Akihiro Tamori and Shuhei Nishiguchi
Department of Hepatology, Osaka City University Graduate School of Medicine, Japan.

ABSTRACT

Some viruses are integrated into host genome. DNA integration of viruses is considered one of the most important steps in the development of several types of human cancers. However, few methods can efficiently detect viral integration sites. Recently developed techniques for polymerase chain reaction (PCR) have facilitated sequencing of the host DNA adjacent to an integrated retroviral provirus. *Alu*-PCR and cassette-ligation-mediated PCR can be used to selectively amplify integrated DNA, provided that a portion of a nearby gene has already been sequenced. Analysis of virus integration can reveal new functions of genes located near integration sites. Such genes may be candidates for new tumor genes or tumor suppressor genes. Hepatitis B virus (HBV) is a major cause of chronic hepatitis and hepatocellular carcinoma (HCC) in humans. HBV DNA is known to be integrated into the human genome of HCC. In this issue, we review recent reports and discuss our findings concerning HBV DNA integration and the pathogenesis of liver cancer.

INTRODUCTION

Over an evolutionary time span, viral infections and the integration of retrovirus genomes or retrotransposons have substantially contributed to the accumulation of repetitive DNA sequences in human genomes. Viruses have evolved highly specialized mechanisms to transport their genome into the nuclei of host cells and fix their genomes permanently by integration into the host genome. Some studies have identified sequences adjacent to viral integration sites and demonstrated the overexpression or disruption of genes near such sites. Several investigations have identified new genes and shed light on the functions of gene-coding protein. Another interesting aspect of viral integration is alteration of the methylation pattern of the host genome [1]. Methylation of promoter regions is closely related to silencing of affected genes, whereas hypomethylation of promoter regions induces gene overexpression. This phenomenon suggests that the integration of foreign DNA can induce important functional consequences extending beyond the site of insertion. Several years ago, the human chromosomal sites of sequence-successive integrated HBV DNA could not be easily identified. However, the Human Genome Project has allowed the location of sequences adjacent to integrated viral DNA to be determined. Protein-coding genes near viral integration sequences can also be found. Now, we can identify viral integration sites in the human genome and study the role of such sites in the pathogenesis of viral infection, especially as related to carcinogenesis. In this chapter, we describe viruses capable of integrating into the human genome, techniques for the detection of integration sites of host DNA, the role of hepatitis B virus integration in hepatocarcinogenesis, and the alteration of DNA methylation pattern by viral integration.

Viruses Capable of Integrating into Human Genome

Integrated viral DNA is detected as a provirus in the host genome. Viral integration is directly or indirectly related to pathogenesis in the host. Some families of viruses can integrate into the genome of target cells. Retrovirus particles contain reverse transcriptase and can integrate into the host genome. Integration is an essential step in the life cycle of retroviruses, resulting in the stable joining of viral cDNA to the host cell chromosomes. Retrovirus particles contain a second enzyme, integrase, which mediates the insertion of viral DNA into essentially random sites in host DNA. These viruses have unique inverted sequences, termed long terminal repeats (LTRs), which contain important regulatory sequences [2]. Viral integration sites of host chromosomes where integration occurs are nonspecific. However, target sequences for viral integration are thought to exist in the host genome. Integration of retroviral DNA may affect host gene expression and disrupt normal cellular functions. The mechanism by which integration sites are chosen is poorly understood, but is apparently influenced by several factors, including DNA sequence and structure, DNA-binding proteins, DNA methylation, and transcription. Integrase, the viral enzyme responsible for catalyzing integration, also plays a key role in controlling the choice of target sites. The integrase domain responsible for target site selection has been mapped to the central core region. A better understanding of the interactions between the target-specifying motif of integrase and the target DNA may foster the development of techniques to manipulate integration into particular chromosomal sites. Another approach to directing integration is to fuse integrase with a sequence-specific DNA-binding protein, which biases integration into the recognition site of the fusion partner in vitro. Successful incorporation of fusion proteins into infectious virions and the identification of proteins optimal suited for fusion with integrase will promote the development of site-specific vectors. Retroviruses are promising tools gene delivery in experimental and therapeutic protocols.

Table: The transformation mechanisms by viruses, which can integrate in host genome (modified table in Principles of virology Second edition, Washington DC, ASM press.)

	Viral genetic information in transformed cells			*Transformation mechanisms*		
Virus	Physical state	Lytic function	Transforming functions	Cell cycle progression	Apoptosis	Immune modulation or stimulation
Adenovirus						
Human subgroup C and A serotype	Integrated into host genome by random recombination	Lost during integration	E1A and E1B gene products	Deregulated; inactivation of pRb family proteins by E1A proteins	Blocked by inactivation of p53 and by E1B 19-kDa protein	Inhibition of transcription of MHC class I genes by subgroup A E1A proteins
Hepadnavirus						
Hepatitis B virus	Episomal; viral sequences are integrated in some hepatocellular carcinoma	None	X protein, Insertional mutagenesis?	Defects in DNA repair; altered transcriptoin of proto-oncogenes and transcriptional regulators	Both pro- and antiapoptotic effects are seen	Continuous tissue damaging due to immune cell attack drives abnormal liver proliferation
Papillomavirus						
Human types 16 and 18	Integrated into host genome by random recombination	Lost during integration	E6 and E7 proteins	Deregulated; inactivation of Rb family proteins by E7 protein	Blocked by E6 protein-induced degradation of p53	None known
Polyomavirus						
Simian virus 40	Integrated into host genome by random recombination	Lost during integration	LT; sT in some circumstances	Deregulated; inactivation of Rb family proteins by LT and stimulation of Mapk	Blocked by inactivation of p53 by LT	None known
Retroviruses						
Transducing (Rous sarcoma virus)	Integrated	None	Virally transduced, cell-derived oncogenes	Loss of tissue, developmental, and cell cycle regulation by oncogenes	None	None known
Nontransducing (Rous-associated viruses 1and 2)	Integrated	None	Insertional muta-genesis, alteration of cellular proto-oncogenes (or their expression) in vicinity of proviral integration site	Loss of tissue, developmental, and cell cycle regulation by inappropriate expression of proto-oncogenes	None	Viral superantigen stimulates T cells and induces proliferation of B cells
Long latency (human T-lymphotropic virus type 1)	Integrated	None	Viral transactivor proteins, Tax	Stimulates T-cell proliferation by induction of Nf-kb and IL-2	None	Abnormal T-cell proliferation is a critical early event, but other currently unknown changes are required

Besides retroviruses, other families of viruses can integrate into the host genome, including adenoviruses, hepadnaviruses, human papillomavirus (HPV) types 16 and 18, and polyomaviruses [3]. Unlike retroviruses, the integration step is not necessary for these viruses to replicate. Hepadnaviruses have the direct repeats, DR1 and DR2, which are not LTRs. Another interesting aspect is that viruses with integration ability are closely related to cell transformation. Transformation mechanisms are summarized in the Table. It is unclear how viral DNA integration into host genome induces cell transformation because the exact locations of viral integration have not been identified in genomes from large numbers of cancers.

Techniques for Detection of the Flanking Human Sequence of Integrated Virus DNA

During the past decade, research on virus integrants and flanking cellular sequences was carried out by the conventional cloning technique of screening individual genomic libraries. This procedure requires large amounts of DNA. Small samples of tissue, such as biopsy specimens, cannot be effectively analyzed. Modified techniques for polymerase chain reaction (PCR) can analyze the human genome in limited amounts of tissue. Several methods can be used to amplify unknown flanking sequences of integrated virus DNA. Inverse PCR, first described in 1988 [4,5], can amplify regions that flank any DNA segment of known sequence, either upstream, downstream, or both. Inverse PCR involves digestion of DNA by restriction enzymes with conveniently located sites, circularization of digested DNA fragments at a very low concentration, which favors the formation of monomers, and amplification of the circularized DNA with the use of primers designed towards unknown flanking sequences, in opposite orientation to that for normal PCR [6]. Other methods based on ligation arc also availablc. Cassctte-ligation mediated PCR has been developed to permit selective and specific amplification of cDNA sequences from total cellular RNA [7]. We use this technique to amplify HBV DNA-integrated human sequences [8] (Figure 1). Cassette-ligation mediated PCR involves the following three steps: (i) digestion of DNA with multiple restriction enzymes, (ii) ligation of cleavage products to double-strand DNA cassettes possessing a corresponding restriction site, and (iii) amplification of cassette-ligated restriction fragments containing a short known viral sequence by PCR, using the specific primer that anneals to the known viral sequence and the cassette primer that anneals to one strand of the cassette. In cassette-ligation mediated PCR, one of the nucleotides at the 5' end of the cassette is dephosphorylated to avoid amplification of false-positive fragments. Other methods are based on the nonspecific annealing of primers. Target gene walking PCR or target gene walking by low stringency PCR uses 'targeted' primers designed from target sequences, 'internal' detection primers for use in oligomer extension assay, which are 'targeted' primers located a short distance internally with respect to the PCR primers, and 'walking' primers, used to hybridize to sequences not known to preexist either upstream or downstream of the target sequence. This procedure does not require a ligation step. However, it uses radioisotopes and amplifies some false-positive fragments. In contrast, simple two-step PCR does not require critical ligation, circularization, or radioisotopes [9]. This technique can amplify the unknown cellular DNA flanking integrated proviruses. Two-step PCR consists of three steps: (i) an initial PCR using a biotinylated provirus-specific primer and a partly degenerate, arbitrary primer (flanking primer) with a fixed 3' end, which will hybridize within a statistically defined range in the cellular DNA flanking the provirus; (ii) purification of the PCR products that originate from the provirus primer, performed with the use of streptavidin-coated magnetic beads; and (iii) a second PCR using the purified initial PCR products as templates and a nested provirus-specific primer together with a biotinylated nondegenerate primer derived from the flanking primer. The flanking primers have GGCCT sequences at the 3' end and 5 or 7 random nucleotides next to this fixed set of 5 nucleotides. Occasionally, provirus-flanking cellular coherent fragments cannot be amplified with the use of flanking primers that contain GGCCT in their 3' end. There are also some difficulties in the design of primer sequences and optimizing the annealing conditions for efficient amplification. Restriction-site PCR (RS-PCR) is a direct method that rapidly retrieves unknown sequences adjacent to a known locus through the use of universal primers [10]. The used restriction site oligonucleotide primer (RSO) incorporates (from 5' to 3') a T7 RNA polymerase promoter recognition site, 10 random nucleotides, and 6 bases specific for a chosen restriction endonuclease site. First-round RS-PCR is performed at a permissive reannealing low temperature, followed by a nested second-round PCR performed at a stringent reannealing temperature. This method can rapidly amplify unknown flanking sequences and is used to detect human sequences adjacent to integrated papillomaviruses in cervical tumors. Finally, we would like to introduce *Alu*-PCR, which can amplify unknown sequences adjacent to integrated virus DNA. *Alu*-PCR was first developed to amplify human genomes in the background of nonhuman genomes such as (lambda)-phages, yeast cells, and human-rodent hybrid cells. *Alu* repeat sequences are short interspersed repeat DNA sequences in mammalian genomes. These repeats are found ubiquitously in human DNA, and 900,000 are believed to exist in the haploid human genome, giving an average distance between copies of about 4 kilobases. To avoid amplification between *Alu* sequences, the used primers have sequences tagged with UTP instead of TTP. This method is reported to have several advantages over other protocols used to amplify unknown flanking sequences [11]. First, the reaction can be performed in a single tube. Second, there is no ligation or dilution step. Third, other methods that require nonspecific annealing of primers or degenerated primers sometimes amplify fragments from inside the known sequence. As compared with inverse PCR, *Alu*-PCR is considered more efficient. However, *Alu* sequences appear to be abundant in certain chromosomal regions and deficient in others. Unknown flanking sequences cannot be amplified for integration sites located far from *Alu* sequences. Difficulty in setting the annealing conditions for tag primers is another drawback of this method.

We summarized techniques used to detect unknown flanking sequences adjacent to virus DNA integration sites. There is no optimal method that can detect every virus integration site. The most efficient method and suitable primer sequences should be selected according to the specific viruses or target organs being studied.

HEPATITIS B VIRUS INTEGRATION AND LIVER CANCER

Hepatitis B virus (HBV), classified as a hepadnavirus, is one of the most important causes of human hepatocellular carcinoma (HCC) [12]. The mechanism of human hepatocarcinogenesis and the role of HBV are not fully understood (Figure 2). Epidemiological evidence has shown that HCC is often detected in chronic liver damage, especially hepatic cirrhosis, with HBV infection. Similar to hepatitis C virus, HBV-induced chronic inflammation is a major factor in hepatocarcinogenesis. Continuous viral inflammation is accompanied by repeated hepatocyte necrosis, followed by regeneration. Such an accelerated cell cycle might be associated with accumulation of genetic errors in the liver. Mutations of mitochondrial DNA (mtDNA) have been extensively investigated in hepatocarcinogenesis associated with viral infection. As compared with control liver tissue, the frequency of mtDNA mutations is markedly increased in both noncancerous and cancerous liver specimens from individuals with HCC [13]. The accumulated amount of mtDNA mutations in HCC tissue reflects the degree of malignancy. The frequency of mtDNA mutations in HCC tissue is greater than that in other types of tumors. These observations suggest that the repeated destruction and regeneration of liver tissue in chronic viral hepatitis leads to the accumulation of mtDNA mutations. The genetic instability that results in the high rate of mtDNA mutations in cancerous liver tissue is also consistent with the multicentric hepatocarcinogenesis detected clinically.

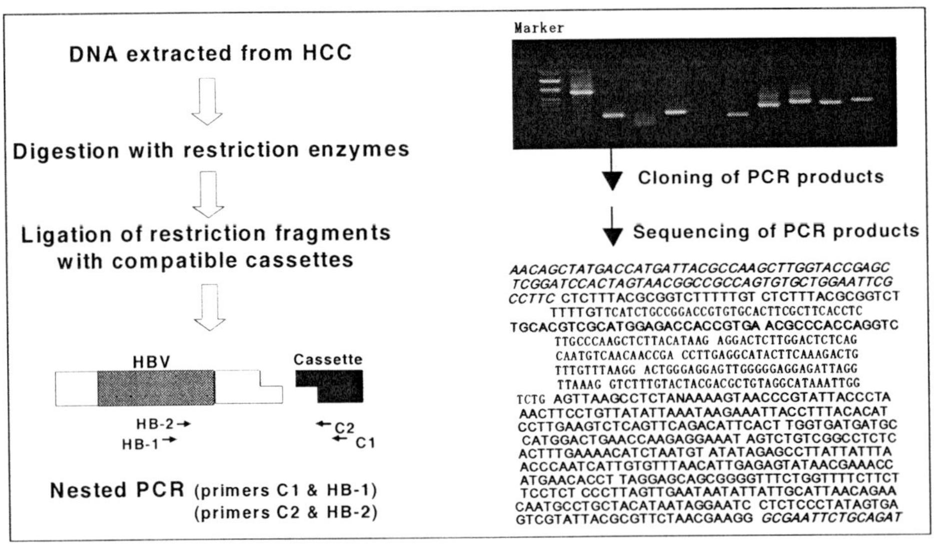

Figure 1. Cassette-ligation-mediated PCR to identify junctions between integrated viral DNA and human DNA and sequence the region on either side of junctions. White bars, human DNA treated with one of three restriction enzymes. Gray bars, HBV DNA. Black bars, cassette DNA. Bars on bottom, amplified PCR products. Small arrows indicate directions of the primers. After ligation of DNA treated with a restriction enzyme to a DNA cassette, nested PCR was done with two sets of primers. PCR products separated by electrophoresis were cut out from gel and were subcloned in vectors. Sequencing the PCR product showed both sides of junctions between the integrated viral DNA and human DNA. Underlined capitals show sequences of primers for PCR, large capitals show a integrated viral sequence, and shadowed capitals shows a flanking host DNA sequence.

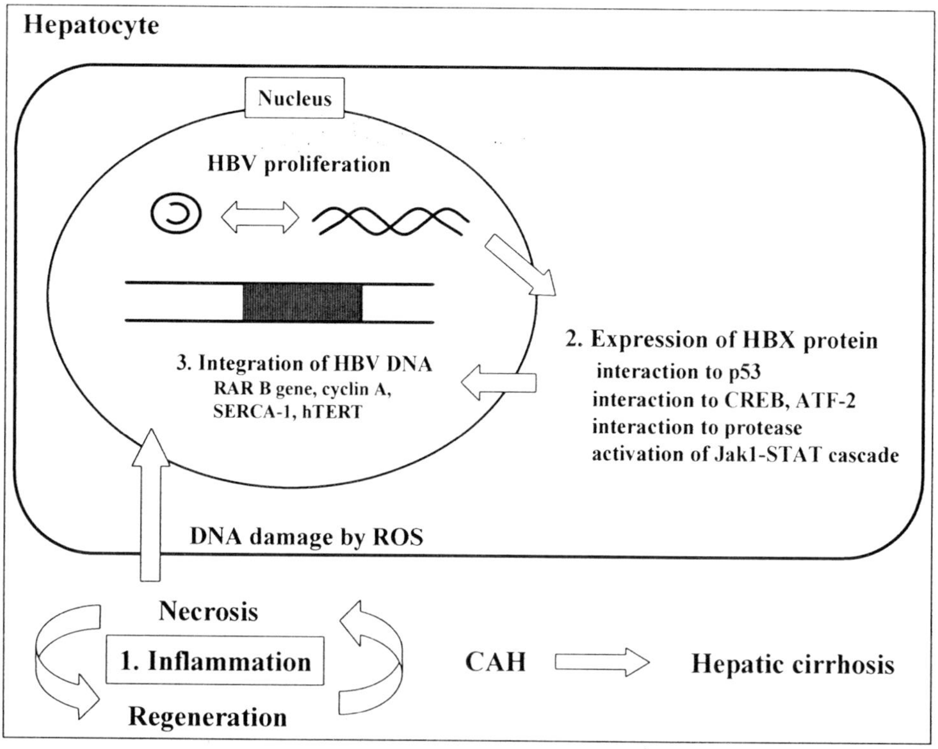

Figure 2. HBV related hepatocarcinogenesis. 1. Chronic liver inflammation by HBV infection. 2. The function of Hepatitis B virus X protein. 3. Insertional mutagenesis by HBV DNA.

HBV has unique functions, some of which promote the transformation of hepatocytes. X protein of HBV, a transcriptional transactivator, plays a role in liver carcinogenesis. HBx-coding plasmids transform NIH 3T3 cells [14] and HBx protein transforms liver cells in transgenic mice [15]. HBx antigens have been detected in many specimens of HCC from patients infected with HBV. HBx protein binds to and inactivates wild-type p53 protein [16]. In HBsAg-positive HCC, associated with a low frequency of p53 gene mutations, p53 protein is inactivated, usually by complexing with HBx protein [17]. HBx protein also represses the transcription of p21 (waf1, cyclin dependent kinase inhibitor) by down-regulating the activity of Sp1 via a p53-independent pathway [18]. In addition, HBx appears to increase levels of insulin-like growth factor I (IGF-I) receptor RNA, a promoter-activity receptor in human hepatocellular carcinoma cell lines [19]. In response to activation of IGF-I receptor by HBx, the mitogenic effects of IGFs (IGF-I and IGF-II) on SNU368 cells may play a role in the pathogenesis of HCC [20]. A recent serial analysis of gene expression profiles in human primary hepatocytes infected with a HBx expression vector confirms that HBx functions as a major regulator in common cellular pathways, which in turn regulate protein synthesis, gene transcription, and protein degradation [21]. However, Henkler et al. did not detect HBx protein in any of 31 specimens of HCC from Chinese patients, most of whom were infected with HBV, on Western blotting, immunoprecipitation, or immunohistochemical staining [22]. Moreover, the roles of HBx protein in liver-cell proliferation and apoptosis remain controversial. In a transgenic mouse model, HBx seems to stimulate the apoptotic turnover of hepatocytes, independently of p53 [23]. In HBV-infected cells, HBx was localized in mitochondria, mapped to amino acids 68-77, which are essential for cell death but lack transactivation function [24]. The ectopic expression of HBx causes loss of mitochondrial membrane potential and subsequently induces mitochondrion-dependent cell death. Apoptosis by HBx expression seems to act against hepatocyte transformation.

Another function of HBV is integration into the human genome of liver cells. HBV DNA is integrated into host DNA in about 90% of liver tumor samples from patients with HBsAg. Some of these patients have chromosomal inversions, translocations, or microdeletions at sites of such integration; subsequently, tumors may develop [25,26]. Genes adjacent to integrated HBV DNA have been identified: in one HCC sample, HBV DNA had been integrated into an exon-like region of the retinoic acid receptor B gene (RAR B) [27], and in another the DNA had been integrated into an intron of the cyclin A2 gene in early-stage HCC [28]. The erythrocytic progenitor cells were transformed by a retrovirus carrying the HBV/RAR B construct. It is speculated that the production of chimeric HBV/RAR B may be involved in hepatocyte transformation. Cyclin A2 plays an important role in checkpoints of the cell cycle. The expression of hybrid HBV/cyclin A2 induces cell transformation. Woodchuck hepatitis virus DNA, which induces liver tumors in woodchucks, was found to be integrated in *c-myc* or *N-myc* in 30% of liver tumors, most of them having increased expression of these genes [29]. However, in humans, HBV DNA is randomly integrated into the genome. Target genes for activation or inhibition by HBV DNA integration remain largely unexplored. Recently, methods for amplifying flanking human sequences of integrated virus DNA have been developed. These techniques have successfully identified some genes successive to integrated HBV DNA. *Alu*-PCR analysis has shown that HBV DNA was integrated into the gene encoding a sarcoendoplasmic calcium ATPase-dependent pump (SERCA1) [30]. The SERCAs are regulators of endoplasmic reticulum calcium stores. Calcium homeostasis has an important role in cell proliferation and viability. HBV integration activated the SERCA1 encoding gene, followed by synthesis of truncated forms of the SERCA protein, which regulates apoptosis.

The use of HBV-Alu PCR has lead to the isolation of new HBV-DNA integration sites [31]. The integrated viral genome altered the expression of key regulatory cellular genes: neurotropic tyrosine receptor kinase 2 (NTRK2) gene, IL-1R-associated kinase 2 (IRAK2) gene, p42 mitogen-activated protein kinase 1 (p42MAPK1) gene, inositol 1,4,5-triphosphate receptor type 2 (IP3R2) gene, inositol 1,4,5-triphosphate receptor (IP3R) type 1 (IP3R1) gene, alpha 2,3 sialyltransferase (ST3GAL VI or SITA) gene, thyroid hormone uncoupling protein (TRUP) gene, EMX2-like gene, and human telomerase reverse transcriptase (hTERT) gene [32]. Another group has reported that the studies of HBV-infected HCC and human papillomavirus (HPV)-infected cervical cancer revealed two HBVs and three HPVs integrated into the hTERT gene. Four of the five integrations occurred within the hTERT promoter and upstream regions, and the other occurred in intron 3 of the hTERT gene. The integration of these viruses was apparently associated with the up-regulation of hTERT expression. The hTERT gene is a telomerase catalytic subunit gene, associated with telomerase activity in human cells. Several transcription factors, including oncogene products (e.g., c-Myc) and tumor suppressor gene products (e.g., WT1 and p53), are able to control hTERT transcription when overexpressed. However, it remains to be resolved whether cancer-associated alteration of these factors is primarily responsible for hTERT activation during carcinogenesis. The hTERT gene may be a target site for viruses frequently associated with human cancers, such as human HPV and HBV. Consequently, virus integration into the promoter region of hTERT activates *cis* hTERT expression. Thus, a variety of cellular and viral oncogenic mechanisms may converge on transcriptional control of the hTERT gene.

We have identified another gene in which HBV DNA is frequently integrated. Integrated HBV DNA was found in the mixed-lineage leukemia (MLL) 2 gene in three HCCs with HBV infection. MLL, a homologue of the *Drosophila trithorax* gene, is involved in translocations in infantile leukemia and amplified in some adult myeloid leukemias [33,34]. MLL2 was identified at chromosome band 19q13.1, closely resembling the MLL gene [35]. MLL2 is amplified in some pancreatic carcinoma and glioblastoma cell lines. This gene was amplified and expressed in three cases of HCC with HBV DNA integration into MLL2.

Taken together, these results suggest that HBV frequently targets cellular genes involved in cell signaling, some of which may be preferential targets for viral integration, followed by the malignant transformation of liver cells.

Methylation Pattern of Human Genome and Integrated Virus DNA

Cyotosine methylation of CpG dinucleotides is an important mechanism of transcriptional regulation in vertebrates. Integrates are frequently methylated *de novo*. Methylation is speculated to be a host defensive mechanism against expression of genes from extrinsic pathogens. This phenomenon was first observed in integrated adenovirus type 12 in hamster tumors [36]. The integrated adenovirus genomes in transformed cells could not be cleaved with methylation-sensitive restriction enzymes such as *HpaII*, whereas the *HpaII* isoschizomer *MspI* cleaved the inserted adenovirus DNA to the same pattern as adenovirus virion DNA. The different results of cleavage with restriction enzymes were ascribed to *de novo* methylation of the integrated DNA. Interestingly, the level of promoter methylation negatively correlated with promoter activity. These results suggest that the sequence-specific methylation of viral promoters leads to the inactivation of these promoters, followed by the down-expression of viral genes. Provirus silencing is followed by DNA methylation of both provirus regulatory regions and adjacent cellular sequences. This CpG methylation is very dense and resistant to the demethylation effects of 5-aza-2'-deoxycytidine. However, little is known about the influence of integrated retrovirus on adjacent chromosomal DNA methylation in the host genome. One report has described the interesting features of a simplified Rous sarcoma virus integrated into a non-transcribed hypermethylated DNA sequence in the Syrian hamster genome [37]. After integration, CpG methylation of this sequence was nearly completely lost, and hypomethylated DNA permitted proviral transcription and hamster cell transformation by the proviral v-src oncogene.

In the human genome, CpG islands in gene promoter regions are targets for methylation and closely related to the expression of the lower stream of genes. Recently, the promoter regions of some tumor suppressor genes were reported to be hypermethylated and their expression inhibited in human cancers. The methylation pattern of some genes can be altered by senescence and viral infection. In HCC, cell cycle regulatory genes are more methylated in patients with hepatitis virus infection than in patients without such infection [38,39]. However, Restriction Landmark Genomic Scanning (RLGS), a two-dimensional system that displays 2000-3000 *Not I* landmark sites in a single gel electrophoresis experiment, has yielded conflicting results. *Not I* landmark spot, i.e., a *Not I* cluster sequence enriched with CpG dinucleotides, has shown high signal intensity in HBV-integrated cell lines or in primary HCCs, but not in HCCs or tumor-cell lines free of HBV integration [40]. Analysis by RLGS has suggested that HBV integration is associated with hypomethylated epigenetic alterations at specific sequences.

It remains controversial which host genome with virus integration is hypermethylated or hypomethylated in human tissue [41,42]. Inconsistent results about alterations in DNA methylation may be related to the approach used in any given study. Except for several cases, virus integration occurs at random sites of the human genome. The methylation pattern appears to be altered at human DNA near sites of viral DNA integration. Many studies have reported that the methylation pattern in the promoter region of common genes is changed in cancers with viral infection. Whether viral integration directly alters the DNA methylation pattern of genes remains a matter of debate. We propose that studies should examine DNA methylation of genes neighboring viral integration sites. Such studies should initially identify viral integration sites in the human genome by analyzing clinical samples. Then, protein-coding sequences and CpG dinucleotides should be looked for near these sites. Methylation-specific sequencing analysis can delineate alterations in DNA methylation associated with integration of viral DNA.

Available evidence thus indicates that by altering methylation patterns in host DNA adjacent to sites of viral integration, some kinds of viruses may influence the expression of human genes in addition to causing infectious disease.

Conclusions

Recently, the retroviral tagged cancer gene database (RTCGD) was developed in mouse hematopoietic tumors with retroviral insertional mutagenesis [43]. RTCGD will make it possible to easily identify common integration sites in the same mouse tumors. Research about interactions between humans and viruses is ongoing. DNA mutations play an important role in allowing some viruses to escape from the human immune system. Other viruses alter the expression of human genes by integrating into the human genome. We have just begun to search for viral integration sites of the human genome by means of modified PCR techniques. Considerable data on integration sites are being compiled and will hopefully elucidate what human sequences on chromosomes are targets for viral integration. Specific sequences of the human genome that are targets for virus integration are also expected to be identified. Single nucleotide polymorphisms (SNPs) in the human genome are important determinants of characteristics such as allergy, drug sensitivity, and immunity against infection [44]. The establishment of target sequences for viral integration will lead to the identification of SNPs for viral integration. Analysis of genes near virus integration sites in human cancers will help to identify unknown tumor genes or tumor suppressor genes. Besides carcinogenesis, unresolved pathogenic mechanisms induced by viral infection might be explained by viral integration followed by altered expression of human genes. In some conditions, viral integration may be closely related to mutations in the human genome. We hope that continued research on viral integration will shed further light on evolution of the human genome.

REFERENCES

[1] Doerfler W, Hohlweg U, et al. (2001) Foreign DNA integration--perturbations of the genome--oncogenesis. *Ann N Y Acad Sci. 945*: 276-288.

[2] Flint SJ, Enquist LW, Racaniello VR, Skaika AM. *Principles of virology*. Second edition, Washington DC, ASM press.

[3] Thorland EC, Myers SL, et al. (2000) Human papillomavirus type 16 integrations in cervical tumors frequently occur in common fragile sites. *Cancer Res. 60*: 5916-5921.

[4] Triglia T, Peterson MG, et al. (1988) A procedure for in vitro amplification of DNA segments that lie outside the boundaries of known sequences. *Nucleic Acids Res. 16*: 8186.

[5] Ochman H, Gerber AS, et al. (1988) Genetic applications of an inverse polymerase chain reaction. *Genetics. 120*: 621-623.

[6] Tsuei DJ, Chen PJ, et al. (1994) Inverse polymerase chain reaction for cloning cellular sequences adjacent to integrated hepatitis B virus DNA in hepatocellular carcinomas. *J Virol Methods. 49*: 269-284.

[7] Isegawa Y, Sheng J, et al. (1992) Selective amplification of cDNA sequence from total RNA by cassette-ligation mediated polymerase chain reaction (PCR): application to sequencing 6.5 kb genome segment of hantavirus strain B-1. *Mol Cell Probes. 6*: 467-475.

[8] Tamori A, Nishiguchi S, et al. (2003) Sequencing of human-viral DNA junctions in hepatocellular carcinoma from patients with HCV and occult HBV infection. *J Med Virol. 69*: 475-481.

[9] Sorensen AB, Duch M, et al. (1993) Amplification and sequence analysis of DNA flanking integrated proviruses by a simple two-step polymerase chain reaction method. *J Virol. 67*: 7118-7124.

[10] Weber KL, Bolander ME, et al. (1998) Rapid acquisition of unknown DNA sequence adjacent to a known segment by multiplex restriction site PCR. *Biotechniques. 25*: 415-419.

[11] Minami M, Poussin K, et al. (1995) A novel PCR technique using Alu-specific primers to identify unknown flanking sequences from the human genome. *Genomics. 29*: 403-408.

[12] Brechot C, Gozuacik D, et al. (2000) Molecular bases for the development of hepatitis B virus (HBV)-related hepatocellular carcinoma (HCC). *Semin Cancer Biol. 10*: 211-231.

[13] Nishikawa M, Nishiguchi S, et al. (2001) Somatic mutation of mitochondrial DNA in cancerous and noncancerous liver tissue in individuals with hepatocellular carcinoma. *Cancer Res. 61*: 1843-1845.

[14] Höhne M, Schaefer S, et al. (1990) Malignant transformation of immortalized transgenic hepatocytes after transfection with hepatitis B virus DNA. *EMBO J. 9*:1137-1145.

[15] Kim CM, Koike K, et al. (1991) HBx gene of hepatitis B virus induces liver cancer in transgenic mice. *Nature. 351*: 317-320.

[16] Feitelson MA, Zhu M, et al. (1993) Hepatitis B x antigen and p53 are associated in vitro and in liver tissues from patients with primary hepatocellular carcinoma. *Oncogene 8*:1109-1117.

[17] Elmore LW, Hancock AR, et al. (1997) Hepatitis B virus X protein and p53 tumor suppressor interactions in the modulation of apoptosis. *Proc Natl Acad Sci USA 94*:14707-14712.

[18] Ahn JY, Chung EY, et al. (2001) Transcriptional repression of p21(waf1) promoter by hepatitis B virus X protein via a p53-independent pathway. *Gene. 275*:163-168.

[19] Kim SO, Park JG, et al. (1996) Increased expression of the insulin-like growth factor I (IGF-I) receptor gene in hepatocellular carcinoma cell lines: implications of IGF-I receptor gene activation by hepatitis B virus X gene product. *Cancer Res. 56*:3831-3836.

[20] d'Arville CN, Nouri-Aria KT, et al. (1991) Regulation of insulin-like growth factor II gene expression by hepatitis B virus in hepatocellular carcinoma. *Hepatology. ;13*:310-315.

[21] Wu CG, Forgues M, et al. (2002) SAGE transcript profiles of normal primary human hepatocytes expressing oncogenic hepatitis B virus X protein. *FASEB J. ;16*:1665-1667.

[22] Henkler F, Waseem N, et al. (1995) Mutant p53 but not hepatitis B virus X protein is present in hepatitis B virus-related human hepatocellular carcinoma. *Cancer Res. 55*:6084-6091.

[23] Terradillos O, Pollicino T, et al. (1998) p53-independent apoptotic effects of the hepatitis B virus HBx protein in vivo and in vitro. *Oncogene. 17*:2115-2123.

[24] Shirakata Y, Koike K. (2003) Hepatitis B virus X protein induces cell death by causing loss of mitochondrial membrane potential. *J Biol Chem. 278*:22071-22078.

[25] Hino O, Shows TB, et al. (1986) Hepatitis B virus integration site in hepatocellular carcinoma 17;18 translocation. *Proc. Natl. Acad. Sci. USA 83*: 8338-8342.

[26] Tokino T, Matsubara K. (1991) Chromosomal site for hepatitis B virus integration in human hepatocellular carcinoma. *J. Virol. 65*: 6761-6764.

[27] Dejean A, Bougueleret L, et al. (1986) Hepatitis B virus DNA integration in a sequence homologous to v-erb-A and steroid receptor genes in a hepatocellular carcinoma. *Nature 322*: 70-72.

[28] Wang J, Chenivesse X, et al. (1990) Hepatitis B virus integration in a cyclin A gene in a hepatocellular carcinoma. *Nature 343*: 555-557.

[29] Fourel G, Trepo C, et al. (1990) Frequent activation of N-myc genes by hepadnavirus insertion in woodchuck liver tumours. *Nature 347*: 294-298.

[30] Chami M, Gozuacik D, et al. (2000) Hepatitis B virus-related insertional mutagenesis implicates SERCA1 gene in the control of apoptosis. *Oncogene. 19*: 2877-2886.

[31] Gozuacik D, Murakami Y, et al. (2001) Identification of human cancer-related genes by naturally occurring Hepatitis B Virus DNA tagging. *Oncogene. 20*:6233-6240.

[32] Ferber MJ, Montoya DP, et al. (2203) Integrations of the hepatitis B virus (HBV) and human papillomavirus (HPV) into the human telomerase reverse transcriptase (hTERT) gene in liver and cervical cancers. *Oncogene. 22*: 3813-3820.

[33] Djabali M, Shelliri L, et al. (1992) A trithorax-like gene is interrupted by chromosome 11q23 translocations in acute leukemia. *Nat Genet, 2*: 113-118.

[34] Ziemin-van der Poel, S, McCabe NR, et al. (1991) Identification of a gene, MLL, that spans the breakpoint in the 11q23 translocations associated with human leukemias. *Proc. Natl. Acad. Sci. USA 88*: 10735-10739.

[35] Fitzgerald KT, Diaz MO. (1999) MLL2: A new mammalian member of the trx/MLL family of genes. *Genomics 59*: 187-192.

[36] Remus R, Kammer C, et al. (1999) Insertion of foreign DNA into an established mammalian genome can alter the methylation of cellular DNA sequences. *J Virol. 73*: 1010-1022.

[37] Hejnar J, Hajkova P, et al. (2001) CpG island protects Rous sarcoma virus-derived vectors integrated into nonpermissive cells from DNA methylation and transcriptional suppression. *Proc Natl Acad Sci U S A. 98*: 565-569.

[38] Shen L, Ahuja N, et al. DNA methylation and environmental exposures in human hepatocellular carcinoma. *J Natl Cancer Inst. 94*:755-761.

[39] Kaneto H, Sasaki S, et al. (2001) Detection of hypermethylation of the p16 (INK4A) gene promoter in chronic hepatitis and cirrhosis associated with hepatitis B or C virus. *Gut. 48*: 372-377.

[40] Nagai H, Baba M, et al. (1999) Isolation of NotI clusters hypomethylated in HBV-integrated hepatocellular carcinomas by two-dimensional electrophoresis. *DNA Res. 6*: 219-225.

[41] Remus R, Kammer C, et al. (1999) Insertion of foreign DNA into an established mammalian genome can alter the methylation of cellular DNA sequences. *J Virol.* 73: 1010-1022.

[42] Pion M, Jordan A, et al. (2003) Transcriptional suppression of in vitro-integrated human immunodeficiency virus type 1 does not correlate with proviral DNA methylation. *J Virol. 77*: 4025-4032.

[43] Akagi K, Suzuki T, et al. (2004) RTCGD: retroviral tagged cancer gene database. Nucleic Acids Res. 32 Database issue:D523-D527.

[44] Shastry BS. (2002) SNP alleles in human disease and evolution. *J Hum Genet. 47*: 561-566.

In: Genome Research Advances
Editor: Alberto P. Bellini, pp. 193-205
ISBN: 978-1-60021-805-7

Chapter V

MICROSATELLITE INSTABILITY IN GASTRIC CANCER

Laura Ottini and Mario Falchetti
Department of Experimental Medicine, University of Rome "La Sapienza", Viale Regina Elena, 324, Rome, 00161, Italy

ABSTRACT

Although gastric cancer (GC) incidence and mortality have recently decreased, GC still represents the fourth most common cancer and the second leading cause of cancer-related death worldwide. GC incidence shows a great geographical variation, with the highest frequencies observed in Eastern Asia and Southern America, and much lower rates in many Western Countries.

The aetiology of the disease and the mechanisms involved in its carcinogenesis are still poorly understood. The association between GC risk and positive family history of the disease suggests the involvement of genetic factors in GC pathogenesis, besides the influence of environmental factors.

GC develops through genomic instability and in particular, instability at simple repetitive sequences, called microsatellite instability (MSI), is observed in about 15% of GCs.

Hypermethylation of the promoter of *hMLH1*, one of the main mismatch repair (MMR) genes, has been found to be responsible for the development of the majority of GCs exhibiting MSI.

MSI is identified by analysing paired tumor and normal DNAs for microsatellite loci. Nowadays, guidelines are established to characterize MSI by using a standard panel of 5 microsatellite markers. Three levels of MSI can be identified: high-level MSI (MSI-H), generally defined as MSI in more than 30% of the standard markers; low-level MSI (MSI-L), when changes are exhibited in less than 30% of the markers and no-MSI (MSS), in the absence of microsatellite alterations. The highly sensitive BAT25 and BAT26 mononucleotide repeats can be used to specifically define the MSI-H phenotype and, recently, the role of MMR protein immunohistochemistry in identify MSI-H GCs has been pointed out.

Simple genetic and/or epigenetic inactivation of MMR genes is not by itself a transforming event and therefore additional genetic changes are believed to be necessary for progression to malignancy. MSI-H GCs progress through mutations in coding repetitive sequences in genes involved in cell growth regulation (*TGFβRII, IGFIIR, RIZ, TCF4, DP2*), in apoptosis (*BAX, BCL10, FAS, CASPASE5, APAF1*) and in DNA repair (*hMSH6, hMSH3, MED1, RAD50, BLM, ATR, MRE11*). In most cases, these mutational events inactivate only one of the two alleles of these genes thus leading to haploinsufficiency. As a consequence, these mutations may tip the balance of regulatory pathways and drive cells to further progress in malignancy.

The occurrence of mutations in specific sets of cancer-related genes confers unique clinico-pathological features to MSI-H GCs. In fact, intestinal histotype, antral location, expanding type, lower prevalence of lymph-node metastases, *H. pylori* seropositivity and improved survival have been associated with MSI-H GCs.

In conclusion, MSI is a promising screening tool because it allows for the identification of a specific GC subset characterized by peculiar molecular alterations and clinico-pathological features. This information may indicate improved therapeutic approaches for patients care.

Introduction

Despite a progressive reduction in incidence and mortality, gastric cancer (GC) represents the fourth most common cancer and the second leading cause of cancer-related death worldwide. GC incidence shows a great geographical variation, with the highest frequencies observed in Eastern Asia and Southern America, and much lower rates in many Western Countries (Parkin *et al.*, 2005). Incidence rises progressively with age, with a peak incidence between 50 and 70 years. In general, countries with higher incidence rates show better survival rates than countries with lower incidence, mainly because of the availability of screening for early detection in high-risk areas. This is particularly true for Japan, where mass screening programs are in place and the detection of early GC has led to a 5-year survival of about 95%. Conversely, Western countries in which early GC detection is rare, such US and Europe, show a lower 5-year survival rate ranging from 10 to 20% (Crew and Neugut 2006).

The aetiology of GC and the mechanisms involved in its carcinogenesis are still poorly understood. It is widely accepted that gastric carcinogenesis is a multifactorial process. The marked geographic variation and time trends suggest that environmental or lifestyle factors are among the major contributors to the GC aetiology. This assumption is confirmed by migratory effect on GC incidence: migrants from high-risk countries show a marked diminution in GC risk when they move to lower-risk areas and decrease in GC risk seems to depend on the age at migration (Parkin *et al.*, 2005).

The association between GC risk and positive family history of the disease suggests the involvement of genetic factors in GC pathogenesis, besides the influence of environmental factors. The risk of GC is increased in first-degree relatives of patients with the disease by approximately two- to threefold (La Vecchia *et al.*, 1992; Palli *et al.*, 1994; Lissowska *et al.*, 1999) and an increased GC risk has been found in the twin of an affected person (Lichtenstein *et al.*, 2000). About 5-10% of GC patients show familial aggregation in absence of a segregating model, biomarkers or germ-line mutation which may indicate a hereditary aetiology (Lynch *et al.*, 2005). It has been suggested that familial clustering of *Helicobacter pylori* infection, blood group A and gastric polyps may contribute to GC risk (Kelley *et al.*, 2003). Interestingly, a model fitting to assess the contribution of hereditary and environmental factors found that inherited genes contributed 28%, shared environmental factors 10% and environmental factors 62% (Lichtenstein *et al.*, 2000).

GENOMIC INSTABILITY AND GASTRIC CANCER

Genomic instability represents a key molecular step that occurs early in the carcinogenesis process and creates a permissive environment for the accumulation of genetic and epigenetic alterations in tumor suppressor genes and oncogenes.

Genomic instability may be caused by at least three different sources of DNA damage: 1) environmental factors, including UV rays from sunlight, ionizing radiations and several genotoxic compounds; 2) free radicals, deriving from the oxidative metabolism or from lipid peroxidation; 3) spontaneous breaks of chemical bonds in DNA (Hoeijmakers, 2001).

Unrepaired DNA damages may be mutational sources. In fact, when mutations target genes that control genomic stability a sort of mutational cascade, involving other genes and eventually conferring proliferative advantage and neoplastic transformation through clonal selection, will occur. Subsequent clonal expansion steps may guide to further tumor evolution (Loeb, 2001).

In gastric carcinogenesis, at least two processes seem to be associated to genomic instability: chromosomal instability (CIN) and microsatellite instability (MSI).

CIN represents the most common type of genomic instability observed in human cancers and CIN phenotype has been reported in at least 60% of gastrointestinal tumors (Choi *et al.*, 1998). CIN is characterized by changes in chromosome copy number (aneuploidy) and alterations in chromosomal regions, including allelic losses, gene deletions and/or amplifications. These alterations may induce oncogenes activation and/or tumor suppressor genes inactivation. Molecular mechanisms underlying CIN are not well elucidated and may include mechanisms involved in regulation of mitotic spindle checkpoints and, in particular, genes that control kinetochore structure and function, centrosome and microtubule formation, chromosome condensation, sister chromatid cohesion and cell cycle checkpoints (Nardone 2003; Grady 2004).

MSI represents a hallmark of the hereditary non-polyposis colorectal cancer (HNPCC) syndrome. It has been reported in colorectal cancer and in GCs developing in the context of HNPCC but also in a subset of sporadic GCs ranging from 25% to 50% (Oda *et al.*, 2005; Bacani *et al.*, 2005). MSI is defined as the presence of replication errors resulting in insertions/deletions of bases within nucleotide repeats, known as microsatellite regions. For their repetitive structure, microsatellites are particularly prone to mismatches during replication or in absence of proficient mismatch repair (MMR) system. Repair of mismatches lies, during replication, on proofreading activity of DNA polymerases and, in post-replicative phase, on MMR system.

Mismatch Repair System

In eukaryotes mismatch repair is initiated when complexes of *E. coli* MutS homologs, either MSH2-MSH6 (MutSα) or MSH2-MSH3 (MutSβ), bind to a mismatch. MutSα is primarily responsible for repairing single base-base and insertion deletions loop (IDL) mismatches; MutSβ is primarily responsible for repairing IDL mismatches containing up to 16 extra nucleotides in one strand. In addition, the two complexes may be responsible for repairing some IDL mismatches, especially those with one extra base (Kunkel and Erie, 2005). The MutSα and MutSβ complexes interact with the proliferating cell nuclear antigen (PCNA), in early stage of MMR and in gap repair synthesis. In initial phase PCNA helps to deliver Muts homologs to the newly replicated DNA containing mismatches, thereby enhancing MMR specificity and efficiency (Schofield and Hsieh, 2003). MutS homologs bind to DNA nonspecifically and bend it in search of a mismatch; upon specific recognition of a mismatch, the complexes undergo a conformational change to an initial recognition complex in which the DNA is kinked; finally, MutSs undergo further conformational changes to the ultimate recognition complex, in which the DNA is unbent with the mismatched base possibly being flipped out. ATP induces formation of a mobile clamp state of MutS that can move away from the mismatch and that this mobile clamp state forms only if the MutS protein is bound to a mismatch.

Eukaryotes also encode multiple *E. coli* MutL homologs that form different heterodimers. MutLα (MLH1-PMS2 in humans) is involved in repairing a wide variety of mismatches. The function of human MutLβ (MLH1-PMS1) is unknown, but human MutLγ (MLH1-MLH3) is thought to participate in repairing a subset of IDLs (Kunkel and Erie, 2005). MutS and MutL proteins interacts by forming multiprotein complexes which are important for communication between the mismatch and the signal for strand discrimination. Probably, MutLα binds duplex DNA with high affinity only when the DNA is long enough to bend, suggesting that mismatched DNA, kinked by MutSα, may be a high-affinity substrate for MutLα. The DNA deformation introduced upon MutS binding to a mismatch could be conserved for use in subsequent steps in the repair pathway. Such complexes could serve to amplify the signal for interaction with the downstream protein. If the strand discrimination signal is close to the mismatch, such complexes may result in direct interaction with the signal. In contrast, if the strand discrimination signal is far, the large MutS-MutL complexes at the mismatch will increase the probability that the downstream proteins will locate the MMR initiation complex.

Last step of MMR process consist of the replication error removal by excision and correct DNA synthesis. Excision is mismatch-dependent, initiates at a nick or a gap and preferentially proceeds along the shortest path to the mismatch, terminating about 150 nucleotides beyond the mismatch. MutSα, MutLα, PCNA, the replication factor C, the ssDNA-binding protein RPA, and EXO1 participate in mismatch excision. MutSα activates and confers high processivity to EXO1, and after the mismatch is excised, MutSα and MutLα suppress EXO1 activity while RPA facilitates its displacement from DNA, resulting in termination after mismatch excision. DNA resynthesis is likely catalyzed by DNA polymerase δ.

In MMR-proficient cells microsatellite sequences are relatively stable, while in MMR-deficient cells they show variations in their sequences. In normal cells, MSI phenotype may be tolerated for a reduced period because DNA damage may induce apoptosis. Abrogation of apoptosis may be necessary but not enough to get neoplastic transformation and non-malignant MSI cells may avoid apoptosis but DNA damage load could be very high that they go to a fast senescence (Janin, 2000). In tumors characterized by MSI phenotype, loss of wild type MMR alleles may occur during the life in individuals with cancer predisposition, particularly in the context of continuously proliferating tissues such as gastrointestinal epithelium. Progression in MSI tumors is characterized by a mutation accumulation process which may last for many years but might be completed in a short period because MMR-defective cells could accumulate mutations at each cell division.

MICROSATELLITE INSTABILITY IN GASTRIC CANCER

Microsatellites are tandem repeats of simple DNA sequences (1-6 basepairs) that are abundant and randomly distributed through out the genome. They are usually polymorphic and are stably inherited from one generation to the next.

MSI is characterized by expansions or contractions in the number of tandem repeats at numerous microsatellite loci. This phenomenon occurs during DNA replication and is thought to be due to slippage of the DNA polymerase complex. The MMR system normally recognize and repairs these replication errors. Thus, MSI appears to be a phenotypic marker of an underlying cellular defect involving MMR. Inactivation of the MMR system due to germ-line defects accounts for the colon cancer family syndrome HNPCC. Somatic inactivation of the MMR system additionally accounts for 15% of sporadic colon cancer. In particular, mutations in the main DNA MMR genes, *hMLH1* and *hMSH2*, and less frequently in *hMSH6*, *hPMS1* and *hPMS2* genes, have been associated with MSI in colon cancer (Gologan *et al.*, 2005). In contrast to colon cancer, sporadic or germ line mutations in MMR repair genes have not been frequently observed in GC. Mutations of *hMLH1* and *hMSH2* genes are relatively rare being reported in about 15% and 12% of GC with MSI phenotype, respectively. However, as in colorectal cancer, *hMLH1* silencing, due to promoter hypermethylation, has been found to be responsible for the development of more than 50% of GCs exhibiting MSI phenotype (Yamamoto *et al.*, 1999; Halling *et al.*, 1999; Fleisher *et al.* 1999; Leung *et al.*, 1999; Pinto *et al.* 2000; Wu *et al.* 2000; Bacani *et al.*, 2005). Thus, in addition to genetic mutations, epigenetic alterations of the MMR system may play a relevant role in GC with MSI by causing gene expression inactivation. In fact, more than 50% and 40% of GC with MSI show lack of hMLH1 and hMSH2 protein expression, respectively (Furlan *et al.* 2002; Lee *et al.*, 2002; Scartozzi *et al.*, 2004; Musulen *et al.* 2004; Rugge *et al.*, 2005).

MSI can be easily identified by analyzing paired tumor and normal DNAs for microsatellite loci. Because the majority of microsatellites demonstrate MSI, a variety of different microsatellite loci, including mono-, di-, tri- and tetranucleotide repeats, have been used over the years to define this tumor phenotype. Since mononucleotide repeats tend to be monomorphic or quasimonomorphic, it has been proposed that MSI testing can be performed with tumor DNA only, without the need to compare to normal DNA from the same patient. However, polymorphic alleles for some of these markers have been identified in certain ethnic groups (Simpson *et al.*, 2001), thus potentially leading to false positive results if only tumor DNA is analyzed. In this context, the polymorphic nature of dinucleotide repeats may be useful for quality control purposes.

A standard panel of microsatellite markers, including mononucleotide (*BAT26* and *BAT25*) and dinucleotide (*D2S123*, *D5S346* and *D17S250*) repeats, has been recommended and guidelines for MSI testing (Bethesda Guidelines) were established in 1998 (Boland *et al.*, 1998). Using the reference panel, three levels of MSI can be identified: high-level MSI (MSI-H), generally defined as MSI in more than 30% of the markers, low-level MSI (MSI-L), when changes are exhibited in less than 30% of the markers and microsatellite stable (MSS), when none of the markers show instability. It has become clear that mononucleotide repeats should be included in any panel of microsatellite markers in order to detect MSI-H tumors. In fact, it has been showed that mononucleotide repeats (e.g. *BAT25* and *BAT26*) are highly sensitive and specific for MSI-H phenotype. A revision of Bethesda Guidelines and criteria have been proposed in 2002 (Umar *et al.*, 2004). As results of this revision, MSI-L was defined as instability limited at dinucleotide loci. By adopting these criteria, we reported a frequency of 17% MSI-H, 14% MSI-L and 69% MSS in a series of GC cases from a high-risk population in Central Italy (Ottini *et al.*, 2004). These frequencies are consistent with frequencies observed in other GC series from Western populations (Bacani *et al.*, 2005).

Finally, it is noteworthy to point out the role exerted by MMR protein immunohistochemistry in identifying MSI-H GCs. In fact, loss of MMR protein, in particular hMLH1 and hMSH2, expression is correlated with an MSI-H phenotype in GC (Bacani *et al.*, 2005; Rugge *et al.*, 2005).

Target Genes in MSI-H Gastric Cancer

Simple genetic and/or epigenetic inactivation of MMR genes is not by itself a transforming event and therefore additional genetic changes are believed to be necessary for progression to malignancy. MSI-H GCs progress through mutations in repetitive sequences in genes involved in cell growth regulation, including *TGFβRII, IGFIIR*, *RIZ, TCF4, DP2*; in apoptosis, including *BAX, BCL 10, FAS, CASPASE5, APAF1* and in DNA repair, including *hMSH6, hMSH3, MED1, RAD50, BLM, ATR, MRE11* (Ottini *et al.*, 1998; Menoyo *et al.*, 2001; Duval *et al.*, 2002; Yamamoto *et al.*, 2002; Ottini *et al.*, 2004). These mutations disrupt gene expression and confer cell growth advantage and clonal perpetuation. In most cases, mutational events inactivate only one of the two alleles of these genes thus leading to haploinsufficiency. As a consequence, these mutations may tip the balance of regulatory pathways and drive cells to further progress in malignancy.

Multiple target genes are simultaneously mutated in the majority of the MSI-H GCs and multiple hits insist on different genes in the same pathway, suggesting that a cumulative effect of haploinsufficiency may be the cause of the alteration or inactivation of a specific pathway. The role of heterozygous mutations is not very clear, although they are frequently observed. Probably some tumors are heterozygous for mutations in mononucleotide repeats, but functionally homozygous because other mutations may inactivate the allele wild-type at mononucleotide runs. In other cases the accumulation of heterozygous mutations may reduce the physiological levels of wild-type proteins acting at different levels in the same pathway of cell homeostasis control. This assumption has been corroborated by the experimental verification of functional consequences of MSI-related mutations in some of the most frequently targeted genes.

TGFβRII is the most frequently altered gene in MSI-H GC with deletion/insertion at the poly(A)10 coding repeat ranging from 35% to 92%. Mutations at *TGFβRII* disrupt TGFβ-induced apoptosis and cell growth arrest in the late G1 phase by impeding ligand-receptor interactions. *TGFβRII* has been shown to be heterozygously or homozygously mutated in MSI-H GCs. Functional studies demonstrated that these MSI-related mutations led to loss of the TGFβRII tumor suppressor function (Duval *et al.*, 2002). Recently, this loss has been explained partly by the sensitivity of *TGFβRII* mRNA to the nonsense-mediated decay (NMD) systems (El-Bchiri *et al.*, 2005).

In the group of genes involved in apoptosis *BAX* is the most frequent MSI target gene. Mutations at the coding poly(G)8 were reported in 15-67% of MSI-H GCs. BAX mediates the apoptotic signaling by p53 in response to DNA damage. The identification of MSI-dependent *BAX* mutations may explain the low incidence of p53 mutations in MSI tumors. Inactivation of wild type *BAX* allele by de-novo frameshift mutations confers strong advantage during tumor clonal evolution and contributes to tumor progression by providing cell survival advantage. Interestingly, survival analyses show that *BAX* mutations are indicators of poor prognosis in MSI-H GCs. Intriguingly, MSI cells carrying *BAX* mutations easily develop resistance to non-steroidal anti-inflammatory drugs (Yamamoto *et al.*, 2002).

In the group of genes involved in DNA repair, mutations at the coding poly(A)8 of *hMSH3* and poly(C)8 of *hMSH6* are the most frequently observed in MSI-H GCs, ranging from 9%-64% and 19%-54%, respectively (Duval *et al.*, 2002). MSI-associated mutations in MMR genes may offer an example in which way a cumulative effect of haploinsufficiency may alter or inactivate a specific pathway by multiple hits insisting on different genes. In fact, MSI-related frameshift mutations at the MMR repair genes, *hMSH6* and *hMSH3*, are due to the mutagenic effect of genetic or epigenetic alterations in primary mutator genes such as *hMLH1* and *hMSH2*. In their turn, mutations in the secondary mutator genes (*hMSH6* and *hMSH3*) may increase the tumor cell genomic instability, accelerating tumor progression. In fact, tumor cells with inactivated *hMLH1* and homozygous *hMSH6* poly(C)8 mutations exhibit a higher mutation frequency and a different spectrum of mutations compared to sister cell clones carrying only inactivated *hMLH1* (Yamamoto *et al.* 2002).

It is noteworthy that genes involved in DNA double strand breaks (DSBs) repair, including *RAD50, BLM, ATR* and *MRE11*, represent frequent mutational targets in MSI-H GCs (Duval *et al.*, 2002; Ottini *et al.*, 2004). Mutations in these genes may constitute an additional pathway of progression from MSI-H to CIN phenotype. It has been shown that mutations at the coding *ATR* poly(A)10 abrogate ATR-dependent DNA damage response by a dominant-negative manner (Lewis *et al.*, 2005), leading to chromosomal instability as an ongoing process, since clones originating from the same genetic background displayed different patterns of chromosomal aberrations (Fang *et al.*, 2004). From the cancer perspective an intact *ATR* or downstream *ATR* targets genes represent a major barrier to the progression of MMR-deficient tumors from MSI-H to CIN phenotype. From the therapeutic perspective, tumors that have crossed the threshold from MSI to CIN might be sensitive to different therapeutic agents.

In addition to alterations in coding mononucleotide repeats, mutations in non-coding repeats are frequently observed in MSI-H GCs. These mutations are unlikely to favour neoplastic growth unless they occur in gene regulatory regions controlling levels of gene expression. Recently, *MRE11* has been indicated as a novel mutational target in MSI-H GCs. Deletions at the poly(T)11, constituting an accessory splicing signal within the 3'-splice acceptor site of intron 4, were associated with a reduced production of wild-type MRE11 transcript and protein (Ottini *et al.*, 2004). Moreover, the impairment of MRE11 protein expression correlated with reduced expression of NBS1 and RAD50 (Giannini *et al.*, 2002) and impaired DSBs repair, through non-homologous end-joining recombination, was related to decreased *RAD50* and *MRE11* genes expression in MMR deficient tumors (Koh *et al.*, 2005). This is of particular interest in the context of gastric carcinogenesis considering that these genes play a relevant role in protecting cells from deleterious DNA-DSBs that might be generated by endogenous and exogenous mutagens. Environmental and dietary carcinogen exposures are known to be implicated in the process of gastric carcinogenesis and suspected risk factors for GC include chemical carcinogens, present in cigarette smoke and in some food items (i.e. polycyclic aromatic hydrocarbons), that produce bulky DNA adducts and DNA-DSBs.

Clinico-Pathological Features in MSI-H Gastric Cancer

The occurrence of mutations in specific sets of cancer-related genes confers unique clinico-pathological features to MSI-H GCs in contrast to MSS and MSI-L GCs and, as recently recommended by the revised Bethesda Guidelines, for clinical purposes, MSS and MSI-L tumors should not be considered distinct from one another.

In GC MSI-H phenotype was shown to be associated with individual clinico-pathological characteristics. MSI-H GCs are more frequently located in the distal (antrum) rather than proximal (body and cardia) regions of the stomach, show the intestinal-type rather than diffuse-type differentiation, are of expanding type, tend to have an abundant lymphoid infiltration and not to give rise to lymph node metastases.

Importantly, MSI-H GC cases show a relatively improved long term survival compared to MSS/MSI-L counterparts (Simpson *et al.*, 2001; Scartozzi *et al.*, 2004; Beghelli *et al.*, 2006). This can be explained by a reduced trend to develop metastases and by assuming that a strong host immune response may limit tumor progression. Alterations in MMR may be responsible for the production, by tumor cells, of abnormal tumor-specific peptides which recruit lymphocytes in the tumor and induce an immune response. Thus, activation of cytotoxic lymphocytes within the tumor may lead to increased apoptotic cell death of neoplastic cells.

Moreover, a positive correlation between the presence of *Helicobacter pylori* (*H. pylori*) sieropositivity and MSI GCs has been observed (Ottini *et al.*, 1997; Simpson *et al.*, 2001;

Scartozzi *et al.*, 2004). Active *H. pylori* infection is present more frequently in individuals carrying MSI GCs, suggesting that *H. pylori* may affect MMR system during the stepwise progression of gastric carcinogenesis. Effectively a decrease of MMR proteins expression has been demonstrated in gastric epithelial cells from individuals infected by *H. pylori* (Kim *et al.*, 2002; Park *et al.*, 2005). Thus, deficient MMR associated with *H. pylori* infection may be a relevant mechanism of mutation accumulation in gastric mucosa during the early stages *H. pylori*-related gastric carcinogenesis.

The role of diet in MSI-related carcinogenesis have been also investigated. The risk of MSI GCs was positively associated with high consumption of red meat and meat sauce and negatively associated with consumption of white meat. A positive association was also seen with total protein and nitrite intake. Risk was especially high among subjects reporting both a positive GC family history and a high consumption of red meat. For MSS tumors, a significant protective effect was associated with frequent consumption of citrus and other fresh fruit, garlic, legumes, vegetables, and olive oil and with high intake of beta-carotene and other antioxidants and sugar (Palli *et al.*, 2001). Finally, both cigarette smoking and alcohol consumption seem to play dominant roles in the development of GC with hypermethylation of the *hMLH1* promoter (Nan *et al.*, 2005).

Conclusions

MSI is a promising screening tool because it allows for the identification of a specific GC subset characterized by peculiar molecular alterations and clinico-pathological features. The results summarized here seem relevant to detect the genetic background probably associated with an increased susceptibility to exogenous carcinogens.

MSI can be used as a marker for the screening of genetic instability and, identifying patients with genetic instability, may help to recognize patients with at risk of GC. Indeed, MSI detection is useful for early diagnosis of GC considering that it can be detected in both gastric adenoma and intestinal metaplasia, which are precancerous lesions associated with well-differentiated GC. In this regard, it is noteworthy that the presence and the extent of MSI evaluated on endoscopic biopsy specimen from GC patients, was shown to provide valuable information for making a preoperative genetic diagnosis of GC (Hong *et al.*, 2005).

Finally, the study of the genetic background of MSI-H GC, will allow the detection of more specific molecular markers useful for improving therapeutic approaches for patients care. Currently, there is no consensus whether adjuvant therapy is differentially beneficial for patients with MSI or MSS GC. However, *in vitro* data suggest that MSI cancers are relatively insensitive to 5-fluorouracil (Carethers *et al.*, 1999). Moreover, a reduced cycloxygenase 2 (COX2) expression was reported in MSI-H tumors (Karnes *et al.*, 1998). This could be particularly relevant considering the use of COX2 inhibitors in preventing gastrointestinal tumors.

References

Introduction

Crew KD, Neugut AI. Epidemiology of gastric cancer. *World J Gastroentol* 2006; 12: 354-362.

Kelley Jon R. and Duggan John M. Gastric cancer epidemiology and risk factors. *Journal of Clinical Epidemiology* 2003; 56: 1–9.

La Vecchia C, Negri E, Franceschi S, et al. A family history and the risk of stomach and colorectal cancer. *Cancer* 1992; 70: 50–55.

Lichtenstein P, Holm NV, Verkasald PK, et al. Environmental and heritable factors in the causation of cancer. Analysis of cohorts of twins from Sweden, Denmark and Finland. *N Engl J Med* 2000; 343: 78–85.

Lissowska J, Groves FD, Sobin LH, et al. Family history and risk of stomach cancer in Warsaw, Poland. *Eur J Cancer Prev* 1999; 8: 223–7.

Lynch HT, Grady W, Suriano G., Huntsman D. Gastric cancer: new genetic developments. *Journal of Surgical Oncology* 2005; 90: 114-133.

Palli D, Galli M, Caporaso NE, et al. Family history and risk of stomach cancer in Italy. *Cancer Epidemiol Biomarkers Prev* 1994; 3: 15–18.

Parkin DM, Bray F, Pisani P. Global cancer statistics, 2002. *CA Cancer J Clin* 2005; 55: 74-108.

Genomic Instability and Gastric Cancer

Bacani J, Zwingerman R, Di Nicola N et al. Tumor microsatellite instability in early onset gastric cancer. *J Mol Diagn.* 2005; 7: 465-477.

Choi SW, Park SW, Lee KY et al. Fractional allelic loss in gastric carcinoma correlates with growth patterns. *Oncogene.* 1998; 17: 2655-2659.

Grady WM. Genomic instability and colon cancer. *Cancer Metastasis Rev* 2004; 23:11-27.

Hoeijmakers J.H. Genome maintenance mechanisms for preventing cancer. *Nature* 2001; 411: 366-374.

Loeb L.A. A mutator phenotype in cancer. *Cancer Research* 2001; 61: 3230-3239.

Nardone G. Review article: molecular basis of gastric carcinogenesis. *Aliment Pharmacol Ther.* 2003; 17: 75-81.

Oda S, Zhao Y, Maehara Y. Microsatellite instability in gastrointestinal tract cancers: a brief update. *Surg Today.* 2005; 35: 1005-1015.

Mismatch Repair System

Janin N. A simple model for carcinogenesis of colorectal cancers with microsatellite instability. *Adv. Cancer Res* 2000; 77: 189-221.

Kunkel TA and Erie DA. DNA mismatch repair. *Annu. Rev. Biochem* 2005; 74: 681–710.

Schofield MJ and Hsieh P. DNA mismatch repair: molecular mechanisms and biological function. *Annu. Rev. Microbiol* 2003; 57: 579-608.

Microsatellite Instability in Gastric Cancer

Bacani J, Zwingerman R, Di Nicola N et al. Tumor microsatellite instability in early onset gastric cancer. *J Mol Diagn* 2005; 7: 465-477.

Boland CR, Thibodeau SN, Hamilton SR et al. A National Cancer Institute Workshop on Microsatellite Instability for cancer detection and familial predisposition: development of international criteria for the determination of microsatellite instability in colorectal cancer. *Cancer Res* 1998; 58: 5248-5257.

Fleisher AS, Esteller M, Wang S et al. Hypermethylation of the hMLH1 gene promoter in human gastric cancers with microsatellite instability. *Cancer Res* 1999; 59: 1090-1095.

Furlan D, Casati B, Cerutti R et al. Genetic progression in sporadic endometrial and gastrointestinal cancers with high microsatellite instability. *J Pathol* 2002; 197: 603-609.

Gologan A, Krasinskas A, Hunt J et al. Performance of the revised Bethesda guidelines for identification of colorectal carcinomas with a high level of microsatellite instability. *Arch Pathol Lab Med* 2005; 129: 1390-1397.

Halling KC, Harper J, Moskaluk CA et al. Origin of microsatellite instability in gastric cancer. *Am J Pathol* 1999; 155: 205-211.

Lee HS, Choi SI, Lee HK et al. Distinct clinical features and outcomes of gastric cancers with microsatellite instability. *Mod Pathol* 2002; 15: 632-640.

Leung SY, Yuen ST, Chung LP et al. hMLH1 promoter methylation and lack of hMLH1 expression in sporadic gastric carcinomas with high-frequency microsatellite instability. *Cancer Res* 1999; 59: 159-164.

Musulen E, Moreno V, Reyes G et al. Standardized approach for microsatellite instability detection in gastric carcinomas. *Hum Pathol* 2004; 35: 335-342.

Ottini L, Falchetti M, Saieva C et al. MRE11 expression is impaired in gastric cancer with microsatellite instability. *Carcinogenesis* 2004; 25: 2337-2343.

Pinto M, Oliveira C, Machado JC et al. MSI-L gastric carcinomas share the hMLH1 methylation status of MSI-H carcinomas but not their clinicopathological profile. *Lab Invest* 2000; 80: 1915-1923.

Rugge M, Bersani G, Bertorelle R et al. Microsatellite instability and gastric non-invasive neoplasia in a high risk population in Cesena, Italy. *J Clin Pathol* 2005; 58: 805-810.

Scartozzi M, Galizia E, Freddari F et al. Molecular biology of sporadic gastric cancer: prognostic indicators and novel therapeutic approaches. *Cancer Treat Rev* 2004; 30: 451-459.

Simpson AJ, Caballero OL, Pena SD. Microsatellite instability as a tool for the classification of gastric cancer. *Trends Mol Med* 2001; 7: 76-80.

Umar A, Boland CR, Terdiman JP et al. Revised Bethesda Guidelines for hereditary nonpolyposis colorectal cancer (Lynch syndrome) and microsatellite instability. *J Natl Cancer Inst* 2004; 96: 261-268.

Wu MS, Lee CW, Shun CT et al. Distinct clinicopathologic and genetic profiles in sporadic gastric cancer with different mutator phenotypes. *Genes Chromosomes Cancer* 2000; 27: 403-411.

Yamamoto H, Perez-Piteira J, Yoshida T et al. Gastric cancers of the microsatellite mutator phenotype display characteristic genetic and clinical features. *Gastroenterology* 1999; 116: 1348-1357.

Target Genes in MSI-H Gastric Cancer

Duval A, Hamelin R. Mutations at coding repeat sequences in mismatch repair-deficient human cancers: toward a new concept of target genes for instability. *Cancer Res* 2002; 62: 2447-2454.

El-Bchiri J, Buhard O, Penard-Lacronique V, Thomas G, Hamelin R, Duval A. Differential nonsense mediated decay of mutated mRNAs in mismatch repair deficient colorectal cancers. *Hum Mol Genet* 2005; 14: 2435-2442.

Fang Y, Tsao CC, Goodman BK, Furumai R, Tirado CA, Abraham RT, Wang XF. ATR functions as a gene dosage-dependent tumor suppressor on a mismatch repair-deficient background. *EMBO J* 2004; 23: 3164-3174.

Giannini G, Ristori E, Cerignoli F, Rinaldi C, Zani M, Viel A, Ottini L, Crescenzi M, Martinotti S, Bignami M, Frati L, Screpanti I, Gulino A. Human MRE11 is inactivated in mismatch repair-deficient cancers. *EMBO Rep* 2002; 3: 248-254.

Koh KH, Kang HJ, Li LS, Kim NG, You KT, Yang E, Kim H, Kim HJ, Yun CO, Kim KS, Kim H. Impaired nonhomologous end-joining in mismatch repair-deficient colon carcinomas. *Lab Invest* 2005; 85: 1130-1138.

Lewis KA, Mullany S, Thomas B et al. Heterozygous ATR mutations in mismatch repair-deficient cancer cells have functional significance. *Cancer Res* 2005; 65: 7091-7095.

Menoyo A, Alazzouzi H, Espin E et al. Somatic mutations in the DNA damage-response genes ATR and CHK1 in sporadic stomach tumors with microsatellite instability. *Cancer Res* 2001; 61: 7727-7730.

Ottini L, Falchetti M, D'Amico C et al. Mutations at coding mononucleotide repeats in gastric cancer with the microsatellite mutator phenotype. *Oncogene* 1998; 16: 2767-2772.

Ottini L, Falchetti M, Saieva C et al. MRE11 expression is impaired in gastric cancer with microsatellite instability. *Carcinogenesis* 2004; 25: 2337-2343.

Yamamoto H, Imai K, Perucho M. Gastrointestinal cancer of the microsatellite mutator phenotype pathway. *J Gastoenterol* 2002; 37: 153-163.

Clinico-Pathological Features in MSI-H Gastric Cancer

Beghelli S, de Manzoni G, Barbi S, Tomezzoli A, Roviello F, Di Gregorio C, Vindigni C, Bortesi L, Parisi A, Saragoni L, Scarpa A, Moore PS. Microsatellite instability in gastric cancer is associated with better prognosis in only stage II cancers. *Surgery* 2006; 139: 347-356.

Kim JJ, Tao H, Carloni E, Leung WK, Graham DY, Sepulveda AR. Helicobacter pylori impairs DNA mismatch repair in gastric epithelial cells. *Gastroenterology* 2002;123:542-553.

Nan HM, Song YJ, Yun HY, Park JS, Kim H. Effects of dietary intake and genetic factors on hypermethylation of the hMLH1 gene promoter in gastric cancer. *World J Gastroenterol* 2005; 11: 3834-3841.

Ottini L, Palli D, Falchetti M et al. Microsatellite instability in gastric cancer is associated with tumor location and family history in a high-risk population from Tuscany. *Cancer Res* 1997; 57: 4523-4529.

Palli D, Russo A, Ottini L, Masala G, Saieva C, Amorosi A, Cama A, D'Amico C, Falchetti M, Palmirotta R, Decarli A, Mariani Costantini R, Fraumeni JF Jr. Red meat, family history, and increased risk of gastric cancer with microsatellite instability. *Cancer Res* 2001; 61: 5415-5419.

Park DI, Park SH, Kim SH, Kim JW, Cho YK, Kim HJ, Sohn CI, Jeon WK, Kim BI, Cho EY, Kim EJ, Chae SW, Sohn JH, Sung IK, Sepulveda AR, Kim JJ. Effect of Helicobacter pylori infection on the expression of DNA mismatch repair protein. *Helicobacter* 2005; 10: 179-184.

Scartozzi M, Galizia E, Freddari F et al. Molecular biology of sporadic gastric cancer: prognostic indicators and novel therapeutic approaches. *Cancer Treat Rev* 2004; 30: 451-459.

Simpson AJ, Caballero OL, Pena SD. Microsatellite instability as a tool for the classification of gastric cancer. *Trends Mol Med* 2001; 7: 76-80.

Conclusions

Carethers JM, Chauhan DP, Fink D et al. Mismatch repair proficiency and in vitro response to 5-fluorouracil. *Gastroenterology* 1999; 117: 123-31.

Hong SJ, Choi SW, Lee KH et al. Preoperative genetic diagnosis of gastric carcinoma based on chromosomal loss and microsatellite instability. *Int J Cancer* 2005; 113: 249-258.

Karnes WE Jr, Shattuck-Brandt R, Burgart LJ, DuBois RN, Tester DJ, Cunningham JM, Kim CY, McDonnell SK, Schaid DJ, Thibodeau SN. Reduced COX-2 protein in colorectal cancer with defective mismatch repair. *Cancer Res* 1998; 58: 5473-5477.

In: Genome Research Advances
Editor: Alberto P. Bellini, pp. 207-218

ISBN: 978-1-60021-805-7

Chapter VI

GENOMICS IN THE FIGHT AGAINST MYCOBACTERIAL DISEASES

Antoine Danchin
Genetics of Bacterial Genomes, Institut Pasteur, Paris, France.

Genome sequencing was initially viewed by the vast majority of researchers as a technical endeavor, mostly meant to provide us with the blueprint of the Human Genome [1]. Bacterial genetics, however, implied a much deeper vision because there were indications that genomes displayed some sort of organisation related to functional properties of the cell [2]. This latter vision prompted the creation of a consortium for the sequencing of the model bacterium *Bacillus subtilis* genome with the explicit underlying hypothesis that a genome is not simply a collection of genes, but an organised set that would, when understood, tell much about the nature of life itself [3]. As a matter of fact, as unexpected discoveries of genomics were made, it became self-evident that sequencing was just the first step in the new domain of the genetics of genomes. Perhaps the most seminal discovery in genomics was made in 1990 when, with the sequencing of the first yeast chromosome (chromosome III) and the first 100 kb of sequence of *B. subtilis*, it appeared that, contrary to expectations, a considerable number of genes (at the time, half of them) had completely been ignored until then [4,5]. In this context, and despite the world-wide importance of tuberculosis it nevertheless asked for a formidable effort to convince the relevant instances that sequencing the *Mycobacterium tuberculosis* genome should be a major priorityn [6] because sequencing genomes was still perceived as a waste of resources and not a research program *per se* [3]. It was not immediately understood that sequencing genomes is highly cost-effective, as we shall see later on, not only for gene discovery but also in the identification of drug resistance (though cost was high at the time). The genomics approach is however well accepted now [7].

In a book devoted to the fight against tuberculosis and related diseases, it seems essential to stress first the scientific reasons we have to sequence genomes (at the time of this article, approximately 1,500 public genome programs are completed or under way). The point of view I will hold is that of a microbiologist not particularly familiar with *M. tuberculosis*, but promoting the genetics of genomes as a basic tool, a new kind of microscope, for better understanding life processes and in particular those involved in pathogenesis, thereby improving our chances to target diseases with faster and better efficiency.

Biology in Silico

Biomedical knowledge used to be kept in articles and books. However, as early as 1978, it had been clear that computer support was necessary to allow researchers to interpret and store knowledge associated with DNA and protein sequences as continuous sequences of symbols, similar to text written with alphabets. A study undertaken by the Rockefeller University and the European Molecular Biology Laboratory (EMBL) in Heidelberg, Germany, led to the idea of the creation of a computerised library that would collect gene and gene products sequences. It was even proposed that the program initiating the creation of the EMBL could be the sequencing of the *Escherichia coli* genome. The US Department of Energy, that had been created as the sequel of the Atomic Bomb Casualty Commission, was keen to develop a library of human genome sequences, to explore whether the consequences of the use of nuclear weapons had had an impact on the genome of humans. This was at the root of the creation of the first GenBank in Los Alamos [1,3]. After a long period of discussion and demonstration of its world-wide importance, the International Nucleotide Sequence Database (INSD) was set up between the DNA Data Bank of Japan (DDBJ) at the National Institute of Genetics (NIG) in Mishima, Japan, the EMBL outstation at Hinxton, UK, the European Bioinformatics Institute (EBI), and the National Center for Biotechnology Information (NCBI) in Bethesda, USA, collaboratively managing the huge amount of nucleotide sequence data created throughout the world [8].

Since these early days, genome sequencing has added a new dimension to the usual experimental palette we had to study life: in addition to work *in vivo* and *in vitro*, we have to work "*in silico*", to use computers in order to understand life [9]. This is not yet — this is a fairly recent field — entirely accepted by the community of scientists working in biology, but *in silico* experiments have, by themselves, a major role in the creation of new biological knowledge, complementary but with its own value, to traditional *in vivo* and *in vitro* evidence [10]. Because the Human Genome Project had been so explicitly conceived and presented as a gigantic enterprise meant to construct a technological empire to face competition between the USA and Asia, it is fairly easy to understand why a majority of scientists were reluctant to be involved. This was nevertheless a scientific misjudgement overemphasised by the scarcity of genome programs with a purely scientific aim, most of them being just sequenced, with poor annotation or associated experiments. Indeed, why sequence genomes? Comparing genomes between themselves shows in an illuminating way that the genetic program, which specifies the RNA and the proteins, also contains an architectural program linked to the organisation of the cell (and of the organism), and this has enormous consequences in terms of understanding the way cells cope with their environment, live, and die. Molecular Biology had been a major conceptual revolution and Genomics (with its sequels as all kinds of « omics » : transcriptome, proteome, metabolome...) is an even deeper revolution. In a nutshell, life possesses a central property of self-consistency, and experience shows that this is reflected in the genome itself as an integration directly visible in the way genes are organised and expressed. Molecular Biology used to rely on the alphabetic description of the genes. Genomics is now further exploring that same metaphor, in which the genome is understood as providing the complete text of the recipe allowing the construction, the development, the survival of the organisms of interest, in our case the mycobacteria, especially *Mycobacterium tuberculosis*.

The aim of *in silico* analysis of the genome of mycobacteria is then two-fold: on the one hand, it is meant to understand the specific biotopes of that special class of bacteria, as reflected by the genes and their organisation in the genome, and on the other hand, it is meant to identify specific genes or groups of genes that have a particular importance for dissemination, stability, pathogenesis and virulence of the organism. This latter utility is the more important because it is well known, in the case of tuberculosis for instance, that there is a wide difference between the level of incidence and the level of prevalence of *M. tuberculosis* in the human population [11]. In what follows we shall first analyse the global properties of mycobacterial genomes, then try to identify specific features associated to their biotope, to end with targets suggested by genome analyses *in silico*.

BACTERIAL GENOMES WITH EMPHASIS ON MYCOBACTERIA

The distribution of the genes in genomes is commonly perceived as more or less random. As a matter of fact, a superficial observation of the position of orthologous genes in various genomes gives the impression that they can be located anywhere, and certainly not always at the same position in different genomes [12]. However, it is also clear that genes have a real tendency to form clusters: genes coding for related functions, such as those coding for the subunits of an enzyme for example, are most often located side by side, in a same transcription unit. This was the very basis of the concept of operon. This grouping is certainly not random. And this may even sometimes give clues for the identification of an unknown function. This has been illustrated, for example, in the case of RNA degradation [13], a domain that may have still much to say in terms of drug targets (see also below). In *M. tuberculosis*, the neighborhood of the origin of replication is not very different from that in the distant organism *Bacillus subtilis* [14], and obviously extremely similar to that of *Mycobacterium leprae* [15]. However in both mycobacteria, the position and number of ribosomal RNA genes are particular: rather than being located next to the origin of replication, they are placed away from it. This is in line with the long doubling time of the organism, suggesting a specific selection pressure for slow growth. And, as a matter of fact, while these bacteria do not sporulate, they can stay dormant in their hosts for very long periods of time, a feature that has much to say about the treatment of the disease, which needs to be pursued for long period of time, since dormant bacteria are usually unaffected by many chemicals [16].

If one understands what are the genes submitted to this architectural selection pressure (one must add the genes of the core intermediary metabolism) it is necessary to attempt to identify the nature of the selection pressure that maintains them organised with respect to each other. To this aim we need to think that this probably already happened at the origin of the first cells, and persisted in present day organisms despite the very large time which elapsed since then. Mycobacteria form a large taxonomic group inhabiting diverse biotopes. Free-living mycobacteria such as *M. smegmatis* are relatively fast growers, while other pathogens (e.g. *M. bovis*, *M. avium* subsp. paratuberculosis, *M. ulcerans* and *M. leprae*) grow slowly or extremely slowly [17]. Associated with slow growth is the frequent persistence of latent infection and medium-term or long-term resistance in the environment [18].

Mycobacteria have either one or two *rrn* operons, depending on their growth rate. The *rrnA* operon, located downstream of the *murA* gene (suggesting a close association between biomass construction and synthesis of the envelope), is found in all species, with a variable number of promoters [19]. When present the second operon, *rrnB*, is located downstream of the *tyrS* gene and has a different regulation pattern [20]. The general organisation of the genome structure is similar to that of other genomes, with genes grouped into islands according to their functional properties. The case of *M. leprae* is interesting, in that it displays a massive tendency to create pseudogenes, with massive functional degeneracy [21]. This feature is a common property of pathogens, which have often a tendency to loose biochemical functions, while we would intuitively think they should only gain functions. This is the case for example of loss of cadaverine synthesis in *Shigella flexneri* [22], an avatar of *E. coli*, and, in a massive way, of *Yersinia pestis*, an avatar of *Y. pseudotuberculosis* [23,24]. A common feature of gene grouping is seen, for example, in the fact that as in other genomes sulphur metabolism genes appeared to belong to functional islands [25]. This can be seen in the *M. tuberculosis* genome, for example in the region *moaA1-cyp141*, the former belonging to the molybdopterin biosynthesis pathway, while the latter is likely to be a heme-thiolate monooxygenase.

Respiratory tract, bone, nerves, brain, lymphoid tissue and skin are preferred places for the development of these organisms, and even ocular tuberculosis probably exists [26]. In all cases, this requires a specific set of features allowing the organism to escape the immune defense of the host, as well as to target specific tissues. The genomic DNA of the pathogen is thus directly involved in the fact that Toll-like receptors recognise unmethylated CpG-containing DNA and are thought to act as a barrier against the GC-rich genome of *M. tuberculosis* [27]. However, the protective effect of addition of CpG oligodeoxynucleotide in vaccine preparation appears to be, at best, transient [28]. Another prominent property of the genome is the presence of the PE-PGRS and PPE family of genes that make up approximately 10% of the *M. tuberculosis* genome. Their function is still unknown and until know this has precluded investigation in terms of possible drug targets [29], while their role as antigens is being explored for designing new vaccine approaches [30]. Some are however expressed under the negative control of a transcription sigma factor, *sigD* thought to be important for virulence [31]. Because they code for highly repeated proteins they are expected to be involved in some recognition process and *in silico* studies aim at exploring their interaction with the HLA system [30]. Finally, as in other genome sequences, many genes appear to encode functions that are as yet unknown, and while this reflects our ignorance, this will probably be in the future a source of unexpected targets to fight diseases.

MOLECULAR EPIDEMIOLOGY

A first application of genome studies is the exploitation of the knowledge of the sequences from reference strains to explore variation in the population. Several studies have made use of DNA microarrays to investigate the spread of tubercle bacilli in mini-epidemic and outbreaks, and to analyse the transmission dynamics of tuberculosis transmission [32,33]. While this technique, being somewhat expensive at present, is not very much in use yet because tuberculosis usually prevails in poor populations, it has already led to interesting observations. A first analysis of deletions in reference genomes thus demonstrated that there was a correlation between the extent of deletions and the progressive decrease in pulmonary cavitation [34]. A recent study investigated the use of genome level-informed PCR as a new investigational approach for analysis of outbreak-associated *M. tuberculosis* isolates in Leicester (U.K.) [35]. The DNA microarrays in this study comprised almost 4,000 PCR products of predicted Coding DNA Sequences from the genome of the reference strain H37Rv. In order to make comparative studies, PCR labeled with two different fluorescent dyes, one for the reference genome and the other one from the bacterium of interest, directly collected from the patients. Five deletions in the CH strain relative to the H37Rv reference were uncovered and used for real time analysis of the spread of the bacterium. In view of the rapid spreading of DNA arrays in molecular epidemiology [36], they will become a basic tool in the epidemiology of *M. tuber*culosis, complementing Restriction Fragment Length Polymorphism studies, in particular under their recent avatar, using fluorescent labeling [37].

PREVIOUS TARGETS AND RESISTANCE TO ANTIBIOTICS

While the role of antibiotics in the environment is not entirely clear (they could play a role in competing for a given biotope, act as quorum-sensors, or both [38]), it appears that to all the biological processes that could be obvious targets for new drugs already have been associated to at least one antibiotic: translation is inhibited by aminoglycosides (e.g. streptomycin), tetracyclines or chloramphenicol, transcription by inhibitors of type II DNA topoisomerases, DNA gyrase and topoisomerase IV (which are the targets of quinolones) or inhibitors of RNA polymerase (rifampin and the like), envelope synthesis by beta-lactams, cephalosporins, cycloserine and the like, and metabolism of the mycobacteria-specific mycolic acids by isonicotinic acid hydrazide (isoniazid), thiolactomycin, and triclosan. Thiolactomycin targets the beta-ketoacyl-acyl carrier protein (ACP) synthases KasA and KasB, while triclosan inhibits the enoyl-ACP reductase InhA [39]. There is still some controversy about the role of isoniazid, that needs to be activated by the catalase/peroxidase *katG* gene product of the target cell [40]. Some of its action may derive from the production of superoxide [41] or nitric oxide during its metabolism [42]. Pyrazinamide targets the eukaryote-like fatty acid synthase I [43]. The target of ethambutol is on the biosynthesis of arabinogalactan, a cell wall polymer found in mycobacteria but in few other bacteria [44]. In the domain of genome studies, expression profiling using DNA microarrays allowed researchers to better characterise the targets of all these drugs, but we are still at the beginning of such studies [39,45,46].

While we are lacking detailed analysis of the function of some of these antibiotics, resistant strains of *M. tuberculosis* have kept appearing everywhere, and it is now admitted that at least 250,000 persons are infected each year with multidrug-resistant strains [47]. Many types of resistance exist, combining increased efflux of antibiotics [48], mutated ribosomal proteins or ribosomal RNA [49], mutated RNA polymerase [50], DNA gyrases or topoisomerases [51], or mutated KatG protein [52]. The panoply displayed by the organism to resist treatment is however very varied, and sometimes unexpected [53]. The situation is much less dramatic in the case of leprosy, where the available treatment should allow eradication of the disease in the next few years, provided appropriate support is found to reach isolated populations. Other mycobacterial diseases, such as Buruli ulcer (caused by *M. ulcerans*) are on the rise [54], and it is therefore essential to find new drugs and new targets for future development of drugs.

Genomics as a Tool to Identify New Antibiotics Acting against *M. Tuberculosis*

In a rather bleak situation, let us begin with a strong signal of hope, that illustrates the power of genomics. While tuberculosis spreads again, in particular because of its association with HIV, *M. tuberculosis* remained a challenging target for drug discovery, with no new anti-tuberculosis drugs while multidrug-resistant TB is on the rise. This was true until very recently when Andries *et al.* reported the identification of a new class of drugs, diarylquinolines, that target a subunit of the mycobacterial ATP synthase [55]. Interestingly, these authors used *M. smegmatis* as a cell factory to screen for potential anti-TB drugs of different chemical series and in their work, genome studies are an essential step. Promising drug candidates, including a very efficient one, dubbed R207910, were then tested on *M. tuberculosis*. Subsequently, resistant mutants were isolated, and rather than using the former tedious and slow genetic investigation used to identify the locus of the relevant mutations, the authors boldly sequenced the whole genome of the mutants, identified polymorphism, and tested all putative targets. The genomes of one resistant *M. tuberculosis* strain and two resistant *M. smegmatis* strains as well as the parental *M. smegmatis*, were sequenced to near-completion. Point mutations that conferred resistance were identified by comparative analysis of the genome sequences. There was only polymorphism common to all three independent mutants. It was located in gene *atpE*, the c subunit of the membrane proton channel of ATP synthase, showing that R207910 inhibits the proton pump of *M. tuberculosis* ATP synthase [55].

Remarkably, the target and mechanism of action of R207910 are different from those of previously known anti-TB agents. Inhibition of ATP synthase leads to ATP depletion while alteration of the F0 proton channel will alter proton availability, as is happening with antibiotics such as oligomycin acting on mitochondria, leading to decrease in viability. A comparison of the sequences of ATP synthases of different bacteria and of eukaryotic ATP synthase provided a rationale for the specificity of the antibacterial spectrum and, to a lesser extent, the safety profile of R207910 [55]. It seems likely however that this drug will prevent the simultaneous use of other drugs such as some aminoglycosides, since inhibition of ATP synthase inhibits transport and often leads to resistance [56]. In any event, this remarkable study is a signal of hope, triggered and supported by genomics, for the future of the fight against microbial diseases.

Genome studies provide us with other types of putative targets [57]. We shall indulge here in some speculation and list some of them in what follows, but the list is only for reference and limited by our (lack of) imagination. Target definition requires that the target is specific to the pathogen, and sufficiently dissimilar from any protein of the host. In addition, for slow growing bacteria, for which one needs long-term treatment, it is best that the chosen antibiotic has a narrow spectrum of action, so that it is unlikely that resistance will easily spread through the bacteria of the environment.

Mycobacterium tuberculosis, for reasons that are not well understood, has a large collection of adenylyl cyclases, and these might look as promising targets [58,59]. Unfortunately these enzymes belong to the ubiquitous nucleotidyl cyclases class [60], that is present both in animals and in some bacteria [61], apparently precluding their use as drug targets. One must note however that this may not be a final drawback, as illustrated with the case of ATP synthase, but differences between human and *M. tuberculosis* cyclases need to be explored in depth before trying to construct inhibitors [62].

Another pathway, the shikimate pathway has been suggested as an attractive target for the development of antimycobacterial agents because it has been shown to be essential for the viability of *M. tuberculosis*, but absent from mammals. Work is in progress to characterize relevant binding sites that would allow *in silico* prediction of possible docking of classes of chemicals [63].

In the same way, uridylate kinase differs dramatically from its counterpart in eukaryotes: it belongs to the carbamate kinase/aspartokinase family [64] and is located at the surface of the cell, both in Gram-positive and Gram-negative bacteria [65], where it is thought to play a major role in the formation of the cell's envelope, an excellent putative target. Analysis of the genomic neighbourhood [66] of that particular gene has suggested another putative target, the ribosome recycling factor (RRF), coded by the *frr* gene. It seems plausible that ribosome recycling is controlled by UTP availability, with RRF as a sensor [13]. If this is true, then it seems likely that among existing antibiotics with no known target, some might bind to RRF: identifying them would allow one to design new types of drug leads for this possible new target.

Several targets could be chosen in sulphur metabolism. However, even in model bacteria such as *E. coli* and *B. subtilis*, sulphur metabolism is still poorly known [67]. Mycobacteria, nevertheless, use a particular reducing compound, mycothiol, whose synthesis might reveal an interesting target: many genes are involved in its synthesis and use (*ino1*, *mca*, *mscR*, *mshA*, *mshB*, *mshC*, *mshD*, *mtr*) and this provides a harvest of opportunities for drug development [68-71].

Because molecular oxygen oxidizes the ferrous ion rapidly into the ferric form, that is extremely insoluble when hydrated, and despite the ubiquitous presence of iron on the Earth crust, iron availability is extremely limited for all living organisms, except those living in an anaerobic environment. Mycobacteria, as other bacteria, have developed a complex strategy to scavenge iron from their environment and hosts. Comparison between the development of the avirulent *M. smegmatis* and *M. tuberculosis* or *M. avium* in phagosomes has demonstrated that the pathogens have developed means to acquire iron that the non-pathogens fail to implement [72]. Siderophores such as mycobactin are used in the process, which is certainly much more complex than presently understood. This suggests interesting routes to generate new antituberculosis agents that would act on iron scavenging and transport [73,74].

Finally, one of the most interesting feature of mycobacteria is their ability to survive in a latent infection [16], suggesting that they are remarkably stable. Recent work has suggested that bacteria age, as do other living organisms [75]. Many molecular changes concur with such a ubiquitous process, and it is an interesting challenge for *in silico* genome biology to try to identify the features that may allow mycobacteria to remain younger for a much longer time than other bacteria.

Provisional Conclusion

Genomics has already proven its capacity in demonstrating its power in improving our knowledge about what life is. Recent developments, stemming from the mycobacterial Genome Programs have allowed rapid characterisation of unexpected genome properties, and have been used to rapidly identify targets of new promising drugs. It can be expected that these first successes will be followed by many new ones, and, perhaps, improve our capacity to create efficient vaccines that would allow us to act upstream of the establishment of the disease. It remains remarkable that about half of the genes of these organism are still of unknown function: new areas will most probably be uncovered when we begin to understand what they mean.

References

[1] Watson JD, Cook-Deegan RM. Origins of the Human Genome Project. *Faseb J.* 1991;5:8-11.

[2] Zipkas D, Riley M. Proposal concerning mechanism of evolution of the genome of Escherichia coli. *Proc Natl Acad Sci U S A.* 1975;72:1354-1358.

[3] Danchin A. *The Delphic boat. What genomes tell us*. Cambridge, USA: Harvard University Press; 2003.

[4] Glaser P, Kunst F, Arnaud M, et al. *Bacillus subtilis* genome project: cloning and sequencing of the 97 Kb region from 325° to 333°. *Mol Microbiol.* 1993:371-384.

[5] Oliver SG, van der Aart QJ, Agostoni-Carbone ML, et al. The complete DNA sequence of yeast chromosome III. *Nature.* 1992;357:38-46.

[6] Cole ST. Why sequence the genome of Mycobacterium tuberculosis? *Tuber Lung Dis.* 1996;77:486-490.

[7] Ahmed N, Hasnain SE. Genomics of Mycobacterium tuberculosis: old threats and new trends. *Indian J Med Res.* 2004;120:207-212.

[8] Brunak S, Danchin A, Hattori M, et al. Nucleotide sequence database policies. *Science.* 2002;298:1333.

[9] Danchin A, Medigue C, Gascuel O, Soldano H, Henaut A. From data banks to data bases. *Res Microbiol.* 1991;142:913-916.

[10] Iyer LM, Aravind L, Bork P, et al. Quoderat demonstrandum? The mystery of experimental validation of apparently erroneous computational analyses of protein sequences. *Genome Biol.* 2001;2:RESEARCH0051.

[11] Borgdorff MW. New measurable indicator for tuberculosis case detection. *Emerg Infect Dis.* 2004;10:1523-1528.

[12] Rocha EP. Order and disorder in bacterial genomes. *Curr Opin Microbiol.* 2004;7:519-527.

[13] Noria S, Danchin A. Just so genome stories: what does my neighbor tell me? Paper presented at: Uehara, 2002; Tokyo.

[14] Kunst F, Ogasawara N, Moszer I, et al. The complete genome sequence of the gram-positive bacterium Bacillus subtilis. *Nature.* 1997;390:249-256.

[15] Cole ST. Comparative mycobacterial genomics. *Curr Opin Microbiol.* 1998;1:567-571.

[16] Cardona PJ, Ruiz-Manzano J. On the nature of Mycobacterium tuberculosis-latent bacilli. *Eur Respir J.* 2004;24:1044-1051.

[17] Wayne G, Kubica G. The mycobacteria. In: Sneath P, Mair N, Sharpe M, eds. *Bergey's Manual of Systematic Bacteriology.* Baltimore, MD: Williams and Wilkins; 1986:1435–1457.

[18] Walther BA, Ewald PW. Pathogen survival in the external environment and the evolution of virulence. *Biol Rev Camb Philos Soc.* 2004;79:849-869.

[19] Gonzalez-y-Merchand JA, Garcia MJ, Gonzalez-Rico S, Colston MJ, Cox RA. Strategies used by pathogenic and nonpathogenic mycobacteria to synthesize rRNA. *J Bacteriol.* 1997;179:6949-6958.

[20] Menendez MC, Garcia MJ, Navarro MC, et al. Characterization of an rRNA operon (rrnB) of Mycobacterium fortuitum and other mycobacterial species: implications for the classification of mycobacteria. *J Bacteriol.* 2002;184:1078-1088.

[21] Cole ST, Eiglmeier K, Parkhill J, et al. Massive gene decay in the leprosy bacillus. *Nature.* 2001;409:1007-1011.

[22] Maurelli AT, Fernandez RE, Bloch CA, Rode CK, Fasano A. "Black holes" and bacterial pathogenicity: a large genomic deletion that enhances the virulence of Shigella spp. and enteroinvasive Escherichia coli. *Proc Natl Acad Sci U S A.* 1998;95:3943-3948.

[23] Parkhill J, Wren BW, Thomson NR, et al. Genome sequence of Yersinia pestis, the causative agent of plague. *Nature.* 2001;413:523-527.

[24] Buchrieser C, Rusniok C, Frangeul L, et al. The 102-kilobase pgm locus of Yersinia pestis: sequence analysis and comparison of selected regions among different Yersinia pestis and Yersinia pseudotuberculosis strains. *Infect Immun.* 1999;67:4851-4861.

[25] Rocha EP, Sekowska A, Danchin A. Sulphur islands in the Escherichia coli genome: markers of the cell's architecture? *FEBS Lett.* 2000;476:8-11.

[26] Bodaghi B, LeHoang P. Ocular tuberculosis. *Curr Opin Ophthalmol.* 2000;11:443-448.

[27] Basu S, Fenton MJ. Toll-like receptors: function and roles in lung disease. *Am J Physiol Lung Cell Mol Physiol.* 2004;286:L887-892.

[28] Hsieh MJ, Junqueira-Kipnis AP, Hoeffer A, Turner OC, Orme IM. Incorporation of CpG oligodeoxynucleotide fails to enhance the protective efficacy of a subunit vaccine against Mycobacterium tuberculosis. *Vaccine.* 2004;22:655-659.

[29] Sharma K, Chopra P, Singh Y. Recent advances towards identification of new drug targets for Mycobacterium tuberculosis. *Expert Opin Ther Targets.* 2004;8:79-93.

[30] Chaitra MG, Hariharaputran S, Chandra NR, Shaila MS, Nayak R. Defining putative T cell epitopes from PE and PPE families of proteins of Mycobacterium tuberculosis with vaccine potential. *Vaccine.* 2005;23:1265-1272.

[31] Raman S, Hazra R, Dascher CC, Husson RN. Transcription regulation by the Mycobacterium tuberculosis alternative sigma factor SigD and its role in virulence. *J Bacteriol.* 2004;186:6605-6616.

[32] Narayanan S. Molecular epidemiology of tuberculosis. *Indian J Med Res.* 2004;120:233-247.

[33] Narayanan S, Das S, Garg R, et al. Molecular epidemiology of tuberculosis in a rural area of high prevalence in South India: implications for disease control and prevention. *J Clin Microbiol.* 2002;40:4785-4788.

[34] Kato-Maeda M, Rhee JT, Gingeras TR, et al. Comparing genomes within the species Mycobacterium tuberculosis. *Genome Res.* 2001;11:547-554.

[35] Rajakumar K, Shafi J, Smith RJ, et al. Use of genome level-informed PCR as a new investigational approach for analysis of outbreak-associated Mycobacterium tuberculosis isolates. *J Clin Microbiol.* 2004;42:1890-1896.

[36] Anthony RM, Brown TJ, French GL. DNA array technology and diagnostic microbiology. *Expert Rev Mol Diagn.* 2001;1:30-38.

[37] Ahmed N, Alam M, Rao KR, et al. Molecular genotyping of a large, multicentric collection of tubercle bacilli indicates geographical partitioning of strain variation and has implications for global epidemiology of Mycobacterium tuberculosis. *J Clin Microbiol.* 2004;42:3240-3247.

[38] Goh EB, Yim G, Tsui W, McClure J, Surette MG, Davies J. Transcriptional modulation of bacterial gene expression by subinhibitory concentrations of antibiotics. *Proc Natl Acad Sci U S A.* 2002;99:17025-17030.

[39] Betts JC, McLaren A, Lennon MG, et al. Signature gene expression profiles discriminate between isoniazid-, thiolactomycin-, and triclosan-treated Mycobacterium tuberculosis. *Antimicrob Agents Chemother.* 2003;47:2903-2913.

[40] Zhang Y, Heym B, Allen B, Young D, Cole S. The catalase-peroxidase gene and isoniazid resistance of Mycobacterium tuberculosis. *Nature.* 1992;358:591-593.

[41] Wang JY, Burger RM, Drlica K. Role of superoxide in catalase-peroxidase-mediated isoniazid action against mycobacteria. *Antimicrob Agents Chemother.* 1998;42:709-711.

[42] Timmins GS, Master S, Rusnak F, Deretic V. Requirements for nitric oxide generation from isoniazid activation in vitro and inhibition of mycobacterial respiration in vivo. *J Bacteriol.* 2004;186:5427-5431.

[43] Zimhony O, Vilcheze C, Jacobs WR, Jr. Characterization of Mycobacterium smegmatis expressing the Mycobacterium tuberculosis fatty acid synthase I (fas1) gene. *J Bacteriol.* 2004;186:4051-4055.

[44] Mikusova K, Slayden RA, Besra GS, Brennan PJ. Biogenesis of the mycobacterial cell wall and the site of action of ethambutol. *Antimicrob Agents Chemother.* 1995;39:2484-2489.

[45] Kendall SL, Rison SC, Movahedzadeh F, Frita R, Stoker NG. What do microarrays really tell us about M. tuberculosis? *Trends Microbiol.* 2004;12:537-544.

[46] Marques MA, Espinosa BJ, Xavier da Silveira EK, et al. Continued proteomic analysis of Mycobacterium leprae subcellular fractions. *Proteomics.* 2004;4:2942-2953.

[47] Drobniewski F, Balabanova Y, Coker R. Clinical features, diagnosis, and management of multiple drug-resistant tuberculosis since 2002. *Curr Opin Pulm Med.* 2004;10:211-217.

[48] Siddiqi N, Das R, Pathak N, et al. Mycobacterium tuberculosis isolate with a distinct genomic identity overexpresses a tap-like efflux pump. *Infection.* 2004;32:109-111.

[49] Honore N, Cole ST. Streptomycin resistance in mycobacteria. *Antimicrob Agents Chemother.* 1994;38:238-242.

[50] Telenti A, Imboden P, Marchesi F, et al. Detection of rifampicin-resistance mutations in Mycobacterium tuberculosis. *Lancet.* 1993;341:647-650.

[51] Jacobs MR. Fluoroquinolones as chemotherapeutics against mycobacterial infections. *Curr Pharm Des.* 2004;10:3213-3220.

[52] Mo L, Zhang W, Wang J, et al. Three-dimensional model and molecular mechanism of Mycobacterium tuberculosis catalase-peroxidase (KatG) and isoniazid-resistant KatG mutants. *Microb Drug Resist.* 2004;10:269-279.

[53] Ramaswamy S, Musser JM. Molecular genetic basis of antimicrobial agent resistance in Mycobacterium tuberculosis: 1998 update. *Tuber Lung Dis.* 1998;79:3-29.

[54] Wagner D, Young LS. Nontuberculous mycobacterial infections: a clinical review. *Infection.* 2004;32:257-270.

[55] Andries K, Verhasselt P, Guillemont J, et al. A diarylquinoline drug active on the ATP synthase of Mycobacterium tuberculosis. *Science.* 2005;307:223-227.

[56] Humbert R, Altendorf K. Defective gamma subunit of ATP synthase (F1F0) from Escherichia coli leads to resistance to aminoglycoside antibiotics. *J Bacteriol.* 1989;171:1435-1444.

[57] Cole ST, Alzari PM. Microbiology. TB--a new target, a new drug. *Science.* 2005;307:214-215.

[58] McCue LA, McDonough KA, Lawrence CE. Functional classification of cNMP-binding proteins and nucleotide cyclases with implications for novel regulatory pathways in Mycobacterium tuberculosis. *Genome Res.* 2000;10:204-219.

[59] Reddy SK, Kamireddi M, Dhanireddy K, Young L, Davis A, Reddy PT. Eukaryotic-like adenylyl cyclases in Mycobacterium tuberculosis H37Rv: cloning and characterization. *J Biol Chem.* 2001;276:35141-35149.

[60] Barzu O, Danchin A. Adenylyl cyclases: a heterogeneous class of ATP-utilizing enzymes. *Prog Nucleic Acid Res Mol Biol.* 1994;49:241-283.

[61] Beuve A, Boesten B, Crasnier M, Danchin A, O'Gara F. *Rhizobium meliloti* adenylate cyclase is related to eucaryotic adenylate and guanylate cyclases. *J Bacteriol.* 1990;172:2614-2621.

[62] Sinha SC, Wetterer M, Sprang SR, Schultz JE, Linder JU. Origin of asymmetry in adenylyl cyclases: structures of Mycobacterium tuberculosis Rv1900c. *Embo J.* 2005.

[63] Pereira JH, de Oliveira JS, Canduri F, et al. Interaction of shikimic acid with shikimate kinase. *Biochem Biophys Res Commun.* 2004;325:10-17.

[64] Serina L, Blondin C, Krin E, et al. Escherichia coli UMP-kinase, a member of the aspartokinase family, is a hexamer regulated by guanine nucleotides and UTP. *Biochemistry.* 1995;34:5066-5074.

[65] Gagyi C, Ionescu M, Gounon P, Sakamoto H, Rousselle JC, Laurent-Winter C. Identification and immunochemical location of UMP kinase from Bacillus subtilis. *Curr Microbiol.* 2004;48:62-67.

[66] Nitschke P, Guerdoux-Jamet P, Chiapello H, et al. Indigo: a World-Wide-Web review of genomes and gene functions. *FEMS Microbiol Rev.* 1998;22:207-227.

[67] Sekowska A, Kung HF, Danchin A. Sulfur metabolism in Escherichia coli and related bacteria: facts and fiction. *J Mol Microbiol Biotechnol.* 2000;2:145-177.

[68] Sareen D, Newton GL, Fahey RC, Buchmeier NA. Mycothiol is essential for growth of Mycobacterium tuberculosis Erdman. *J Bacteriol.* 2003;185:6736-6740.

[69] Jaeger T, Budde H, Flohe L, et al. Multiple thioredoxin-mediated routes to detoxify hydroperoxides in Mycobacterium tuberculosis. *Arch Biochem Biophys.* 2004;423:182-191.

[70] McCarthy AA, Peterson NA, Knijff R, Baker EN. Crystal structure of MshB from Mycobacterium tuberculosis, a deacetylase involved in mycothiol biosynthesis. *J Mol Biol.* 2004;335:1131-1141.

[71] Fetterolf B, Bewley CA. Synthesis of a bromotyrosine-derived natural product inhibitor of mycothiol-S-conjugate amidase. *Bioorg Med Chem Lett.* 2004;14:3785-3788.

[72] Wagner D, Maser J, Lai B, et al. Elemental analysis of Mycobacterium avium-, Mycobacterium tuberculosis-, and Mycobacterium smegmatis-containing phagosomes indicates pathogen-induced microenvironments within the host cell's endosomal system. *J Immunol.* 2005;174:1491-1500.

[73] Holmes MA, Paulsene W, Jide X, Ratledge C, Strong RK. Siderocalin (Lcn 2) also binds carboxymycobactins, potentially defending against mycobacterial infections through iron sequestration. *Structure (Camb).* 2005;13:29-41.

[74] Ratledge C. Iron, mycobacteria and tuberculosis. *Tuberculosis (Edinb).* 2004;84:110-130.

[75] Stewart EJ, Madden R, Paul G, Taddei F. Aging and Death in an Organism That Reproduces by Morphologically Symmetric Division. *PLoS Biol.* 2005;3:e45.

In: Genome Research Advances
Editor: Alberto P. Bellini, pp. 219-235
ISBN: 978-1-60021-805-7

Chapter VII

Genomic Instability of the APC Gene Found in Glioblastoma

Nives Pecina-Slaus[1,2,*], Vili Beros[3], Tamara Nikuseva Martic[1,2] and Floriana Bulic-Jakus[2]

[1]Laboratory of Neurooncology, Croatian Institute for Brain Research, School of Medicine University of Zagreb, HR-10000 Zagreb, Croatia;
[2]Department of Biology, School of Medicine, University of Zagreb, HR-10000 Zagreb, Croatia;
[3]Department of Neurosurgery, University Hospital "Sisters of Charity", 10000 Zagreb, Croatia.

Abstract

The etiology and pathogenesis of tumors of the central nervous system are still inadequately explained. This study analyzes tumor suppressor gene—adenomatous polyposis coli (APC) in 28 patients with glioblastoma, the most aggressive form of glial tumors. APC protein has a structural role in adherens junctions, but also plays a signaling role as a negative regulator of the wnt pathway. Our interest in the APC gene stemmed principally from the findings that the wild-type APC protein is highly expressed in the central nervous system, and upon the finding that it is critically involved in particular syndromes, among which brain tumors play a significant role. Glioblastoma samples were tested for gene instability by PCR/loss of heterozygosity using RFLP method. Two polymorphic markers were used: an Rsa I polymorphic site in exon 11, and an Msp I polymorphic site in exon 15. The results of our analysis for both markers showed allelic loss of the APC gene in 40% of our sample out of 25 heterozygous patients (informativeness 89%). Another 20% of samples demonstrated allelic imbalance of the APC allele in tumor tissue. Altogether, there were 15 samples (60%) demonstrating

* Correspondence concerning this article should be addressed to Nives Pecina-Slaus, Laboratory of Neurooncology, Croatian Institute for Brain Research, School of Medicine University of Zagreb, Salata 3, HR-10000 Zagreb, Croatia. e-mail: nina@mef.hr; tel. +385 1 46 21 140; fax: +385 1 45 50 744; +385 1 49 20 050; +385 1 45 96 942.

instability of this tumor suppressor gene. Despite increasing knowledge on glioma biology and genetics, the prognostic tools for glioblastoma still need improvement. Our findings on genomic instability of the APC gene may contribute to a better understanding of glioblastoma genetic profile and could be used as a prognostic marker of disease evolution and progression.

Keywords: glioblastoma, adenomatous polyposis coli gene (APC), loss of heterozygosity, wnt signaling pathway.

INTRODUCTION

The genome of tumor cells is affected by two crucial types of changes: the accumulation of somatic mutations and the acquisition of genomic instability. The usual incidence rate of spontaneous somatic mutations is much lower compared to the rates of genetic changes observed in tumor cells and this increased frequency is the result of genomic instability that characterizes such cells. It is now believed that the increase of frequency of genetic changes is to be at least 100 times. Although it is usually assumed that the genetic instability is a later event in tumor progression, the question whether it may be an early causative event [1] in the formation of a specific tumor type is still speculative.

The well-known two-hit hypothesis for tumor suppressor genes in hereditary cancer, that has nowadays transcended to the multi-hit hypothesis, describes that as many as seven distinct and sequential mutations are needed for the development of a malignant tumor, whereas fewer could give rise to benign precursors. Kinzler and Vogelstein [2] investigating colo-rectal carcinoma provided evidence for this hypothesis and showed that the mutations could accumulate over a span of years to decades. On the other hand a body of evidence indicates that the mutation rate of normal human cells is much too low to account for the accumulation of this number of genetic alterations over a single individual's life time [3]. We know that multiple genetic changes are required for carcinogenesis and the inadequate incidence rate observed during a human lifetime seeks explanation. The answer to this discrepancy is given in the so-called mutator hypothesis. A mutator gene is a gene whose mutation increases the level of genetic changes of the individual's genome [4]. The mutator gene can be any gene, but usually comes from the group of genes whose products are involved in DNA repair mechanisms or genes whose products control the fidelity of DNA replication. Mutations of those genes are referred to as mutator mutations, and cancer cells are said to exhibit a mutator phenotype with the increased rate of genomic instability.

There are two categories of genomic instability observed in human cancers: microsatellite instability (MIN) and chromosomal instability (CIN). MIN involves the defects in the repair of short mismatches in the replication/repair machinery [5] resulting from the slippage of DNA polymerase as it synthesizes segments of short repeats. The consequential instability observed in a tumor is sometimes termed as RER+ tumor phenotype from Replication Error Positive phenotype. RER-positive tumor samples can be visualized on high resolution gels after electrophoresis techniques as samples with bands on different positions in comparison to bands of autologous normal tissue. CIN occurs at the chromosome level and is caused by chromosomal mis-segregation due to improper alignment of chromosomes in mitosis. The resulting daughter cells comprise aneuploidy and allelic imbalance. Besides instability at the whole chromosomal level, there are other forms of CIN. Translocations, segmental duplications and deletions, as well as gene amplifications, can also be considered a form of CIN, although mechanisms that cause such alterations are distinct from partitioning at mitosis [1,6].

Our study analyzed genomic instability in the brain tumor glioblastoma, with regard to the role of the tumor suppressor gene—adenomatous polyposis coli (APC). Protein product of this gene is a component of the adherens junction. Its central part binds to E-cadherin, the most important molecule of cell-cell adhesion, via beta-catenin molecule [6,7]. APC is illustrative of multiple roles that certain tumor suppressors play in a cell. Besides its structural role in cellular architecture, the protein product of the APC gene is also one of the key players of the wnt signaling pathway. The wnt/wingless pathway was first discovered in mouse and *Drosophila* and is one of the most interesting signal transductions, in which key components have multiple functions. In vertebrate cells, it is named after Wnt proteins, a family of highly conserved secreted signaling molecules that regulate cell-to-cell interactions during embryogenesis. Insights into the mechanisms of Wnt action have emerged from several systems: genetics in *Drosophila* and *Caenorhabditis elegans*; biochemistry in cell culture; and ectopic gene expression in *Xenopus* embryos. Many Wnt genes in the mouse have been mutated, leading to very specific developmental defects. As currently understood, Wnt proteins bind to receptors of the Frizzled family on the cell surface. Through several cytoplasmic relay components, the signal is transduced to beta-catenin, which then enters the nucleus to activate transcription of Wnt target genes [9,10]. Although the main signaling molecule of the pathway is beta-catenin, APC is a critical component of the beta-catenin destruction machinery heading to the proteasome, and acts as a negative regulator of the wnt pathway. When wnt ligand is absent, beta-catenin binds to APC protein and is being destroyed. Another tantalizing compartment of the wnt signaling pathway lies in downstream transcriptional activation. In response to wnt signaling, or under the circumstances of mutated APC, beta-catenin is stabilized, accumulates in the cytoplasm and enters the nucleus, where it finds a partner, a member of the DNA binding protein family LEF/TCF (lymphoid enhancer factor-T cell factor). Together they activate new gene expression programs. Target genes for beta-catenin/TCF encode c-MYC and cyclin D1 proteins [11], explaining why constitutive activation of the wnt pathway can lead to cancer. The schematic illustration of key components of the wnt signaling is shown in figure 1.

It has been well documented that wnt genes, together with other components of wnt signaling pathway, are implicated in cancer [10], especially in neoplasms of epithelial origin. Mutations in the adenomatous polyposis coli gene are responsible for familial adenomatous polyposis and the majority of sporadic colorectal cancers. The APC gene (chromosome 5q21) [12] is organized in 16 translated exons and encodes a 2843 amino acid protein that is expressed in specific (frequently post-mitotic) epithelial and mesenchymal cells of several fetal and adult human tissues. The final protein product of APC gene with a mass of 311.8 kDa shows little similarity to other known proteins. There are three known APC protein isoforms that arise through a complicated pattern of alternative splicing. The conventional APC splice isoform is present in many cell types, while other splice forms are present only in differentiated cells [13]. These alternative splice forms exclude exon 1 but include alternate exons 5 'of the missing exon 1 [14]. Two of the splice variants, BS (brain-specific) and 0.3 (exon 0.3) contain in-frame translation initiation codons, and the protein isoforms lack exon 1 encoded homodimerization domain. It is interesting that these protein isoforms are expressed in neuronal cells and other tissues with nondividing cell populations. Carson et al. [15] demonstrated that all isoforms cause similar reduction in beta-catenin activity and show no difference in the subcellular localization, and concluded that different isoforms are not trancriptionally different but rather involve unidentified protein functions of the specific domains. The majority of identified mutations of the APC gene are frameshift mutations [16] that almost always lead to a truncated protein product, devoid of C-terminal peptide sequences. A number of tumors, including colorectal, gastric, lung, hepatocellular, renal and laringeal tumors demonstrate loss of heterozygosity (LOH) of this gene [2,17-19]. It has been shown that tumor suppressors are differentially expressed during normal embryonic development [20] and that specific developmental pathways are often inappropriately regulated during tumorigenesis [21,22].

Our interest in elucidating the role of tumor suppressor gene—adenomatous polyposis coli (APC) stemmed principally from the findings that wild-type APC protein is expressed in the central nervous system [23,24] and that there are strong indications that wild-type APC protein is critically involved in initiation of neuronal differentiation [25]. Moreover, new knowledge on wnt signaling shows that wnt proteins regulate critical developmental processes of normal brain development [26-30]. Processes that include cellular adhesion and synaptic rearrangements require the expression of molecular components of the wnt pathway [31]. As we have already pointed out, the wnt pathway has repeatedly been implicated in tumorigenesis and lately in brain tumorigenesis as well [32]. These findings inspired us to focus our interest on the key player of the wnt signaling, APC tumor suppressor gene. The majority of brain tumors arise sporadically except those that are part of defined clinical syndromes. APC protein is critically involved in particular syndromes, like the Tourcot's syndrome, which includes the development of primary brain tumors such as medulloblastomas and gliomas [33,34]. Cytoplasmically associated to E-cadherin is beta-catenin, yet another important molecule in brain tumorigenesis. Activating mutations of beta-catenin gene are found in a variety of tumors suggesting that beta-catenin gene acts as an oncogene. In 2003. beta-catenin was identified as a critical factor for dendritic morphogenesis [35]. Mutations of this gene have been reported in sporadic medulloblastoma [36] and recently this molecule was proposed prognostic factor for medulloblastoma [37].

The etiology and pathogenesis of the tumors of the central nervous system are still inadequately explained. Our knowledge on genetic background of specific histopathological type of brain tumor still needs to be elucidated although great progress has been achieved along with the increasing advances in molecular genetics. Glioblastoma is the most aggressive form of glial brain tumor. The majority of glioblastomas arise without clinical or histological evidence of less malignant lesion and this kind of glioblastoma is categorized as primary or *de novo* glioblastoma. They manifest in older patients (mean age 55 years) after a short clinical history. In contrast, secondary glioblastomas develop more slowly by malignant progression from diffuse or anaplastic astrocytoma and they manifest in younger patients (mean age 40 years) [38]. At present specific genetic alterations found to be associated with glioblastoma are emerging [39-41]. Both types of glioblastoma have disrupted p53 and Rb1 pathways, although in different ways. Primary glioblastomas have amplification of the 12q14 region encompassing the CDK4 and MDM2 genes which results in their over-expression and the dysfunction of both pathways. In addition to the genetic abnormalities involving p53 and Rb1 pathways, glioblastomas exhibit loss of alleles from 10q in over 90% of cases. This genetic region encompasses PTEN tumor suppressor gene. Moreover, amplification of the epidermal growth factor receptor (EGFR) gene has been reported in about 35% of glioblastoma [40,42,43]. Secondary glioblastomas generally have no wild type p53 due to loss of one allele and mutation of the other allele and also loss of functional Rb1 pathway in a similar manner. Nevertheless, the genetic basis of primary as well as reoccurring glioblastoma is unclear, and much work is still required to determine the final list of genes involved, as well as the point in time of their activation/inactivation. In this chapter we offered a new candidate to fill in the puzzle of genetic basis of glioblastoma. Novel genes and molecular candidates involved in mechanisms of brain tumor formation will offer improvement in comprehension, diagnosis and treatment of this disease.

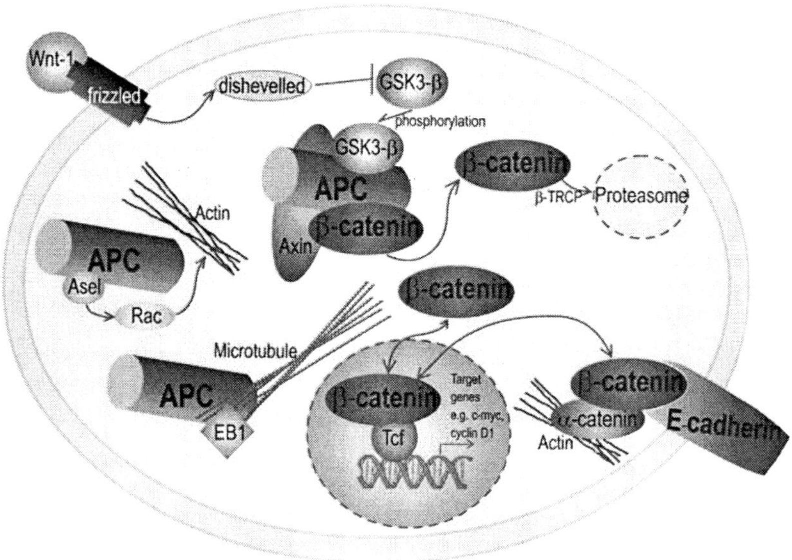

Figure 1. Wnt signaling pathway demonstrating its key components: frizzled (receptor of WNT proteins; GSK3-beta (glycogen-synthase-kinase 3-beta); Adenomatous polyposis coli (APC); beta-catenin. E-cadherin mediated adhesion is also shown since it modulates wnt signaling by sequestering cytoplasmic beta-catenin.

MATERIALS AND METHODS

Twenty-eight glioblastoma samples together with 28 autologous blood tissues were collected from the Department of Neurosurgery, University Hospital "Sisters of Charity", Zagreb, Croatia. Using the magnetic resonance imaging (MRI) tumor lesions were found in different cerebral regions (predominantly temporal and parietal region), with the surrounding zone of perifocal oedema (table 1). During the operative procedure the tumor was maximally reduced using a microneurosurgical technique. The patients had no family history of brain tumors and did not undergo chemotherapy or radiotherapy prior to surgery. Collected tumor tissues were frozen in liquid nitrogen and transported to the laboratory, where they were immediately transferred at -75°C. The peripheral blood samples were processed immediately. The histopathological classification was glioblastoma. The local Ethical Committee approved our study and patients gave their informed consent.

Table 1. Genetic changes of the APC tumor suppressor gene, anatomical location of glioblastoma and the duration of symptoms.

Patient No.	APC gene exon11/Rsa I	APC gene exon15/MspI	Localization	Symptom duration/months
1[1]	heterozygous	**A.I.**	FTP left	6
2	homozygous b	**A.I.**	FTP left	2
3	**A.I.**	heterozygous	P left	1
4	**A.I.**	heterozygous	TP right	7
5	**A.I.**	heterozygous	T right	0,5
6	heterozygous	heterozygous	FTP right	0,5
7	heterozygous	heterozygous	T left	1
8	heterozygous	heterozygous	FTP left	0,5
9	heterozygous	heterozygous	FT left	0,5
10	heterozygous	heterozygous	P right	1
11[2]	homozygous b	homozygous d	P right	2
12	homozygous a	heterozygous	bifrontal	1
13	homozygous a	heterozygous	FPO right	3
14	homozygous a	heterozygous	FTP left	2
15	homozygous a	heterozygous	T right	3
16	homozygous a	heterozygous	T left	2
17	homozygous b	heterozygous	TO right	2
18[1]	**LOH**	heterozygous	P left	12
19	heterozygous	homozygous c	T left	10
20[3]	heterozygous	homozygous c	T right	18
21	**LOH**	homozygous c	P left	1
22[4]	**LOH**	homozygous c	TO right	3
23	**LOH**	**LOH**	TP left	1
24	**LOH**	**LOH**	T left	3
25[3]	**LOH**	**LOH**	FTleft	2
26	**LOH**	ND	P left	2
27	homozygous b	ND	FP right	2
28	homozygous a	ND	FP left	0,5

Heterozygous = without LOH or A.I; a = both alleles 133 bp; b = both alleles have restriction sites (85+48 bp); c = both alleles 500 bp; d = both alleles have restriction sites (2 x 250); A.I. = allelic imbalance; [1]from diffuse astrocytoma (GII); [2]from anaplastic astrocytoma (GIII); [3]from oligoastrocytoma; [4]from oligodendroglioma; FTP= frontotemporoparietal region; P = parietal region ; TP = temporoparietal region ; T = temporal region; FPO = frontoparietooccipital region; TO = temporooccipital region; FT = frontotemporal region; FP= frontoparietal region; ND = not determined.

DNA Extraction

Tumor sample for DNA isolation was the part of obvious tumor mass evaluated by the neurosurgeon and based on macroscopic appearance and tissue color, density, and consistency on gross section. The sample was also evaluated for the percentage of tumor cells by pathologist and consisted of more than 85% of tumor cells.

Approximately 0.5 g of tumor tissue was homogenized with 1 ml extraction buffer (10 mM Tris HCl, pH 8.0; 0.1 M EDTA, pH 8.0; 0.5% sodium dodecyl sulfate) and incubated with proteinase K (100 μg/ml; Sigma, USA; overnight at 37°C). Phenol chloroform extraction and ethanol precipitation followed.

Blood sample was used to extract lymphocyte DNA. Five ml of blood was lysed with 7 ml distilled water and centrifuged (15 min/5000 *g*). The pellet was then processed as for DNA extraction from the tissue samples.

Polymerase Chain Reaction

The optimal reaction mixture (25 μl) for APC's exon 11 amplification was: 20 pmol of each primer (5'-GGACTACAGGCCATTGCAGAA-3' and 5'-GGCTACATCTCCAAAAGT CAA-3'), 200 μM of each dNTP, 200-400 ng template DNA, 1 μl (0.5 U) of Taq polymerase, 1 mM $MgCl_2$, 5 μl 10 X reaction buffer II (500 mM KCl, 100 mM Tris-HCl, pH 8.3). Polymerase chain reaction (PCR) conditions were: initial denaturation, 4 min/95°C; denaturation, 1 min/94°C; annealing, 2 min/58°C; extension, 1.5 min/72°C; 35 cycles.

To amplify fragment of the exon 15 of the APC gene we used in a volume of 25 μl: 5 pmol of each primer (5-ATGATGTTGACCTTTCCAGGG-3 and 5-CTTTTTTGGCAT TGCGGAGCT-3), 200 ng template DNA, 2.5 μl 10X buffer II, 1.5 mM $MgCl_2$, 25 μM of each dNTP, 0.2 μl (5U/μl) of Taq polymerase. PCR conditions were: initial denaturation, 5 min/95°C; denaturation, 30 s/95°C; annealing, 30 s/57°C; extension, 45 s/72°C; final extension, 72°C/7 min; 30 cycles. The PCR products for both markers were analyzed on 2% agarose gels.

Loss of Heterozygosity

Loss of heterozygosity (LOH) of the APC gene was detected on the basis of restriction fragment length polymorphism (RFLP) of the PCR products. Two different polymorphisms were investigated. One is an Rsa I polymorphic site in exon 11, and the other is an Msp I polymorphic site in exon 15. PCR amplification of exon 11 generated a 133- bp fragment that is cleaved to 85- and 48- bp fragments by Rsa I restriction if the polymorphic site is present, and remains uncleaved if the site is absent [44]. The amplified fragment of exon 15 is 550 bp long and is cleaved with the Msp I restriction endonuclease to two 250 bp fragments if the restriction site is present. LOH/Rsa I was demonstrated only in informative (heterozygous) persons when the tumor DNA showed loss of either the single uncut band (133 bp) or of the two cut bands (85+48 bp) compared to autologous blood tissue. For Msp I polymorphism heterozygous patients demonstrated two bands (550+250 bp), while LOH was shown when either band was missing in comparison to the autologous blood tissue. Samples that demonstrated quantitatively weaker allelic band in tumor tissue than in normal blood DNA were described as samples with allelic imbalance (A.I.).

PCR aliquots (10-15 μl) were digested with 6 U Rsa I (Gibco, USA; 12 h at 37°C) and with 6 U Msp I (AGS, Germany, overnight at 37°C) and were electrophoresed on Spreadex gels EL 300 in the SEA 2000 submarine electrophoresis apparatus (Elchrom scientific, Switzerland) at 120V. Temperature of the running buffer was kept constant at 55°C. The samples with LOHs were additionaly electrophoresed on 15% polyacrylamide gels stained with silver.

RESULTS

From 28 glioblastoma samples 79% appeared to have developed *de novo* , while six samples developed from previously diagnosed tumors of lower malignancy grade, two from diffuse astrocytoma (GII), one from anaplastic astrocytoma (GIII), two from oligoastrocytoma and one from oligodendroglioma. The histopathological diagnosis of glioblastoma (malignancy grade IV) was based on recognition of poorly differentiated astrocytic tumor cells with brisk mytotic activity and marked nuclear atypia. Differentiated elements were intermingled with bizzare multinucleated tumor giant cells. Prominent microvascular hyperplasia, in which proliferating blood vessels come to be lined by cells heaped up in disorderly fashion and transformed into glomeruloid or solid tufts (figure 2), and necrosis were also recognized as diagnostic features. The age of patients varied from 31 to 77 years (mean age = 59.1), and there were 15 females and 13 males. Tumor localization and the duration of symptoms are shown in table 1.

Genomic Instabilities

The Rsa I polymorphic site in APC's exon 11 was analyzed first. From 28 glioblastoma samples analyzed, 18 (64.3%) were informative for this polymorphism. For Msp I polymorphism in APC's exon 15, we found 20 heterozygous patients (71.4%). The results of our analysis showed 7 samples with LOH/exon 11 of the APC gene (39%) when tumor DNA was compared to autologous constitutive blood DNA. Another three samples demonstrated allelic imbalance (quantitatively weaker allelic band in tumor tissue than in normal blood DNA). Genetic changes of exon 15 demonstrated 3 samples with LOHs (15%) and another two samples with allelic imbalance.

When summing the informativeness provided by both markers, our sample showed (25/28) 89% of heterozygosity. The results of our analysis for both markers showed allelic loss of the APC gene in 10 glioblastoma samples out of 25 heterozygous patients (40%). Another 20% of samples demonstrated allelic imbalance of the APC allele in tumor tissue. Altogether, there were 15 samples (60%) demonstrating instability of this tumor suppressor gene. Three glioblastoma samples had both LOH at exon 11 and 15, while four demonstrated

LOH exclusively on exon 11, and another three A.I. were detected in this genetic area. Two A.I. found on exon 15 did not show losses on exon 11. Genomic instabilities of tumor suppressor gene investigated and the polymorphic status for both markers are summarized in table 1. LOHs of the APC gene that both markers revealed are shown in Figure 3 A (exon 11) and B (exon 15).

When we correlated our molecular findings with the demographic variables we observed that genomic instability of the APC gene was confined to younger patients.

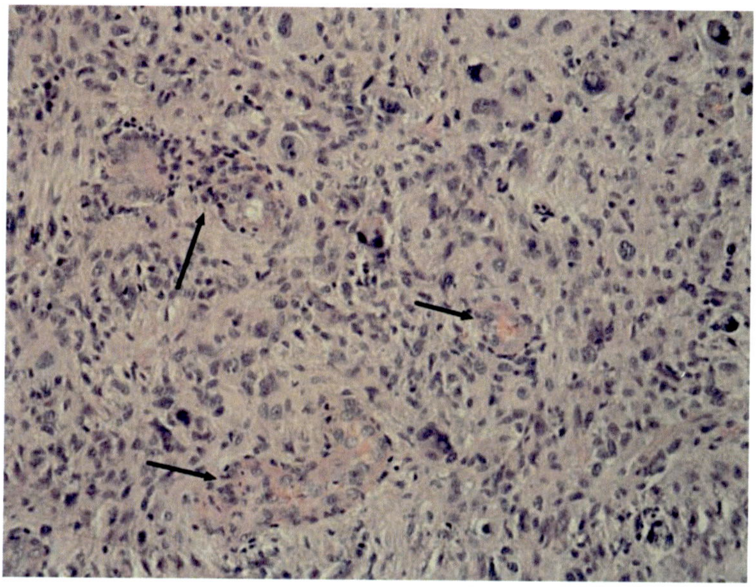

Figure 2. Glioblastoma multiforme (HE, x 200). Dense cellularity, striking pleomorphism, brisk mytotic activity and prominent microvascular proliferation (arrows).

Discussion

The mechanisms of brain tumor initiation and progression have not yet been completely investigated and elucidated. As with other tumors, the formation of brain tumors is the result of multiple consecutive genetic changes that represent a critical factor in tumor evolution. No consistent genetic abnormalities have been detected that would indicate the genetic profile of glioblastoma [40]. Histopathologically, an unambiguous distinction of glioma subtypes has remained elusive, but they clearly evolve through different gentic pathways. It also remains to be shown whether these subtypes differ significantly with respect to the molecular genetic alterations underlying the oncogenesis and progression of glioblastoma. With this in mind, we investigated a new candidate gene, tumor suppressor gene APC, in a set of twenty eight glioblastomas. Although APC has been thought of primarily as a colon-specific tumor suppressor gene, its association with certain brain tumors and its expression in the CNS suggests that it performs important functions in these tissues also.

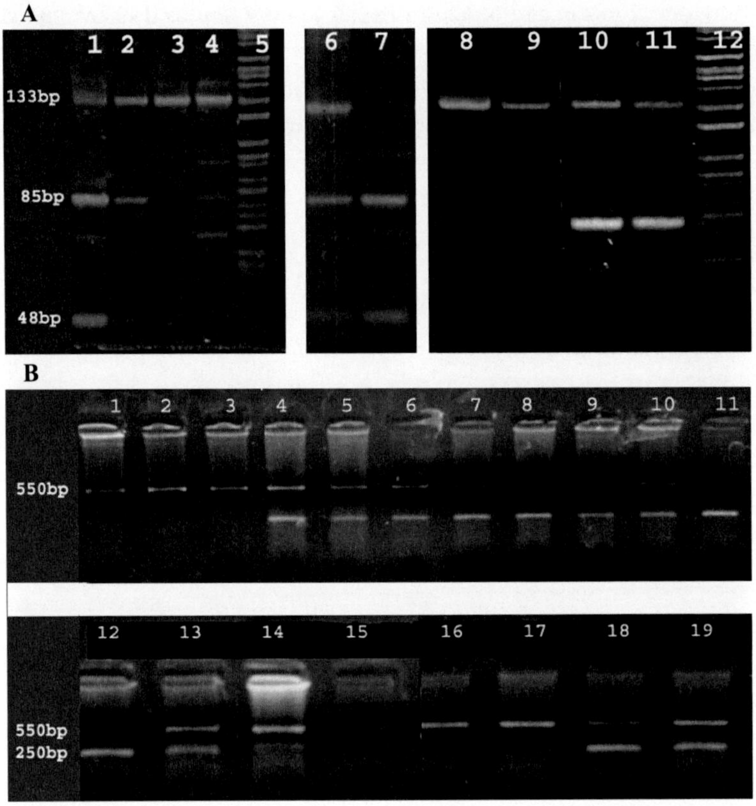

Figure 3. A. Loss of heterozygosity of APC gene in two patients with glioblastoma. Exon 11/RsaI/RFLP is demonstrated. Lanes 1, 2- heterozygous sample (tumor and blood); lane 3- LOH in glioblastoma patient (the digested/cut allele is missing); lane 4- informative blood sample of the same patient showing undigested/uncut allele 133 bp and cut allele (85+48 bp); lanes 5, 12 - standard M3 (Elchrom scientific); lane 6- informative blood sample; lane 7- LOH in the corresponding glioblastoma sample (uncut allele is missing); lanes 8, 9 – uninformative homozygous sample showing uncut allele (tumor and blood), lane 10- informative blood sample; lane 11- allelic imbalance in the corresponding glioblastoma (uncut allele is weaker, 48 bp fragments are not shown). B. Loss of heterozygosity of APC gene in 3 patients with glioblastoma. Exon 15/MspI/RFLP is demonstrated. Lane 1 - standard DNA/50 bp ladder; lanes 2, 3, 16, 17 – uninformative homozygous samples showing uncut alleles of 550 bp (tumors and blood); lanes 4, 5, 6, 7 - heterozygous samples, both alleles, cut and uncut, are visible; lanes 8, 10, 13, 14, 19 - informative blood samples; lane 9- allelic imbalance in the corresponding glioblastoma (uncut allele is weaker); lane 11, 12- LOHs in the corresponding glioblastomas, uncut alleles are missing; lane 15 – LOH in the corresponding glioblastoma cut allele is missing; lane 18 – allelic imbalance in the glioblastoma corresponding to the blood in lane 19 (uncut allele is weaker).

Genes involved in formation and acquisition of the full metastatic potential of specific tumor are not only those responsible for cell proliferation and survival, but also genes responsible for the control of cell adhesion and cell motility [45]. It is now apparent that tumor malignancy can, in certain aspects, be explained by alterations in the adhesive properties of neoplastic cells. One of the most important hallmarks of malignant gliomas is their invasive behavior. Despite modern diagnostics and treatment the median survival time does not exceed 15 months which is in accordance to extreme invasive characteristics of glioblastoma. Demuth and Berens in their review paper [46] state that glioblastoma recur predominantly within 1 cm of the resection cavity mainly due to the fact that at the time of surgery, cells from the bulk tumor have already invaded normal brain tissue. The problem of new rise of the malignant cells left behind the surgical resection of diffuse gliomas is one of the main disadvantages of therapeutic decision-making. The culprit of the highly invasive phenotype of human gliomas is thought to be associated to the cadherin group of adhesion molecules. Classical cadherins such as E-type and N-type are involved in forming both adherens and synaptic junctions in the nervous system. Shimamura and Takeichi [47] found that E-cadherin is transiently expressed in restricted regions of the mouse embryonic and adult brain. Although E-cadherin molecule is a well known and almost universal suppressor of invasion, little is known on the role of cell-cell adhesion in astrocytes and its alteration in migrating and invasive glioblastomas. In 2002. [48] Perego and co-workers demonstrated that disorganization of cadherin mediated junction, in which APC is included, is required to promote migration and invasiveness in glioblastoma cell lines.

The results of the analysis reported in this chapter suggested that APC plays an important role in the mechanisms of development and progression of this specific tumor of the central nervous system. Our previous investigations are also in favor of this presumption. We showed that neuroendocrine tumor pheochromocytoma exhibits microsatellite instabilities of the E-cadherin gene, which is the main adherens junction member molecule and an indirect member of the wnt signaling [49]. Preliminary analysis on oligoastocytoma showed gross deletion of the APC gene [50]. which is also indicative of involvement of this tumor suppressor gene in the formation and development of central nervous system tumors.

The analysis described in the present chapter showed that 60% of glioblastoma samples demonstrated LOH and allelic imbalance of the APC gene, which indicated that gross deletions of APC are part of genetic profile of these tumors. Allelic imbalances, quantitatively weaker allelic bands in tumor tissue than in normal blood, that we observed in 20% of our sample probably was not the result of normal tissue DNA interference. Because each glioblastoma sample was maximally reduced during the operative procedure using a fine microneurosurgical technique, our samples tested for genomic instabilities consisted of 90% of tumor cells. The explanation for allelic imbalances is that not all tumor cells of the sample underwent this sort of alteration. Behaviorally, gliomas can be viewed as consisting of two discreet subpopulations of cells, the proliferative cells at the tumor core, and cells invading the brain parenchyma. Thus, the gene expression profile of the tumor core may not necessarily depict the profile of genes active in the invading rim. Demuth and Berens [46] identified genes differentially expressed in invasive glioma cells and illustrated the differing biology of the invasive cells in contrast to the tumor cells at the core. Findings on different glioblastoma cellular components could explain for the portion of glioblastoma that did not show instabilities of APC gene. The relatively high number of allelic losses of the APC gene in our sample may be attributed to random variation in tumors, but in our opinion the frequency of APC's genomic instability in our sample was too high to be random. The observed high frequency led us to conclude that gross deletion of the APC gene that our glioblastoma samples comprised are an important event in the mechanisms of glioblastoma formation. Unfortunately, it was still not clear whether those changes represent an initiation event or come along the path of glioblastoma progression.

In our study genetic alterations were more frequently confined to exon 11 of the APC gene. Three glioblastoma samples had both LOH at exon 11 and 15, which could indicate that the deletion is extensive and goes all the way to exon 15. The alterations that were uniquely found on exon 15 could indicate that this genetic area is affected, and the same applies to the alterations found exclusively on exon 11. Exon 11 is a much smaller exon than exon 15, the biggest exon of the APC gene which encompasses 3/4 of its coding region (6577 bp). We could suspect that losses of exon 11 would not have such severe consequences at the protein level as would losses of exon 15. The alterations of the APC gene found in our sample were not correlated with sex of the patients. Nevertheless, we observed that genomic instability of the APC gene was confined to younger patients. The glioblastomas we considered primary because the diagnosis of glioblastoma was made at the first biopsy, without clinical or histopathologic evidence of a less malignant precursor lesion. Patients with glioblastoma de novo are older at diagnosis, compared to patients with secondary glioblastoma, and the mean age of our patient group (=59.1) was somewhat higher than the mean age reported in literature of 55 years.

The clearest example of wnt signaling involvement in brain tumor formation is the critical role of the tumor suppressor gene APC in the Tourcot's syndrome. Association of polyposis of the colon with a CNS malignancy is represented with this syndrome. Genetically predisposed patients are younger at diagnosis of glioblastoma compared to patients with sporadic disease. The relative risk of brain tumor formation is 23 times greater for Tourcot's syndrome patients with APC-mutation when compared to the general population between the ages of 0 and 29. Hamilton et al, [33] reported 14 families with Turcot's syndrome in whom patients with medulloblastoma had mutations in the APC gene, whereas patients with glioblastoma multiforme had germline mutations in one of the mismatch-repair genes. Mutation reports by Mori et al. [51] indicated lack of mutations in neuroepithelial tumors they investigated (glioblastomas included), but presence of germ line mutations of APC gene in 3 Turcot's syndrome patients. Other authors found germline mutations of APC gene in Tourcot's syndrome patients, but failed to detect them in the primary astrocytoma [52], or glioblastoma multiforme [53]. These results suggest that APC gene is associated with pathogenesis of Turcot's syndrome, but that at least one other gene is responsible for the genesis of neuroepithelial tumors in the CNS.

Somatic mutations of the APC gene were reported predominantly in sporadic medulloblastoma [34], and are much less frequent in glioblastoma. Nevertheless, Steigerwald and co-workers [13] found base change mutations in APC gene in two of 23 sporadic glioblastomas examined and a heterozygous G to A transition at position 148 of exon 0.3 in cell line SW 1088 from a human astrocytoma. This base change affects the amino acid sequence of the APC gene. One of the sporadic glioblastoma they investigated had an additional G following a string of seven Gs located at position 34 of exon 0.1 , but the effect on amino acid sequence could not be predicted. Microarray technology has been applied to the genome of glioblastoma, because it allows a large number of hybridizations to be performed simultaneously and was used for array comparative genomic hybridization by Roversi et al. [54]. Twenty-five primary glioma cell lines were investigated for gross and subtle genetic changes involved in glioma initiation and progression at the whole genome level. In this extensive study the authors have found among many other candidate chromosomal regions, recurrent losses of 5q22.2-q23.3 region in 10 glioblastoma cell lines (IV WHO grade) they analyzed, which is in accordance with our results.

So far little is known about the events that are involved in tumor initiation and early development of glioblastoma. Different subtypes of gliomas differ significantly in their age distribution, growth potential, tendency for progression and clinical course.

The work that we described in this chapter elucidated the potential contribution of genomic instability of the APC tumor suppressor gene, to malignant transformation of the specific tumors of the central nervous system. The functional consequences of the observed instabilities that we found at the genetic level, would need to be confirmed at the protein level by analyzing changes of APC's protein expression. We can speculate that wild-type APC protein is going to be absent in glioblastoma cells by inactivation of both alleles of the APC gene. One allele could be deleted, and the remaining one might have suffered another type of mutation, probably point mutation or a small deletion. On the other hand, only one allele may have suffered gross deletion, leaving the other one to do the job of wild type protein expression. It is therefore unclear whether the clinical evolution of glioblastoma is due to quantitative reduction or complete loss of APC protein product. Whether genetic losses we found have consequences at the protein level is left for future studies.

Nevertheless, our study offers the APC tumor suppressor gene as a potential candidate involved in the mechanisms of glioblastoma development. We consider our finding a specific but relevant contribution to understanding the pathophysiological mechanisms of brain tumor formation.

Future Perspectives

It would be interesting to assess the consequence that the instabilities of the APC gene might have at the protein level. Future studies need to confirm the functional changes of the APC molecule by analyzing its protein expression or changes at the APC's mRNA levels.

The question whether microsatellite genetic instabilities, that indicate involvement of mismatch repair, have a role in glioblastoma development and progression is also very tempting and should be considered for future work.

Further investigation regarding this gene on a bigger sample as well as employment of additional markers of the APC gene should be performed in the future to confirm our suggestions.

It would also be relevant to obtain data on wnt target genes, such as c-myc, cyclin D1 in future studies.

Acknowledgements

I would like to thank Prof. Dr. sc. Ivica Kostovic for critically reading the manuscript and for his constructive suggestions. This work was supported by grant 108-1081870-1905 and 0108049 from Ministry of Science Sports and Education, Republic of Croatia.

REFERENCES

[1] Bignold LP. Embryonic reversions and lineage infidelities in tumour cells: genome-based models and role of genetic instability. *Int J Exp Pathol* 2005, 86, 67-79.

[2] Kinzler KW, Vogelstein B. Lessons from hereditary colorectal cancer. *Cell* 1996, 87, 159-170.

[3] Prochownik EV. Functional and physical communication between oncoproteins and tumor suppressors. *Cell Mol Life Sci* 2005, 62, 2438-2459.

[4] Beckman RA, Loeb LA. Genetic instability in cancer: theory and experiment. *Semin Cancer Biol* 2005, 15, 423-435.

[5] Ionov Y, Peinado MA, Malkhosyan S, Shibata D, Perucho M. Ubiquitous somatic mutations in simple repeated sequences reveal a new mechanism for colonic carcinogenesis. *Nature* 1993, 363, 558-561.

[6] Gagos S, Irminger-Finger I. Chromosome instability in neoplasia: chaotic roots to continuous growth. *Int J Biochem Cell Biol* 2005, 37, 1014-1033.

[7] Rubinfeld B, Souza B, Albert I, Muller O, Chamberlain SH, Masiarz FR *et al.* Association of the APC gene product with beta-catenin. *Science* 1993, 262, 1743-1745.

[8] Pecina-Slaus N. Tumor suppresor gene E-cadherin and its role in normal and malignant cells. *Cancer Cell Int* 2003, E3, 17 (http://www.cancerci.com/content/3/1/17).

[9] Morin PJ, Sparks AB, Korinek V, Barker N, Clevers H, Vogelstein B, *et al.* Activation of beta-catenin-Tcf signaling in colon cancer by mutations in β-catenin or APC. *Science* 1997, 275,1787-1790.

[10] Peifer M, Polakis P. Wnt signaling in oncogenesis and embriogenesis - a look outside the nucleus. *Science* 2000, 287, 1606-1609.

[11] He TC, Sparks AB, Rago C, Hermeking H, Zawel L, da Costa LT, *et al.* Identification of c-MYC as a target of the APC pathway. *Science* 1998, 281, 1509-1512.

[12] Groden J, Thliveris A, Samowity W, Carlson M, Gelbert L, Albertsen H, *et al.* Identification and characterization of the familial adenomatous polyposis coli gene. *Cell* 1991, 66, 589-600.

[13] Steigerwald K, Santoro IM, Kordich JJ, Gismondi V, Trzepacz C, Badiali M, *et al.* A distinct splice form of APC is highly expressed in neurons but not commonly mutated in neuroepithelial tumours. *J Med Genet* 2001, 38, 257-262.

[14] Bardos J, Sulekova Z, Ballhausen WG. Novel exon connections of the brain- specific (BS) exon of the adenomatous polyposis coli gene. *Int J Cancer* 1997, 73,137-142.

[15] Carson DJ, Santoro IM, Groden J. Isoforms of the APC tumor suppressor and their ability to inhibit cell growth and tumorigenicity. *Oncogene* 2004, 23, 7144-7148.

[16] Sieber OM, Tomlison IP, Lamlum H. The adenomatous polyposis coli (APC) tumour suppressor - genetics, function and disease. *Mol Med Today* 2000, 6, 462-469.

[17] Pecina-Slaus N, Pavelic K, Pavelic J. Loss of heterozygosity and protein expression of APC gene in renal cell carcinomas. *J Mol Med* 1999, 77, 446-453.

[18] Pecina-Slaus N, Gall-Troselj K, Slaus M, Radic K, Nikuseva-Martic T, Pavelic K. Genetic changes of the E-cadherin and APC tumour suppressor genes in clear cell renal cell carcinoma. *Pathology* 2004, 36, 145-151.

[19] Pecina-Slaus N, Kljaic M, Nikuseva-Martic T. Loss of heterozygosity of APC and CDH1 genes in laryngeal squamous cell carcinoma. *Pathol Res Pract* 2005, 201, 557-563.

[20] Maric S, Bulic-Jakus F, Jezek D, Juric-Lekic G, Kos M, Vlahovic M. Expression of the proliferating cell nuclear antigen and protein products of tumour suppressor genes in the human foetal testis. *Andrologia* 2004, 36, 24-30.

[21] DiCiommo D, Gallie BL, Bremner R. Retinoblastoma: the disease, gene and protein provide critical leads to understand cancer. *Sem Canc Biol* 2000, 10, 255-269.

[22] Kato N, Shibuya H, Fukase M, Tamura G, Motoyama T. Involvement of adenomatous polyposis coli (APC) gene in testicular yolk sac tumor of infants. *Hum Path* 2006, 37, 48-53.

[23] Brakeman JS, Gu SH, Wang XB, Dolin G, Baraban JM. Neuronal localization of the Adenomatous polyposis coli tumor suppressor protein. *Neuroscience* 1999, 91, 661-672.

[24] Wedgwood S, Lam WK, Pinchin KM, Markham AF, Cartwright EJ, Coletta PL. Characterization of a brain-selective transcript of the Adenomatous polyposis coli tumor suppressor gene. *Mamm Genome* 2000, 11, 1150-1153.

[25] Dobashi Y, Katayama K, Kawai M, Akiyama T, Kameya T. APC protein is required for initiation of neuronal differentiation in rat pheochromocytoma PC12 cells. *Biochem Biophys Res Commun* 2000, 279, 685-691.

[26] Joyner AL. Engrailed, Wnt and Pax genes regulate midbrain-hindbrain development. *Trends Genet* 1996, 12, 15-20.

[27] Patapoutian A, Reichardt LF. Roles of wnt proteins in neural development maintenance. *Curr Opin Neurobiol* 2000,10, 392-399.

[28] Judas M, Milosevic JN, Rasin MR, Heffer-Lauc M, Kostovic I. Complex Patterns and Simple Architects: Molecular Guidance Cues for Developing Axonal Pathways in the Telencephalon. In: Kostovic I, editor. *Guidance Cues in the Developing brain*. Berlin Heidelberg: Springer-Verlag; 2003; p.1-32.

[29] Lie DC, Colamarino SA, Song HJ, Desire L, Mira H, Consiglio A, *et al.* Wnt signalling regulates adult hippocampal neurogenesis. *Nature* 2005, 437, 1370-1375.

[30] Pozniak CD, Pleasure SJ. A tale of two signals: Wnt and Hedgehog in dentate neurogenesis. Sci. STKE 2006 (319), pe5. [www.stke.org/cgi/content/full/sigtrans;2006/319/pe5]

[31] Hall AC, Lucas FR, Salinas PC. Axonal remodeling and synaptic differentiation in the cerebellum is regulated by WNT-7a signaling. *Cell* 2000, 100, 525-535.

[32] Howng SL, Wu CH, Cheng TS, Sy WD, Lin PC, Wang C, *et al.* Differential expression of Wnt genes, beta-catenin and E-cadherin in human brain tumors. *Cancer Lett* 2002, 183, 95-101.

[33] Hamilton SR, Liu B, Parsons RE, Papadopoulos N, Jen J, Powell SM, *et al.* The molecular basis of Turcot's syndrome. *N Engl J Med* 1995, 332, 839-840.

[34] Koch A, Waha A, Tonn JC, Sorensen N, Berthold F, Wolter M, *et al.* Somatic mutations of WNT/wingless signaling pathway components in primitive neuroectodermal tumors. *Int J Cancer* 2001, 93, 445-449.

[35] Yu X., Malenka R.C. Beta-catenin is critical for dendritic morphogenesis. *Nature Neurosci* 2003, 6, 1169-1177.

[36] Yokota N, Nishizawa S, Ohta S, Date H, Sugimura H, Namba H, *et al.* Role of wnt pathway in medulloblastoma oncogenesis. *Int J Cancer* 2002, 101,198-201.

[37] Ellison DW, Onilude OE, Lindsey JC, Lusher ME, Weston CL, Taylor RE, *et al.* United Kingdom Children's Cancer Study Group Brain Tumour Committee. Beta-catenin status predicts a favorable outcome in childhood medulloblastoma: the United Kingdom Children's Cancer Study Group Brain Tumour Committee. *J Clin Oncol* 2005, 23, 7951-7957.

[38] Kleihues P, Louis DN, Scheithauer BW, Rorke LB, Reifenberger G, Burger PC, *et al.* The WHO classification of tumors of the nervous system. *J Neuropathol Exp Neurol* 2002, 61, 215-225.

[39] Holland E. Gliomagenesis: Genetic alterations and mouse models. *Nat Rev Genet* 2001, 2, 120-129.

[40] Collins VP. Brain tumours: classification and genes. *J Neurol Neurosurg Psychiatry* 2004, 75 (Suppl 2), ii2-11.

[41] Fukushima T, Favereaux A, Huang H, Shimizu T, Yonekawa Y, Nakazato Y, *et al.* Genetic alterations in primary glioblastomas in Japan. *J Neuropathol Exp Neurol.* 2006, 65, 12-18.

[42] Ekstrand AJ, James CD, Cavenee WK, Selinger B, Pettersson RF, Collins VP. Genes for epidermal growth factor receptor, transforming growth factor alpha, and epidermal growth factor and their expression in human gliomas in vivo. *Cancer Res* 1991, 51, 2164-2172.

[43] Behin A, Hoang-Xuan K, Carpentier AF, Delattre JY. Primary brain tumours in adults. *Lancet* 2003, 361, 323-331.

[44] Pecina-Slaus N, Slaus M. Genetic polymorphism in exon 11 of the APC tumor suppressor gene in a Croatian sample. *Homo* 2000, 51, 151-155.

[45] Gumbiner BM. Cell adhesion: The molecular basis of tissue architecture and morphogenesis. *Cell* 1996, 84, 345-357.

[46] Demuth T, Berens ME. Molecular mechanisms of glioma cell migration and invasion. *J Neurooncol* 2004,70, 217-228.

[47] Shimamura K, Takeichi M. Local and transient expression of E-cadherin involved in mouse embryonic brain morphogenesis. *Development* 1992, 116, 1011-1019.

[48] Perego C, Vanoni C, Massari S, Raimondi A, Pola S, Cattaneo MG, *et al.* Invasive behaviour of glioblastoma cell lines is associated with altered organisation of the cadherin-catenin adhesion system. *J Cell Sci* 2002, 115, 3331-3340.

[49] Pecina-Slaus N, Nikuseva-Martic T, Gall-Troselj K, Radic K, Hrascan R. Replication Error-positive Samples Found in Pheochromocytomas. *In vivo* 2005, 19, 359-366.

[50] Pecina-Slaus N, Beros V, Houra K, Cupic H. Loss of heterozygosity of the APC gene found in a single case of oligoastrocytoma. *J Neurooncol* 2006, 78, 213-215.

[51] Mori T, Nagase H, Horii A, Miyoshi Y, Shimano T, Nakatsuru S, *et al.* Germ-line and somatic mutations of the APC gene in patients with Turcot syndrome and analysis of APC mutations in brain tumors. *Genes Chromosomes Cancer.* 1994, 9,168-172.

[52] Barel D, Cohen IJ, Mor C, Stern S, Shapiro R, Shomrat R, *et al.* Mutations of the adenomatous polyposis coli and p53 genes in a child with Turcot's syndrome. *Cancer Lett* 1998, 132,119-125.

[53] Suzui M, Yoshimi N, Hara A, Morishita Y, Tanaka T, Mori H. Genetic alterations in a patient with Turcot's syndrome. *Pathol Int* 1998, 48, 126-133.

[54] Roversi G, Pfundt R, Moroni RF, Magnani I, van Reijmersdal S, Pollo B *et al.* Identification of novel genomic markers related to progression to glioblastoma through genomic profiling of 25 primary glioma cell lines. *Oncogene* 2006, 25, 1571-1583.

INDEX

A

D

E

F

G

H

N

O

P

Q

R

S

T

U

V

W

X

Y